Earth Observation and Monitoring: Techniques and Challenges

Earth Observation and Monitoring: Techniques and Challenges

Edited by Oscar Barker

SYRAWOOD PUBLISHING HOUSE
New York

Published by Syrawood Publishing House,
750 Third Avenue, 9th Floor,
New York, NY 10017, USA
www.syrawoodpublishinghouse.com

Earth Observation and Monitoring: Techniques and Challenges
Edited by Oscar Barker

International Standard Book Number: 978-1-64740-400-0 (Hardback)

Cataloging-in-publication Data

Earth observation and monitoring : techniques and challenges / edited by Oscar Barker.
p. cm.
Includes bibliographical references and index.
ISBN 978-1-64740-400-0
1. Earth (Planet)--Observations. 2. Earth sciences--Remote sensing.
3. Remote sensing. 4. Geographic information systems. I. Barker, Oscar.
QE33.2.R4 .E27 2023
550.28--dc23

TABLE OF CONTENTS

PREFACE

The process of gathering information about the physical, chemical and biological systems of the Earth is called Earth observation. It uses remote sensing technology to continuously monitor land, water, environment and atmosphere. Earth observation plays an instrumental role in assessing and mitigating negative impacts across multiple areas such as agriculture, biodiversity, climate, disasters, ecosystems, energy, health, water and weather. In the recent years, various developments have taken place in satellite remote sensor technology and retrieval algorithms. These advancements have improved climate research, and observations related to land, oceans and atmosphere. Earth observation and monitoring can also help in evaluating and preventing disasters, climate change, and biodiversity loss. This book provides comprehensive insights into the subject of Earth observation and monitoring. It presents researches and studies performed by experts across the globe. Students, researchers, experts, and all associated with geography, geology, and Earth sciences will benefit alike from this book.

Various studies have approached the subject by analyzing it with a single perspective, but the present book provides diverse methodologies and techniques to address this field. This book contains theories and applications needed for understanding the subject from different perspectives. The aim is to keep the readers informed about the progresses in the field; therefore, the contributions were carefully examined to compile novel researches by specialists from across the globe.

Indeed, the job of the editor is the most crucial and challenging in compiling all chapters into a single book. In the end, I would extend my sincere thanks to the chapter authors for their profound work. I am also thankful for the support provided by my family and colleagues during the compilation of this book.

Editor

1

Combining SAR and Optical Earth Observation with Hydraulic Simulation for Flood Mapping and Impact Assessment

Emmanouil Psomiadis [1], Michalis Diakakis [2] and Konstantinos X. Soulis [1,*]

[1] Department of Natural Resources Management and Agricultural Engineering, Agricultural University of Athens, 75 Iera Odos st., 11855 Athens, Greece; mpsomiadis@aua.gr

[2] Faculty of Geology and Geoenvironment, National and Kapodistrian University of Athens, 15784 Panepistimioupoli Zografou, 10679 Athens, Greece; diakakism@geol.uoa.gr

* Correspondence: soco@aua.gr

Abstract: Timely mapping, measuring and impact assessment of flood events are crucial for the coordination of flood relief efforts and the elaboration of flood management and risk mitigation plans. However, this task is often challenging and time consuming with traditional land-based techniques. In this study, Sentinel-1 radar and Landsat images were utilized in collaboration with hydraulic modelling to obtain flood characteristics and land use/cover (LULC), and to assess flood impact in agricultural areas. Furthermore, indirect estimation of the recurrence interval of a flood event in a poorly gauged catchment was attempted by combining remote sensing (RS) and hydraulic modelling. To this end, a major flood event that occurred in Sperchios river catchment, in Central Greece, which is characterized by extensive farming activity was used as a case study. The synergistic usage of multitemporal RS products and hydraulic modelling has allowed the estimation of flood characteristics, such as extent, inundation depth, peak discharge, recurrence interval and inundation duration, providing valuable information for flood impact estimation and the future examination of flood hazard in poorly gauged basins. The capabilities of the ESA Sentinel-1 mission, which provides improved spatial and temporal analysis, allowing thus the mapping of the extent and temporal dynamics of flood events more accurately and independently from the weather conditions, were also highlighted. Both radar and optical data processing methods, i.e., thresholding, image differencing and water index calculation, provided similar and satisfactory results. Conclusively, multitemporal RS data and hydraulic modelling, with the selected techniques, can provide timely and useful flood observations during and right after flood disasters, applicable in a large part of the world where instrumental hydrological data are scarce and when an apace survey of the condition and information about temporal dynamics in the influenced region is crucial. However, future missions that will reduce further revisiting times will be valuable in this endeavor.

Keywords: floods; remote sensing; hydraulic modelling; agriculture; flood impact; Sperchios catchment

---

1. Introduction

Floods are considered to be one of the most common, recurring and catastrophic types of natural hazard [1,2]. Over the last few decades, they have been documented as a significant threat to human life, including a noteworthy number of fatalities as well as extended economic damages [3–9]. Across the Mediterranean region, most commonly, flooding results from intense rainfall within a short duration of time, constituting a serious risk for human life, infrastructures and agricultural land. Small- to medium-sized catchments are most frequently prone to flooding, often intensified by their intrinsic features, such as geological and geomorphological characteristics (low permeability and steep slopes), drainage network features, watershed texture and geometry, land use/cover, along with extreme meteorological forcing [10–12].

Timely mapping, monitoring and impact assessment of flood events are vital for the coordination of flood relief efforts and the elaboration of flood management, risk mitigation plans and many prevention initiatives. This knowledge is usually hard to obtain based on conventional approaches due to flood temporal and spatial characteristics [13].

Conventional observational networks often face difficulties in flood monitoring due to the temporal and spatial scales of flood occurrence in the region [14,15]. Traditional approaches to flood delineation are mostly based on field inspections and ground observations as well as airborne surveillance. However, in incidents that extend over a large area, such methods can be impractical, time-consuming, expensive and sometimes dangerous for the personnel and the equipment involved. Moreover, timely remote sensing (RS) monitoring of flood events can be problematic due to the usual inhibitive cloud cover (cc) [16]. During the last few decades, Earth Observation (EO) passive and active multitemporal satellite systems and RS practices offer timely and efficient tools for examining and delineating flooding over large areas. Their continuous measurements offer valuable information on inundated areas, which is useful for the effective response contributing likewise to damage assessment and hazard management [16,17].

Active microwave radar systems can be useful for such purposes since they can provide high-resolution and frequent day and night observations of the Earth's surface under practically any weather conditions since they can penetrate clouds and heavy rainfall [18,19].

In the framework of the Copernicus Earth Observation (EO) program, the European Space Agency (ESA) carried out an innovative satellite program named Copernicus, which comprises the constellation of the Sentinel satellites. Each mission of Sentinels aims towards a distinct sector of Earth observation (land, ocean and atmosphere), and its products are utilized in numerous applications. Sentinel-1 (S1) is a Synthetic Aperture Radar (SAR) operation system, that provides a continuous dataset of radar images, independent of the time of day and the weather conditions, enabling flood events monitoring and mapping [20]. The contribution of SAR sensors, having a C band sensor (wavelength $\approx$ 5.6 cm), rises from the sensibility of the backscatter signal to calm water surfaces, that typically employs the distinct properties of water in the microwave spectrum. Specifically, calm flood water surfaces work as mirrors and reflect incoming radiation away from the sensor, which normally results in low backscatter measurements [21]. Additionally, the S1 SAR mission incorporates improved spatial resolution (10 m), a 12-day repeat cycle using two identical satellites (S1A and S1B) with a 180° orbital phasing difference enhancing repeat cycle to 6-days, and various polarization modes that have enhanced the observing potential, making it ideal for flood monitoring [22,23]. Polarization is a key component in flood detection, as has been shown in numerous former surveys [24]. The suitable selection of horizontal (H) or vertical (V), single (VV or HH) or dual (HH + HV or VH + VV) polarity plays a decisive role in the enhanced distinction of backscatter values between dry and wet areas [24]. VV polarized products are greatly affected by surface roughness conditions, but in many cases, they still deliver adequate results for the mapping of the flooded area, because this specific polarization identifies clearly the partially submerged features, which can be valuable in flood damage estimation [23,25]. The major drawbacks of SAR data are related to a complicated mixture of several surface characteristics, which act negatively on the flood detection, such as the incidence angle, relative water height, vegetation type, density and orientation, and wind consequences, which can cause opposite backscatter responses for similar situations [13].

Passive multispectral instruments record the earth's surface reflected or emitted energy in the visible and infrared part of the electromagnetic spectrum. The high-resolution sensors can offer good detection of the flooded areas in cases that clouds, floating vegetation and trees do not conceal the water surface [13,18,26]. Satellite datasets recorded by the passive sensors, such as those onboard the Landsat-7 (L7) and Landsat -8 (L8), provide unique data of the land surface, appropriate for land use/cover (LULC) classification and delineation of floods, due to their exceptional spatial, spectral and temporal resolution [27,28]. Numerous studies utilizing EO information with different techniques have been conducted worldwide for the detailed classification of LULC [29–32] and the mapping of inundated areas [13,17,33,34] along with their impact on agricultural land [20,35,36]. Moreover, utilizing multispectral optical data, several different remotely sensed water indices have been developed for the detection of open surface water [37–39]. The modified Normalized Difference Water Index (MNDWI) provided by Xu [37], is an improved form of the simple Normalized Difference Water Index (NDWI) provided by McFeeters [39] and is capable of delineating water surfaces using green and near-infrared (NIR) spectral bands of Landsat [37,40], while at the same time effectively removing the land noise from built-up, vegetation and soil cover [37,38].

Apart from the valuable information on the flood extent, flood depth, discharge and inundation duration are among the most critical issues in flood impact assessment and deliver crucial inputs for land use planning, insurance claims management and a better knowledge of basin hydrological conditions [41]. In contrast to the far-reaching amount of literature regarding the flood extent mapping, only a few surveys emphasize the assessment of flood duration, such as those of O'Hara et al. [20], Bhatt et al. [42], Rahman et al. [43] and Rättich et al. [41]. Nowadays, the availability of a great number of multitemporal satellite data and the very high temporal revisit period ensures the appropriate monitoring and analysis of an affected area [16,41,44]. Moreover, water depth is considered essential for flood impact monitoring and an accurate digital elevation model (DEM) is the most efficient tool for a better assessment of these impacts. Especially, in a flat plain area, the precision of flood assessment directly depends on the analysis of the DEM [13,45]. A high-resolution DEM is crucial to extract the details of the topography, acquire the water surface elevations, and simulate flood inundation and depth [13,46,47].

RS-derived geospatial products, such as Digital Elevation Models (DEM) and Geographical Information System (GIS) analysis, can also deliver valuable input for hydraulic modelling, with high accuracy spatial data of elevation. Numerical hydrodynamic models are an important tool that capitalizes these produces for understanding flood flows and planning flood risk management. Hydrologic Engineering Center's River Analysis System (HEC-RAS) is considered to be one of the most suitable packages for creating and evaluating flood simulations, due to its improved capability to model natural rivers and canals [48–50]. HEC-RAS has the capability of one- and two-dimensional simulations and is used to simulate the floodplain area [48,51]. Especially in the case of 1-D modelling, there are widespread applications that have demonstrated its capability in simulating water flow. In the field of floods, while RS applications are most active in the post-flood period (e.g., flood mapping), and hydraulic modelling is more useful in pre-flood hazard assessment (e.g., simulating flood flows), there has been literature that couples both approaches in certain applications.

In data-poor regions and given the difficulty of instrumental observations in flash flood-prone basins, there is a need to exploit emerging remote sensing technologies to enrich instrumental data and improve the understanding of critical flood metrics (i.e., peak discharge, flood extent and depth, return period) in the course of post-flood investigations.

In this context, the principal objectives of this study are: (a) Examine whether the coupling of RS data (using different methods) and hydraulic modelling (using different flow scenarios) has the potential to delineate inundated areas and contribute to estimating key flood metrics in poorly gauged basins, such as peak discharge and recurrence intervals of the obtained flood extent and depth; (b) Explore the potential to assess the impact on the crops of the affected area through the estimation of the inundation duration of the flood event using multitemporal RS data. This analysis exploited a flood event that took place from 31 January to 1 February 2015, following intense and heavy rainfall that occurred in the Sperchios river basin, in Central Greece.

2. Study Area—Background

The Sperchios River basin lies between 38°44′ and 39°05′N, and 21°50′ and 22°45′E, covering an area of approximately 1823 km^2, originating from Mount Timfristos (2327 m) in the Fthiotida Prefecture in central Greece. Having a west to east direction, the river traverses the Sperchios Valley, between Mount Oiti and the western part of Mount Othrys and then outflows into the sea at the Maliakos Gulf [52,53]. The Sperchios river develops a complex meander track about 30 km long, at the coastal part of its floodplain. (Figure 1a).

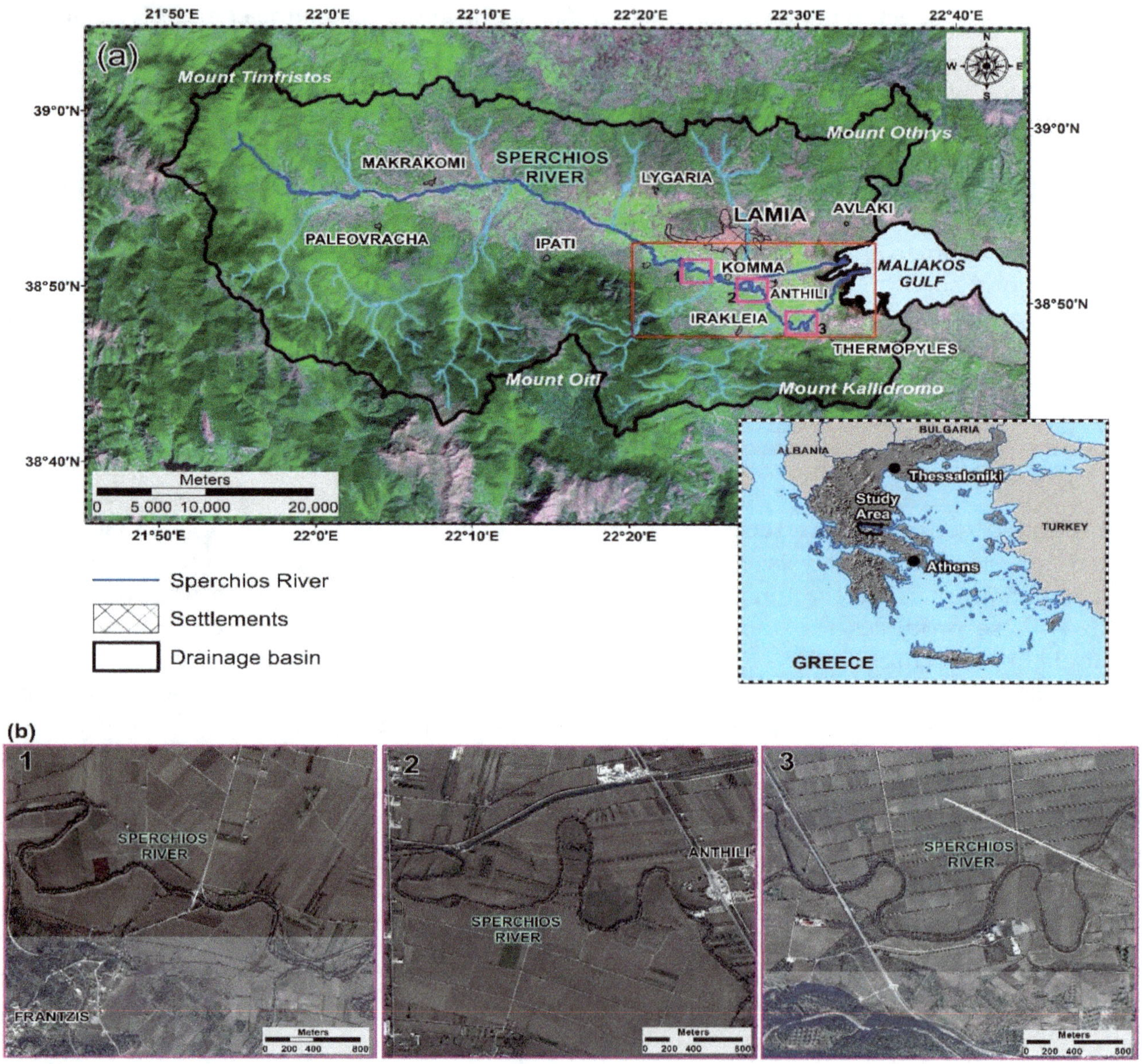

Figure 1. (**a**) The location of the Sperchios river catchment in Central Greece (background image: Landsat 8, 457 as RGB), the red rectangular line displays the river's intense meandering area, (**b**) Very high resolution frames at the meandering area using orthophoto images (the frames correspond to the pink squares with the same numbers, scale 1:10,000).

The Sperchios floodplain is an area dominated both in terms of land use and local economy by agricultural activities. According to the Hellenic Statistical Authority (ELSTAT 2011), the Prefecture of Fthiotida has a total of 169,542 residents with a large percentage of them lying at the eastern coastal part of the catchment, at the city of Lamia and several other smaller villages. Moreover, the main road and railway network, as well as critical facilities and industrial units, lie in the area (Figure 1b).

The Sperchios river catchment is a graben, controlled by major NW-SE and E-W trending faults [54]. The main lithological formations include flysch (south-southwest and northwest part), dolomites and limestones (south and southeast part), an ophiolitic complex in a shale-chert formation (north and north-eastern part) and Neogene and Quaternary unconsolidated deposits (central plain area) (Figure 2) [52,53,55].

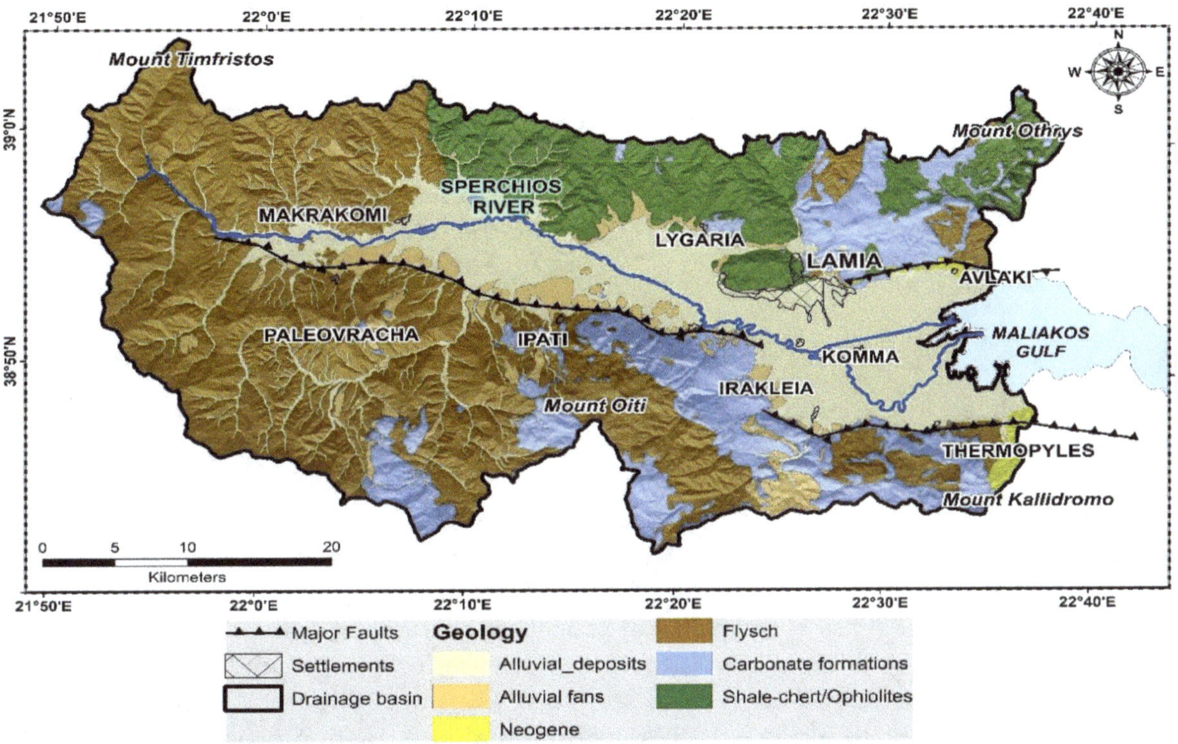

Figure 2. The simplified geological status of the area, including faults and fracture systems (using as a background the hillshade relief of the basin).

Catchment Characteristics and Flood Susceptibility

Flooding vulnerability in the basin is intensified by its topography, geological background and drainage network characteristics.

The steep slopes, especially in the southern area, where at a relatively small distance, an altitude difference of 300 m can be observed, and the dense drainage network contributes to a very fast hydrological response and short concentration times. Furthermore, the western and southwestern part of the Sperchios catchment is dominated by impermeable rocks (flysch), leading to a dense drainage network and increased runoff volumes.

As is reported in more detail by Psomiadis et al. [54], who presented a comprehensive analysis of the morphometric characteristics of the main sub-catchments of Sperchios watershed and their effect on flow routing, the drainage network follows the tectonic movements, the geology, and the morphometry, and consequently, the stream branches in the south are shorter and denser, while in the north they are longer and sparser. These morphometric characteristics result in very high flow velocities, especially at the southern part of the Sperchios river watershed. The torrential hydrological conditions (rapid flow and high runoff volumes over a short period) during storm events combined with the strong meandering formations of the eastern plain part of the Sperchios riverbed, which significantly slow down water flow results in the inability of the river to effectively drain these vast amounts of water towards its outlet at the Maliakos Gulf, provoking an increase in water level and, ultimately, overtopping, and, by extension, frequent flash-flood events.

Additionally, the intense agricultural activity, the high population concentration, and the huge infrastructure projects, including the main road (National Highway and E65) and railway network constructions that cluster at the eastern coastal area make this area quite vulnerable to flood hazards [11,53,54].

The Sperchios river basin has suffered in the past many intense and catastrophic floods. Based on historical records and the literature, many noteworthy events can be identified, with the most significant being those of 1889, 1939, 1954, 1984, 1987, 1994, 1997, 2001, 2003 and 2012 [10,11,53,56]. The areas frequently flooded are roughly 20,000 to 100,000 ha including mostly cropland and villages located near the Sperchios riverbed. Typically, these areas are covered with water for around four to five days excluding some areas with locally lower altitudes, which can be inundated for eight to ten days.

The studied flood event occurred from January 31 to February 1 of 2015, following high amounts of rainfall, which led to rapid flow (because of the abovementioned morphological and geological characteristics) of large water quantities at a short time into the river, flooding a large area near its outlet.

3. Materials and Methods

3.1. Datasets

The dataset utilized in the present research contains satellite-born images, topographical data, and statistical archives. For the multi-temporal monitoring of the flood event, all the available optical and SAR images of Sentinel-1, Landsat 7 and Landsat 8 were acquired. Unfortunately, Sentinel-2A has been set in orbit since 23 June 2015, four months after the flood occurrence.

Five Sentinel-1 Level-1 Ground Range Detected (GRD) SAR images were acquired for the flood extent and the inundation duration mapping (Table 1, Figure 3a,b). The images have been acquired through the Sentinels Open Access Hub and were processed using the Sentinel's Application Platform (SNAP) [57], an open-source common architecture software, offered by the ESA. Each image has a coverage range of 250 km with a 10×10 m spatial resolution. The data were selected with the Interferometric Wide (IW) swath mode, which supports dual polarization products (VV + VH), using the Terrain Observation with Progressive Scans SAR [58].

Table 1. The selected Sentinel-1 and Landsat-7/8 satellite datasets.

Satellite/Path-Row or Relative Orbit	Acquisition Date	Use
Sentinel-1 C-SAR/102	21 January 2015	Pre-Flood (reference image)/Flood monitoring
Sentinel-1 C-SAR/175	02 February 2015	Post-Flood/Flood monitoring
Sentinel-1 C-SAR/7	03 February 2015	Post-Flood/Flood monitoring
Sentinel-1 C-SAR/80	08 February 2015	Post-Flood/Inundation duration monitoring
Sentinel-1 C-SAR/7	15 February 2015	Post-Flood/Inundation duration monitoring
Sentinel-1 C-SAR/175	26 February 2015	Post-Flood/Inundation duration monitoring
Landsat-8 OLI/184033	20 December 2014	LULC Classification/Permanent water delineation
Landsat-7 ETM+/184033	13 January 2015	Permanent water delineation
Landsat-7 ETM+/183033	07 February 2015	Flood monitoring (29% cc)
Landsat-7 ETM+/184033	14 February 2015	Post-Flood/Inundation duration monitoring (21% cc)
Landsat-8 OLI/183033	15 February 2015	Post-Flood/Inundation duration monitoring
Landsat-8 OLI/184033	11 April 2015	LULC Classification/Winter crops
Landsat-8 OLI/184033	14 June 2015	LULC Classification/Winter crops
Landsat-8 OLI/184033	18 September 2015	LULC Classification/Summer crops

Moreover, regarding the optical data, eight images of L7/8 Level 1 geometrically corrected to the UTM 34N WGS84 ellipsoid with standard terrain correction applied were obtained free of charge (Path/Row: 184/033 and 183/033) through the United States Geological Survey portal [59] (Table 1, Figure 3). The optical data were processed only for permanent water bodies and flood duration mapping, and not for the flood monitoring, since there were no cloud-free images available during the flood event. Additionally, the multispectral images were used for the classification of LULC and the delineation of the existing crops at the winter period where the flood event occurred.

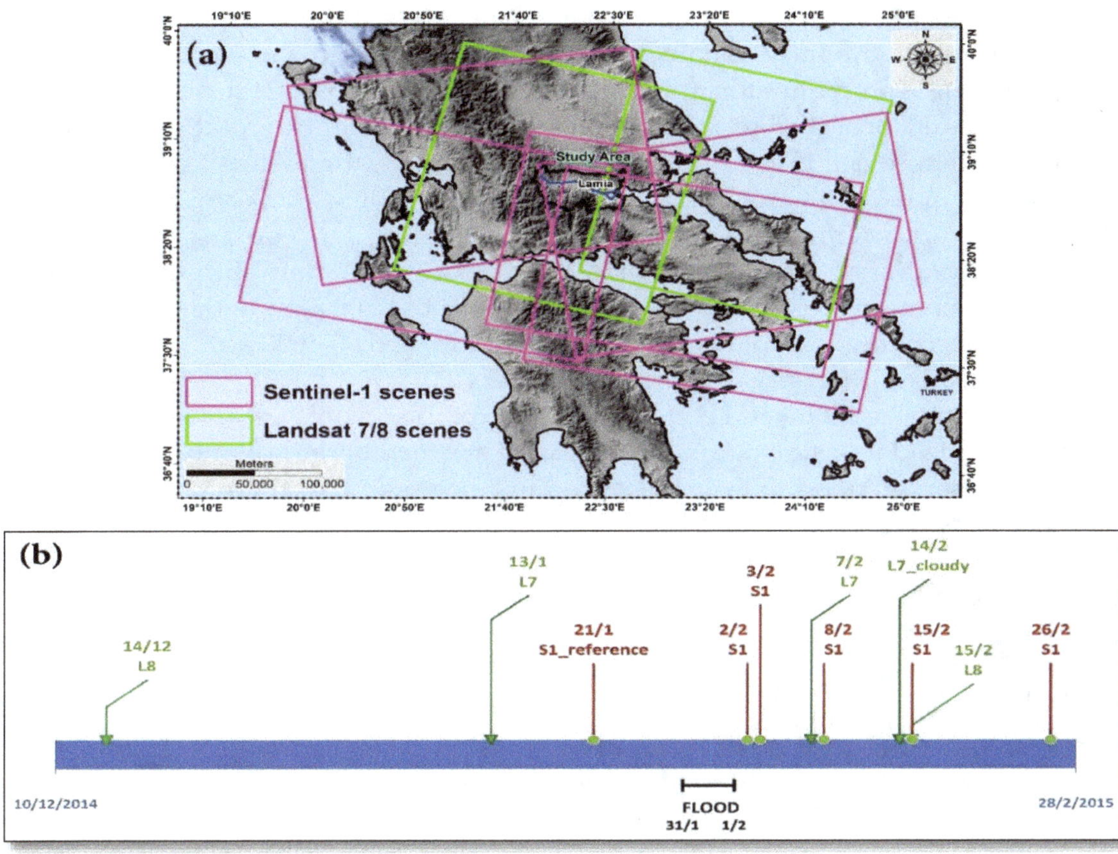

Figure 3. (**a**) Spatial and (**b**) Temporal distribution of the Sentinel-1 (relative orbits 7, 80, 102 and 175) and Landsat 7/8 (path/row 184/033 and 183/033) datasets.

For the cross-checking of the permanent water areas (small lakes, ponds or marshes), the Global Surface Explorer database (https://globalsurface-water.appspot.com/) of the European Commission's Joint Research Centre (ECJRC) was used. This water dataset was derived from the entire inventory of the Landsat satellites and for this case, we used the maximum water extent product, which concentrates information about the detected water surfaces throughout a period of 32 years.

The land cover classification was supplemented with the farmers' declarations and Greek Land Parcel Identification System (LPIS) data obtained by the Integrated Administration and Control System of the Greek Payment and Control Agency for Guidance and Guarantee Community Aid (OPEKEPE) (https://www.opekepe.gr/en/).

Furthermore, geomorphological data from the Hellenic Military Geographical Service (HMGS) topographic maps were used to mine different types of info-layers such as rivers, roads, railway tracks, etc., and geological maps of the Geological and Mineral Exploration Institute for the extraction of the geological information. A very detailed DEM (5 m) derived from the stereo pairs of very high accuracy aerial images of the National Cadastre and Mapping Agency S.A., were obtained to be used as an input in the HEC-RAS processing. Finally, statistical data concerning the sub-municipalities, the areas and the kind of crops affected by the flood episode provided by the Greek Agricultural Insurance Organization (ELGA), were processed and compared with the results of this study.

The satellite image processing was accomplished utilizing ENVI (v.5.5) and SNAP (v.7.0) software (L3Harris Geospatial Solutions, Pearl East Circle Boulder Co, Boulder, CO, USA and European Space Agency, respectively), while the GIS-based analysis was made utilizing ArcGIS (version 10.7, Environmental Systems Research Institute-ESRI, Redlands, CA, USA).

3.2. *Flood Event—February 2015*

The meteorologically unstable period with extreme rainfalls of the studied event concerns the period from 23 January to 15 February 2015. Especially, on 31 January, the meteorological station of Makrakomi, provided the most reliable data, as it is located in the western, upstream part of the basin (Figure 1a). The main storm started at 6:00 a.m. on the 31st of January and finished twenty-two hours later, on February 1st. In this length of time, the total height of rainfall was 59 mm when the average height of rain output for January in the area is about 110 mm. In total, for the entire period under study, the meteorological stations of Makrakomi and Lamia (at the city of Lamia) recorded 101.6 mm and 45.2 mm, respectively [60]. According to the meteorological data analysis of the National Observatory of Athens (NOA), the main rainfall peak took place on 31 January, with a maximum recording of 7.0 mm during 9:00–12:00 a.m. at the meteorological station of Makrakomi (3 h recording, cumulative) and 5.4 mm during 6:00–9:00 a.m. at the station of Lamia (http://www.iersd.noa.gr/GR/services1_gr.html#weather4).

Evaluating the analyzed data and considering the hydrometeorological features of the basin [61], the event is characterized as a storm rain episode with a return period of only 2–5 years, which is in contrast with the observed flood. Nevertheless, it must be noted that there is a considerable level of uncertainty regarding the spatial variability of rainfall due to the lack of meteorological stations at the higher parts of the watershed. For this reason, in this study, indirect estimation of the recurrence interval of the obtained flood extent and depth was attempted by comparing it with different flood flow scenarios simulated through hydraulic modelling.

3.3. *Methodology*

The basic steps of the methodology applied are presented in the flowchart of Figure 4, and comprehensively analyzed in the following sections.

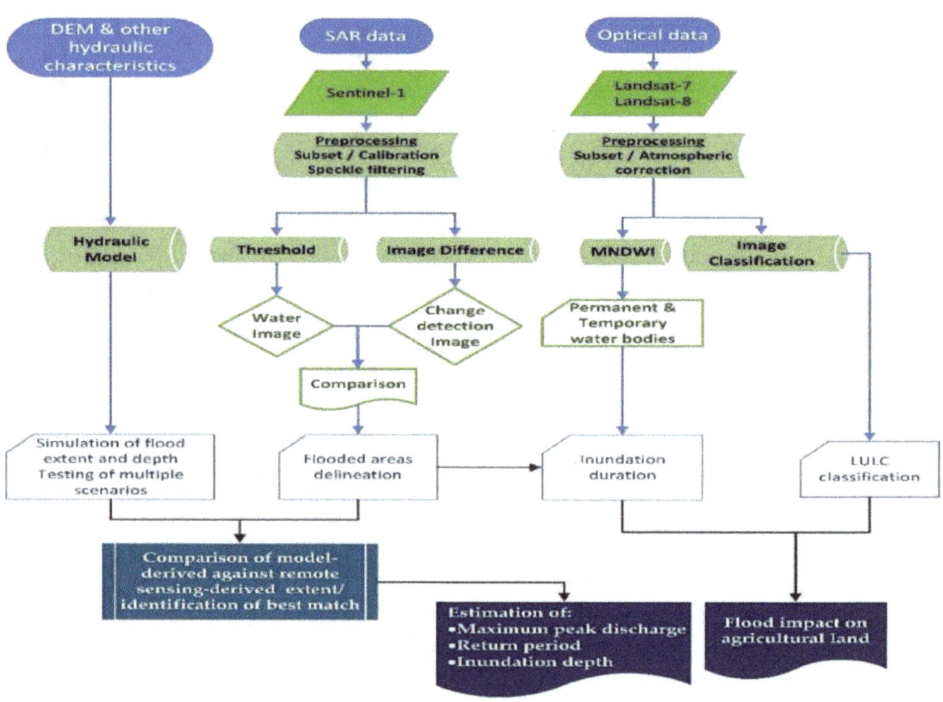

Figure 4. Methodology flowchart.

3.3.1. Optical Data Processing

Optical data acquired by Landsat satellites (7 and 8) have related sensor observation characteristics, due to their similar technical and orbital characteristics. Nevertheless, some basic pre-processing steps are always necessary for each image to enhance the quality and accuracy of the finally used products. For instance, the interfering atmosphere among the satellite and the Earth's surface can reduce the quality of the acquired data [13]. As atmospheric conditions can differ both spatially and temporally, standardized atmospheric models—such as that provided by the Quick Atmospheric Model (QUAC) of ENVI—are essential for the atmospheric correction of the images [62]. The final preprocessing phase of the L7 ETM+ image includes the gap-filling that derives from the scan line corrector (SLC) failure. Therefore, the Landsat gap-fill plugin of ENVI was applied to recover the data loss.

3.3.2. LULC Classification

Land cover classification based on Landsat images has been applied for the recognition of different LULC categories using a variety of classification methods applied on EO data [63]. Landsat 8 OLI data, that are primarily characterized by enhanced spectral, spatial, and radiometric resolution, were utilized for LULC mapping [64–67]. Four cloud-free, geometrically corrected images, of a one-year growing period (2014–2015, one image for every season), were obtained, to distinguish the different crop types of the area. The images were acquired via the United States Geological Survey portal (https://earthexplorer.usgs.gov/) [13,68,69].

Then, the LULC map created from the supervised classification of Landsat 8 scenes, utilizing the Maximum Likelihood classification technique using 440 ground control samples, which were collected from the farmers' declared parcels (OPEKEPE, Figure 5), field survey and very high-resolution orthophoto maps. This very high accuracy geospatial dataset of the parcels was used to update the accuracy of the classified LULC map in the cultivated area. Then, post-classification filtering was performed to remove high-frequency deviations and misclassified pixels.

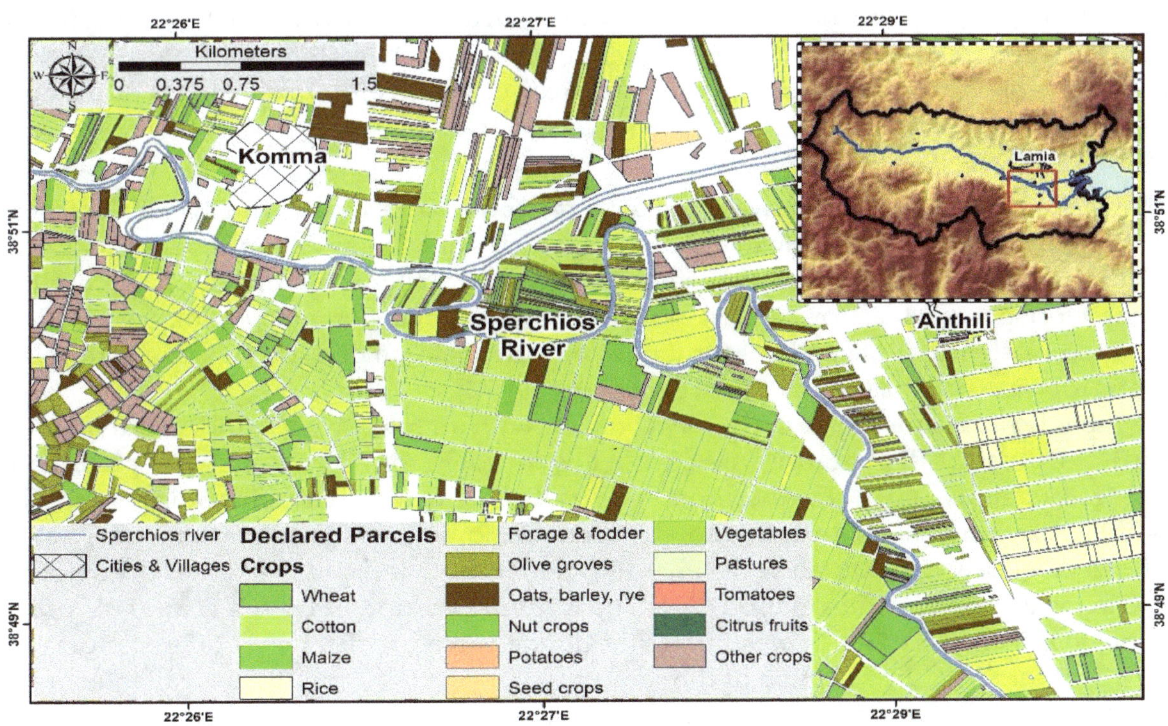

Figure 5. The farmers' declared parcels used for the classification of crop types.

The data samples were divided into two datasets, one for training (70%) and one for validation (30%). Utilizing the training dataset, and twelve classes, a two-step classification process was applied. The initial step comprised an initial broad distinction between general land cover types (water, forest, cultivations, built-up classes, etc.). In the second step, the above general land cover types were additionally classified into summer and winter arable crops, and different kinds of forest categories (evergreen, deciduous, etc.) Finally, post-classification filtering was performed to remove high-frequency deviations and misclassified pixels.

3.3.3. Permanent Water Bodies, Flood Delineation and Inundation Duration Detection

The basic processing step for the precise delineation of flood extent concerns the calculation of a water index using the multispectral optical data. The Modified Normalized Difference Water Index (MNDWI) enhances open water characteristics while suppressing noise from built-up land, vegetation, and soil. The calculation of MNDWI and the associated spectral bands is presented in Equation (1).

$$\text{MNDWI} = (p_{\text{GREEN}} - p_{\text{SWIR}})/(p_{\text{GREEN}} + p_{\text{SWIR}}) \quad (1)$$

where p_{Green} and p_{SWIR} are the reflectance of the green and Short Wave Infrared (SWIR) bands that in case of L8 are band 3 (green) and band 6 (SWIR), while for L7 are band 2 and band 5, respectively.

The MNDWI was calculated for two images (20/12/14 and 13/01/15) before the flood event in order to distinguish the permanent water bodies (small lakes, ponds or marshes) of the area and exclude them from the detected flooded or post-flood inundated areas afterwards. Likewise, it was applied to four images (6/2, 7/2, 14/2 and 15/2) after the flood to detect the inundation duration period and to estimate the possible impact on the crops of the affected areas.

For the delineation and isolation of the water surfaces from the surrounding land cover, the establishment of a suitable threshold is needed. The appropriate adjustment of the threshold, which is descriptive of the conditions that temporally exist in the area under study, can improve the extraction results of the water index [70,71]. The mining of the water parts was made utilizing the raster calculator operation of ArcGIS (script: Con (“raster”> = a, “raster”,a); where a is the threshold point that must be selected for the separation of water surfaces), knowing that MNDW index values for water bodies are greater than 0.2 while all the other surface characteristics are <0.2. Thus, a water cover (WC) polygon is acquired showing only the areas of water, either permanent or inundated.

3.3.4. Sentinel-1 SAR Data Processing

Radar systems transmit microwave electromagnetic radiation to the Earth’s surface and a receiver collecting and strengthening at certain frequencies and polarization modes the echoes of the backscattered signal (backscattering coefficient σ^0) after interacting with the objects depending on the surface characteristics, such as roughness, moisture, vegetation cover, morphology etc. [23]

One of the most crucial parameters for the suitable delineation of a flooded area is the selection of the most appropriate polarization of the radar waves [16,72].

The available radar images for this study come up with an Intensity VV and VH dual polarization. Noise restricts the use of VH because water has a lower radar cross-section in cross-polarization than in co-polarized channels (VV). Furthermore, generally for water surface mapping, co-polarization (VV) has a 0° phase difference and therefore is preferred over cross-polarization (VH), which has a phase difference evenly distributed between −180° and +180°, and thus the latter does not comprise considerable target-specific information [73]. The implementation of both VH and VV polarization revealed that the VH polarization overestimates the flooded areas, while the VV offers a more reasonable depiction [74]. Moreover, the selection of the same polarization (VV) for all the available images offers better results at the following interpretation procedure [11].

The data preprocessing for the radar images followed specific steps using SNAP. The initial step concerns the subset of the images, which reduces the size of the data [75]. Then, a calibration of the images was carried out, comparing the instrument’s measuring accuracy to a known standard. Subsequently, a speckle filter was applied. The suitable selection of the speckle filter is crucial for the delineation of the flooded area. Speckle shows up as a grainy “salt and pepper” texture in an image [28,76,77], due to the random constructive and destructive interference from the multiple scattering returns (individual reflectors) that take place within each resolution cell. Speckle is a type of noise which reduces the quality of an image and makes interpretation more difficult. Thus, it is commonly necessary to diminish speckle impact before interpretation and analysis. The SNAP program provides several types of filters and window sizes that were applied in the present study, such as Lee (3 × 3, 5 × 5 and 7 × 7), Lee Sigma (5 × 5, 7 × 7), Refined Lee and Gamma Map (3 × 3, 5 × 5 and 7 × 7). The Lee filter using a window of 7 × 7 was finally selected since it was verified as the most appropriate for better delineation of the flooded area, avoiding also the overestimation of the water surface that could provide an unfiltered image [23]. In the literature review, lots of studies have also concluded that the Lee filter provides great results, using statistical parameters, such as the mean and standard deviation with a determined window size considering different factors for smoothing.

Ultimately, the σ^0 outputs were terrain corrected by applying the "Range Doppler Terrain Correction" algorithm with the SRTM 1 arc-sec DEM and resampled from 10 to 30 m according to the terrain model resolution and fitting thus to the same resolution of Landsat datasets [11].

Moreover, it must be noted that since the incidence angle variation of the different relative orbits of S1 was small, varying from 34.6 at the near range to 46.1 at the far range, the temporal range of the images was short (1 month) and due to the fact that the study area is relatively flat (plain area), the incidence angle correction was skipped [78].

3.3.5. Flood Monitoring and Inundation Duration Detection

With the goal of detecting the flooded area, two different methods were examined. The first method used was the calibration threshold or binarization technique that uses only the "flood image" and sets a threshold value of radar backscatter signal which is followed by binary band arithmetic to separate whether a given raster pixel is flooded or not [74,79,80]. Thus, in the thresholding technique, the suitable threshold value of the filtered backscatter coefficient was specified manually (ranging from 2.14×10^{-2} to 2.32×10^{-2} sigma value for the processed images) by analyzing the logarithmic histogram of the σ^0. The histogram demonstrates a peak of different magnitude. Low backscatter values are related to the water, and high backscatter values represent the non-water class. In some spots of the image (especially at the south part), the low backscatter values are associated with terrain shadows in the mountainous area which are not related to the flooded areas and were not taken into account. To make the image binarization, band arithmetic was used, putting the logical value "true" for values less than the chosen threshold and "false" for the higher values, creating thus the final "Water" image.

The change detection method deals with the separation of water from non-water areas by comparing one "reference image" with one "flood image", making use of a change detection technique. Thus, the pair of S1 SAR images acquired for the flood mapping comprises an image during the flood, and another one before the flood event, which here is called "Flood Image-FI", and "Reference Image-RI", respectively. This method compares the same area at two distinct points in time, before (not flooded) and throughout the flood event. The result of this image-to-image difference comparison is a grayscale image composed of single-band continuous data, that reflects the change. Brighter spots correspond to increased reflectance, showing the areas that are not covered by water, while darker areas correspond to decreased reflectance, representing areas covered by water.

3.3.6. Hydraulic Model Analysis

The one-dimensional (1D) model (HEC-RAS model) was applied to simulate the flood extent of different discharge scenarios in the main channel of the Sperchios river in the segment between the Kompotades town and the outlet. The 1D models have been considered effective in predicting the flood extent [48,81], which was the objective in this application where hydraulic modelling is coupled with remote sensing techniques to indirectly approximate the reoccurrence interval of the flood event and to estimate the inundation depth.

The terrain was represented by a DEM (resolution 5 m) from which the model extracted a sequence of cross-sections upon which it simulated flood flow and estimated flood extent and depth. The topographic data upon which the cross-section data were based, were derived by the integration of HEC-GeoRAS in a Geographic Information System (GIS) environment. HEC-RAS determines water surface elevation at cross-sections set by the user at representative locations across the floodplain, by a solution of the energy and energy head loss equations and simulates the depth between the cross-sections using conservation of momentum and friction loss evaluation techniques [82,83]. Then, once the water surface is generated, it subtracts the terrain elevation with the elevation of the water surface to determine water depth across the inundated area. According to the 1D model principles, it was assumed that floodwater flows in the longitudinal direction concerning the main channel. One-dimensional models represent the terrain as a sequence of cross-sections.

The model uses the well-established conservation of mass and conservation of momentum equations for simulating 1-D flow [84]. As input to the model, the channel network and the riverbanks, the cross-section geometry extracted by the DEM, the reach lengths, and hydraulic works and infrastructure data were included. Furthermore, Manning roughness coefficients were selected based on the available literature [85–88] using the landcover datasets of CORINE. For boundary conditions, we used different discharge scenarios until we identified the one closer to the extent found by the RS techniques described above. In this way, despite the lack of discharge data for this event, by repeated trials, it was possible to indirectly identify the magnitude of the flood by assuming that the real maximum discharge during the event was the one that was in a maximum degree of agreement among the trials, with the flood extent as described by the RS techniques. The model estimated both the extent and the depth of floodwaters under these peak discharge scenarios and was plotted and compared with the remote sensing technique results in a GIS environment.

4. Results

4.1. LULC Classification

In the present study, the final LULC map, attained by the Maximum Likelihood classification method, was reclassified into 7 main classes, i.e., (a) coniferous and sclerophyllous forest, (b) broad-leaved forest, (c) tree crops (orchards and olives), (d) build-up areas, (e) summer arable crops, (f) winter arable crops and (g) grasslands-sparsely vegetated areas (Figure 6).

The natural vegetation areas (coniferous-sclerophyllous, broad-leaved forest and grasslands-sparsely vegetated areas) cover the 69.54% (1267.73 km^2) of the basin, mainly in the mountainous parts, while the crops occupy primarily the central plain area of the basin. The summer crops occupy the greater part of the plain area covering the 14.35% (261.59 km^2) of the basin and 52.97% of the total cultivated land. The rest of the cultivated area is occupied by winter crops-bare soils (24.54%) and olive orchards (22.49%).

The accuracy assessment of the classified LULC map was achieved utilizing the second validation set. The outcomes were evaluated using overall accuracy (OA), and the Kappa coefficient [89,90]. The classification OA and Kappa coefficient were estimated at 89.4% (118 correctly classified parcels out of 132 reference samples) and 0.82, respectively, indicating a highly accurate result, especially in the plain cultivated area, where the OA was calculated at 94.1% (79 correct out of 84 reference samples).

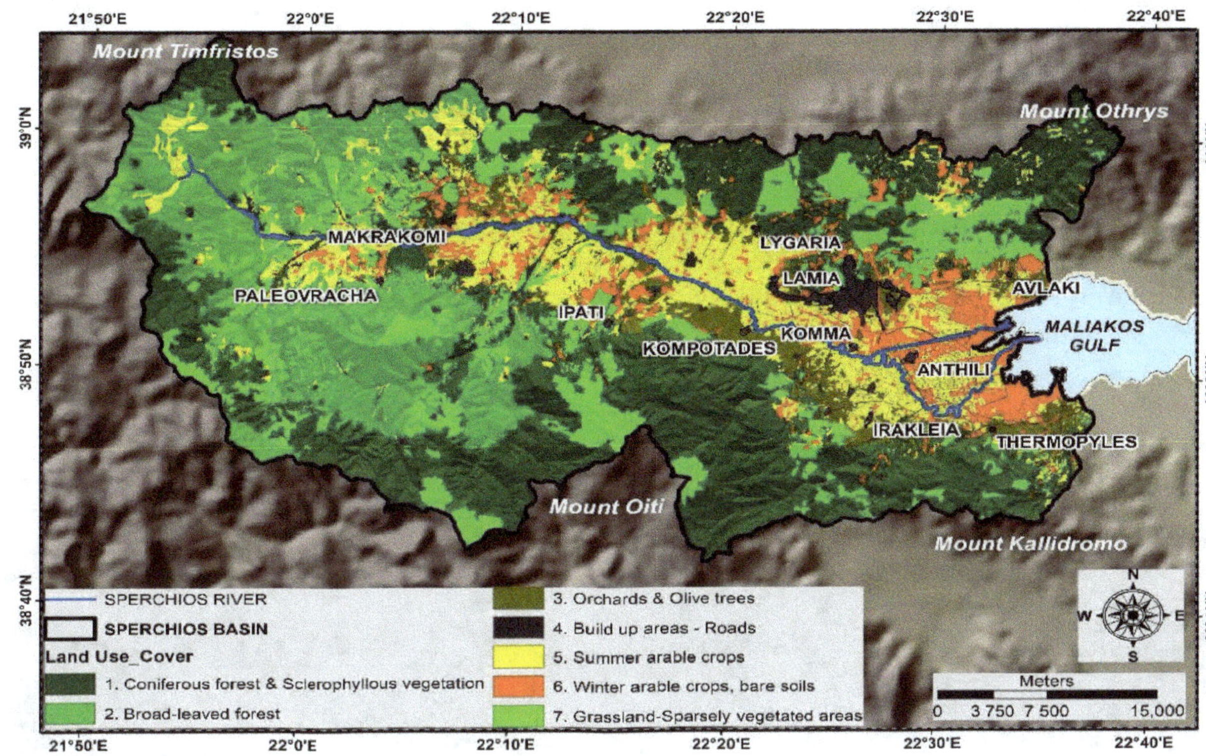

Figure 6. Land use/cover classified map.

4.2. Flooded Area Detection and Flood Duration Monitoring

4.2.1. Permanent Water Detection

For the delineation of the permanent water bodies or marshes in the area, optical data acquired before the flood event were used. The permanent water areas were estimated at 1.93 km^2 (Figure 7a,b). The comparison displays a very high coincide of the two maps, especially at the coastal estuary area, the north branch of the Sperchios river and the small lake situated at the place of the construction of the river proportional divider (rpd) next to the Komma village where the north (new) branch of the Sperchios river initiates. The rpd was created in the first place in 1956 as a flood protection project and rebuilt in 2007 in parallel with the construction of the new National highway and the new National railway that crosses the river close to the village of Anthili (Figure 1).

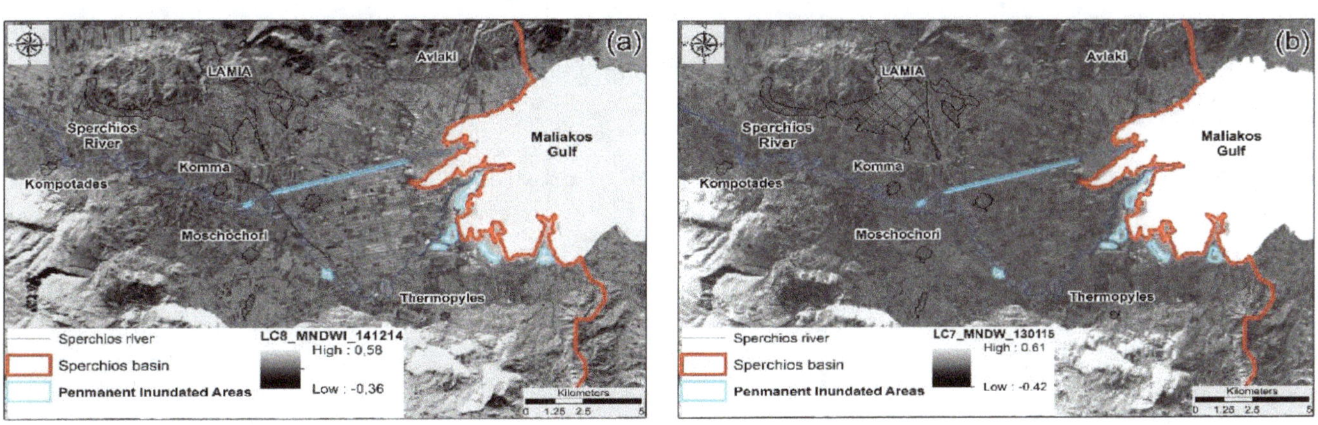

Figure 7. The permanent water area delineation using the Modified Normalized Difference Water Index derived from Landsat 7 and Landsat 8 images before the flood event (**a**) L8_20 December 2014, (**b**) L7_13 January 2015.

4.2.2. Flooded Extent Detection

The flood extent map was extracted by comparing and evaluating the two different methods, threshold calibration and change detection. These methods are two of the most frequently used in active RS to separate flooded area in a radar image [11,17,25,74,91–93].

The first method uses a threshold value of radar backscatter (calculated as a function of the incident angle of the sensor and digital number) [25,26], which is set in decibels (dB), and a binary algorithm is followed to define whether an area is flooded. The result is an image where the flooded area appears in white, while all the other images are presented in black. The total flood extent was calculated as 32 km^2 (Figure 8).

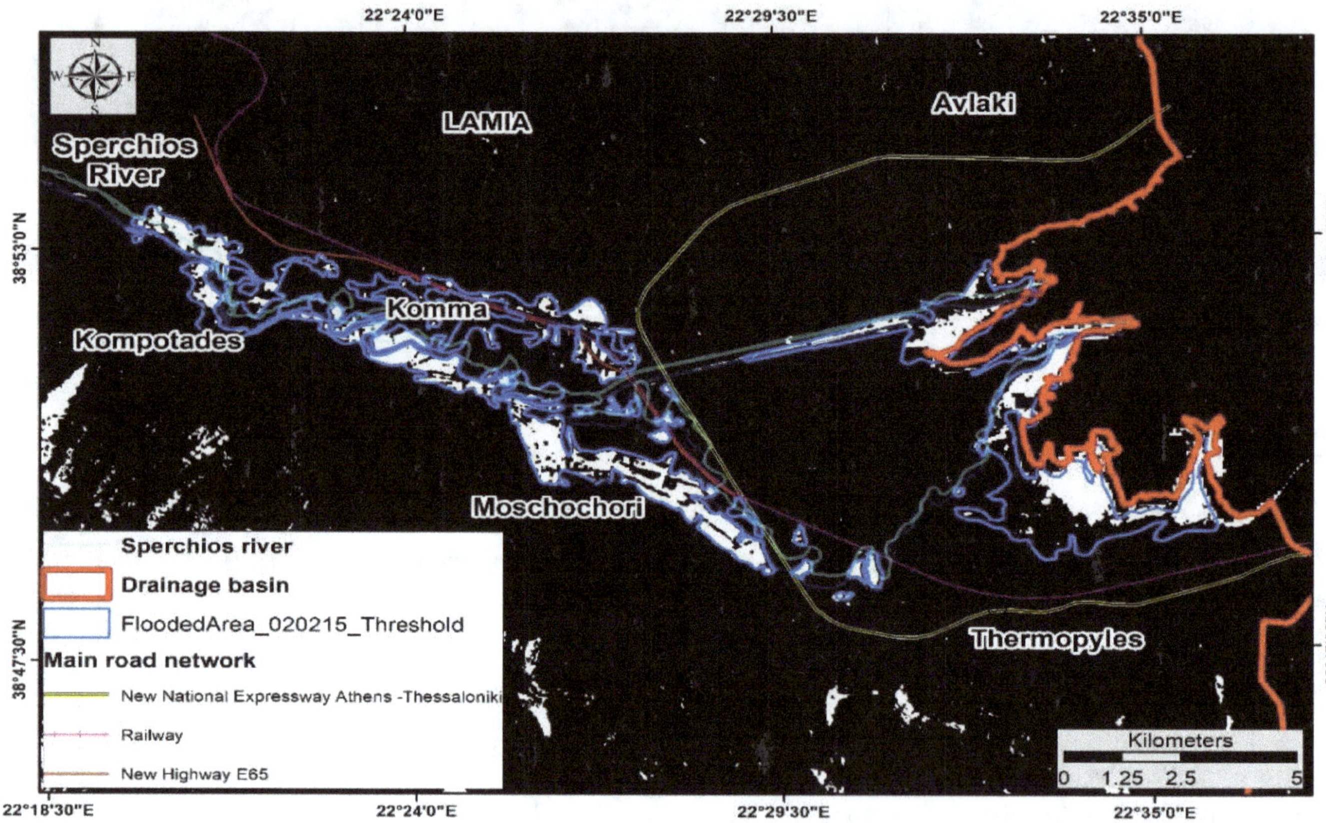

Figure 8. The flooded area mapping utilizing the first method (using threshold) of Flood Image (FI) processing.

The change detection method that uses the image differencing approach estimates the flooded areas by observing the radar backscatter considerable coefficient decline before and after the flood (Figure 9) [26,93]. It delivers a change detection image where the inundated area appears in the dark grey shade, while all the other image parts appear in light grey shades. The total flood extent was calculated as 33.4 km^2 (3340 hectares).

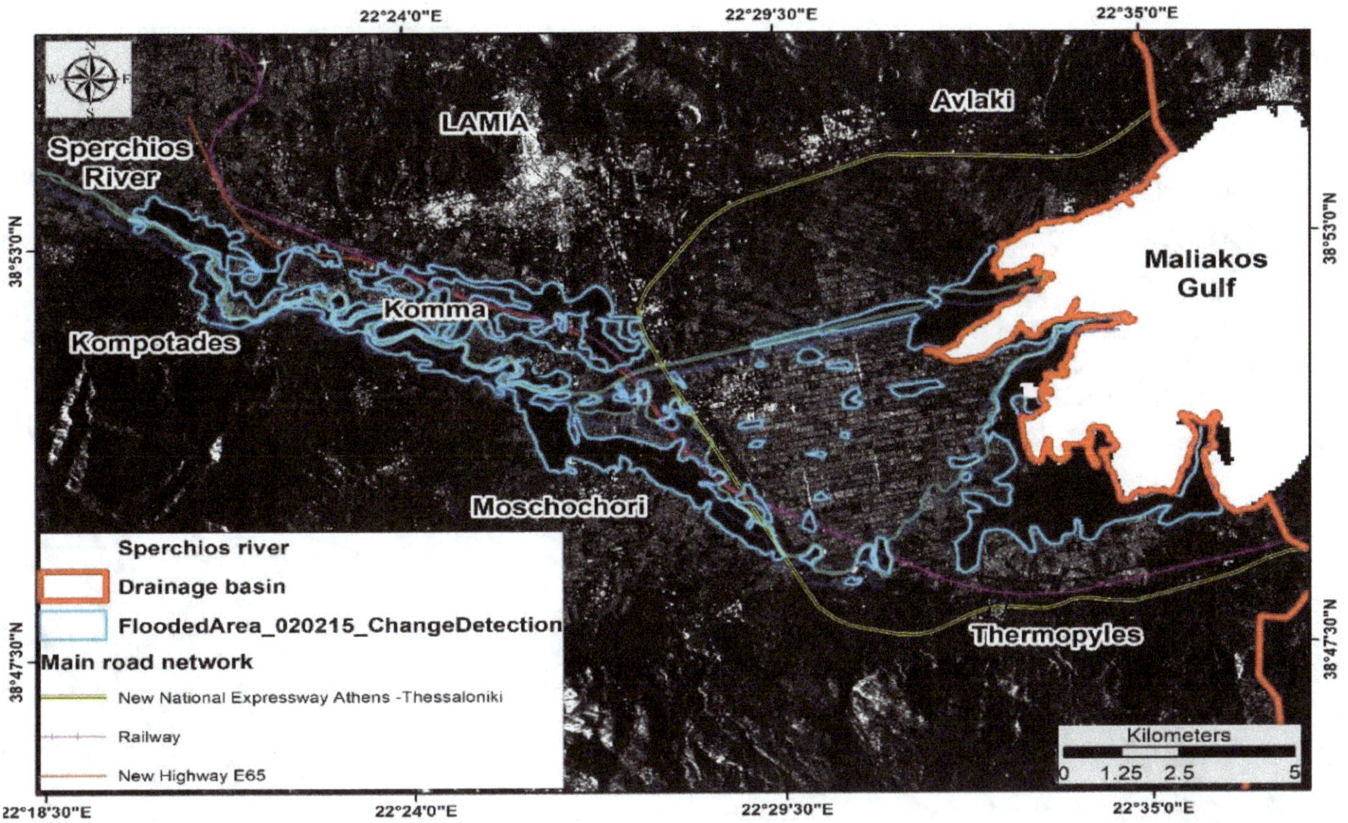

Figure 9. The flooded area mapping utilizing the second method (image difference) of Reference and Flood Images (FI, RI) processing.

The final flooded area was obtained after a thorough analysis and interpretation of the results derived from the two methods, taking into consideration the estimated flooded areas obtained from the image two days after the flood, on 3rd February and subtracting the permanent areas with water.

4.2.3. Inundation Duration

Through the analysis of Landsat images, it was indicated that those acquired on 6th and 14th of February should be excluded since they do not provide reliable results due to the extensive cloud cover. Besides, the low quality of the image acquired on 7th February restricts the proper interpretation (Figure 10). From the evaluation of the rest of the Landsat and the S1 processed images, the inundated areas were estimated approximately equal to 32.62 km^2 on 3rd February (S1), 10.78 km^2 on 8th February (S1), 5.70 km^2 on 15th February (L8, S1) and 4.75 on 26th February (S1) (Figure 10).

The fact that the inundated area on 3rd February is larger than the flooded area on 2nd February reveals that large amounts of precipitation or snowmelt likely took place in the mountainous parts, which along with the concentration time of the basin and the prolonged rainfall period led to the continuous water concentration in the coastal zone for an extended period. The long inundation duration can cause suffocating conditions on existing crops, resulting thus in greater damage.

Total inundation duration was computed for each pixel, by intersecting the four final inundation maps of the four selected dates. Consequently, a combined map was extracted showing the duration of the inundated surfaces (Figure 11). A basic prerequisite for the computation process was that the days without satellite observation between two acquired days were assumed that remained flooded [41]. From this map, it was estimated that the inundated areas one (1) day (3/2/15) after the flood event were 34.1 km^2, in six (6) days (8/2/15) were 9.33 km^2, in thirteen (13) days (15/2/15) were 5.09 km^2 and in twenty-four (24) days 4.01 km^2.

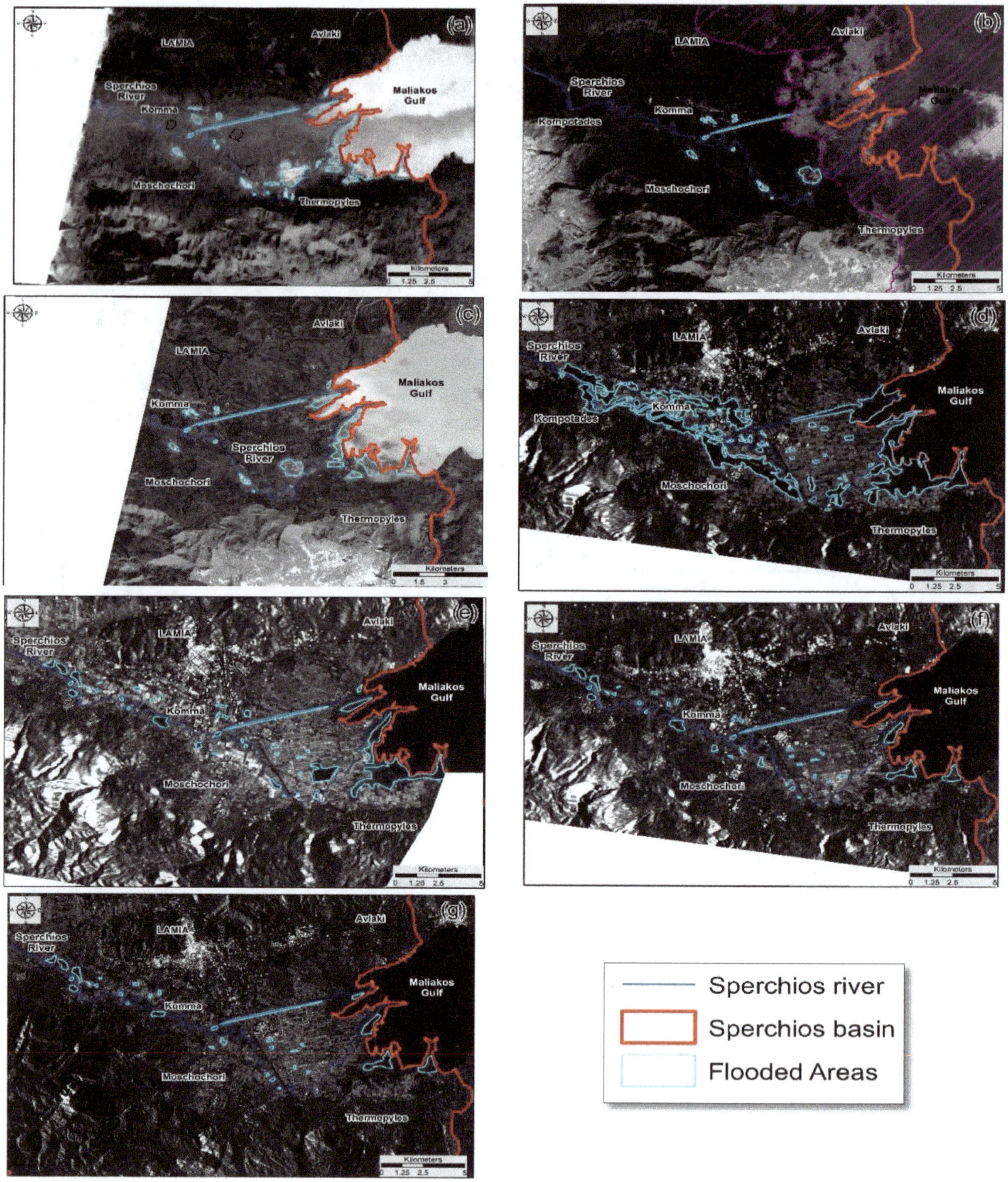

Figure 10. Inundated duration delineation after the flood event using the Modified Normalized Difference Water Index derived from Landsat 7 and Landsat 8 images: (**a**) Landsat7_07 February 2015, (**b**) Landsat7_14 February 2015, (**c**) Landsat8_15 February 2015, as well as Sentinel-1 images: (**d**) S1_03 February 2015, (**e**) S1_08 February 2015, (**f**) S1_15 February 2015 and (**g**) S1_26 February 2015.

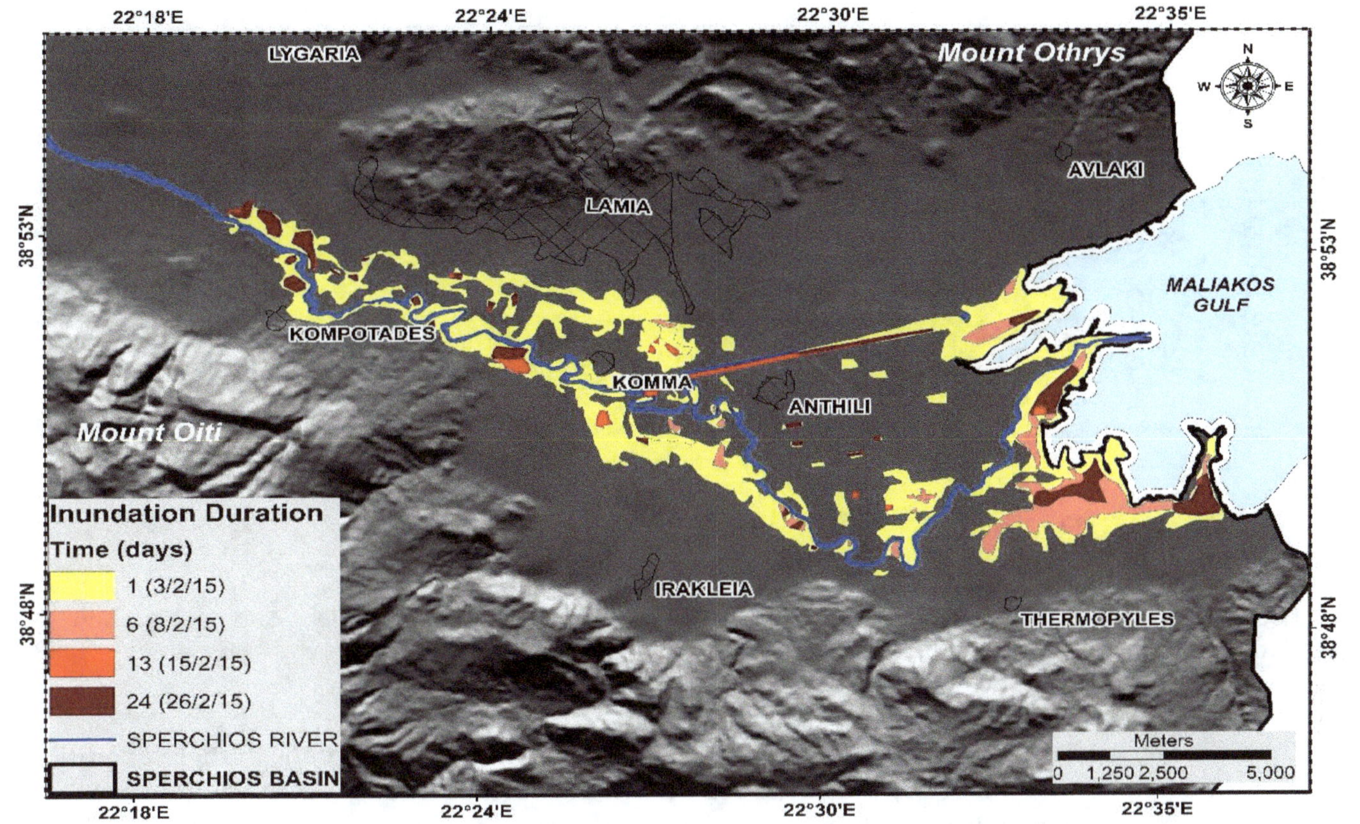

Figure 11. The inundation duration map, showing the inundated areas in one, six, thirteen and twenty-four days after the flood event.

It must be noted that the accuracy of the estimated inundation duration for the various regions does not only depend on the accuracy of the methodology but is also strongly influenced by the satellite data frequency and distribution [41]. This is worsened by satellite crossings that occasionally do not cover the full extent of the study area, potentially miscalculating the full water surface coverage [74].

4.3. Impact on the Cultivated Areas

Aiming to investigate the spatio-temporal flood impacts at fine spatial scales, the overlapping of the temporal recorded inundated areas on the classified land cover map was accomplished. The results show that 56.61% of the LULC flooded were summer crops, 34.57% winter crops and 1.29% olives-orchards, while 7.37% and 0.16% were built-up areas and riparian forests, respectively.

Considering that the flood took place during winter, the most susceptible cultivations to damage were vegetables, forage crops and cereals, which were present in the fields during the flood. Regarding the 11.55 km^2 of the flooded winter crops (mainly cereals, fodder crops and vegetables), 4.09 km^2 (35.41%) were inundated for one day, 3.18 km^2 (27.45%) for six days, 2.26 km^2 (19.57% or 226 ha) for thirteen days and 2.03 km^2 (17.58% or 203 ha) for twenty-four days (Figure 12). A very small part of the flooded area was covered by olive groves and fruit trees (0.14 ha).

According to the Greek Agricultural Insurance Organization (ELGA) statistical data, a total of 77.3 ha of crops were damaged and were compensated and concerned 74 ha of cereals and forage crops and 3.3 ha of vegetables. It should be noted that ELGA regulations dictate that compensations are only paid under specific circumstances, for example when the total damaged production is greater than 20%, the total annual compensation of each parcel does not exceed 80%, etc.

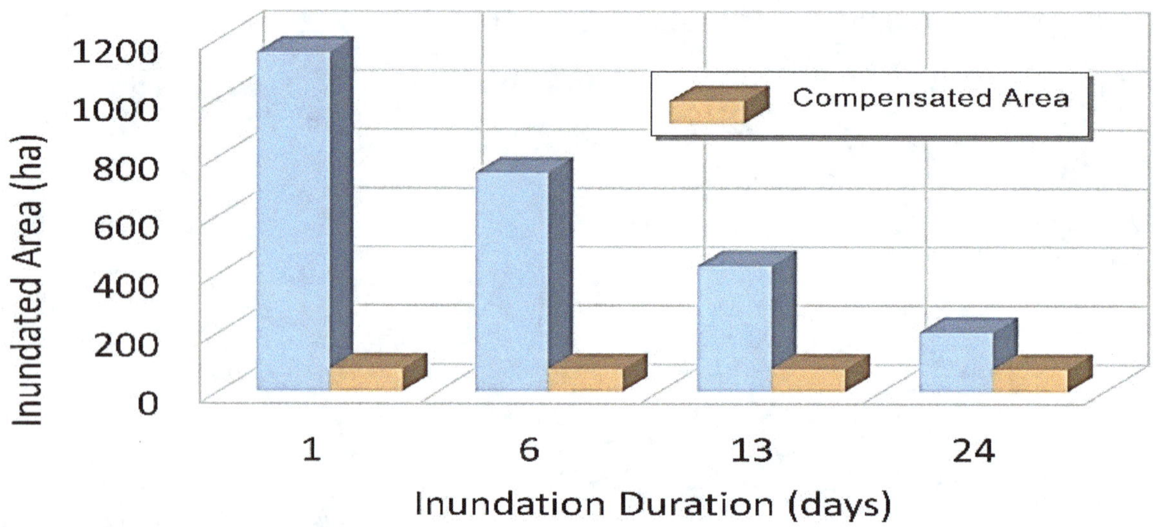

Figure 12. Affected cultivated areas with winter crops, forage crops and vegetables against inundation duration and the compensated area.

Crops are mostly affected by water-logged conditions due to lack of oxygen. In these conditions, the amount of oxygen available to plant tissues below the surface of the water level decreases as plants and microorganisms use up what is available. Furthermore, one of the most important factors in oxygen depletion is the temperature, with higher temperatures leading to higher rates of oxygen depletion [94]. Considering that this flood occurred during winter, it is expected that crops can tolerate flood conditions for a more prolonged period. According to Salamin [95], the expected damage for cereals for various inundation durations during February is 5% for 7 days, 10% for 11 days and 20% for 15 days. In the case of forage crops, it is 5% for 11 days and 10% for 15 days. Accordingly, considering the temporal dynamics of the flood (Figures 11 and 12), the inundated area for more than 15 days obtained with the used methodology is comparable to the area calculated and compensated by ELGA based on field surveys and farmers' declarations. The observed differences are reasonable considering the different sensitivity of each crop to temporal inundation as well as that the expected damage in cereals is near the damage limit covered by ELGO (greater than 20%).

4.4. Hydraulic Model Results and Comparison with Remote Sensing Outcomes

From the testing of different discharge scenarios in the hydraulic model, the discharge value that shows the highest degree of agreement with the satellite-derived flood extent was identified (Figure 13). The discharge value was estimated at 1100 m^3/s, although exact matching of the two extents was impossible in any of the discharge scenarios, probably due to small differences in channel and surface geometry and flow conditions between the real and the simulated world. For the above-estimated peak discharge value, the model calculated the flood depth across the inundated area with maximum values of 5.5 m, as presented in Figure 14.

Based on the discharge data of the area, the discharge value found to be in agreement with the satellite-derived flood extent is estimated to have a return period of approximately eight years [96]. Moreover, the created flood inundation map of the basin using the HEC-RAS model, as compared with the results of the S1 data, demonstrates that the simulation is currently acceptable regarding the flood extent detection predictive ability, giving adequate results in terms of simulating flood events.

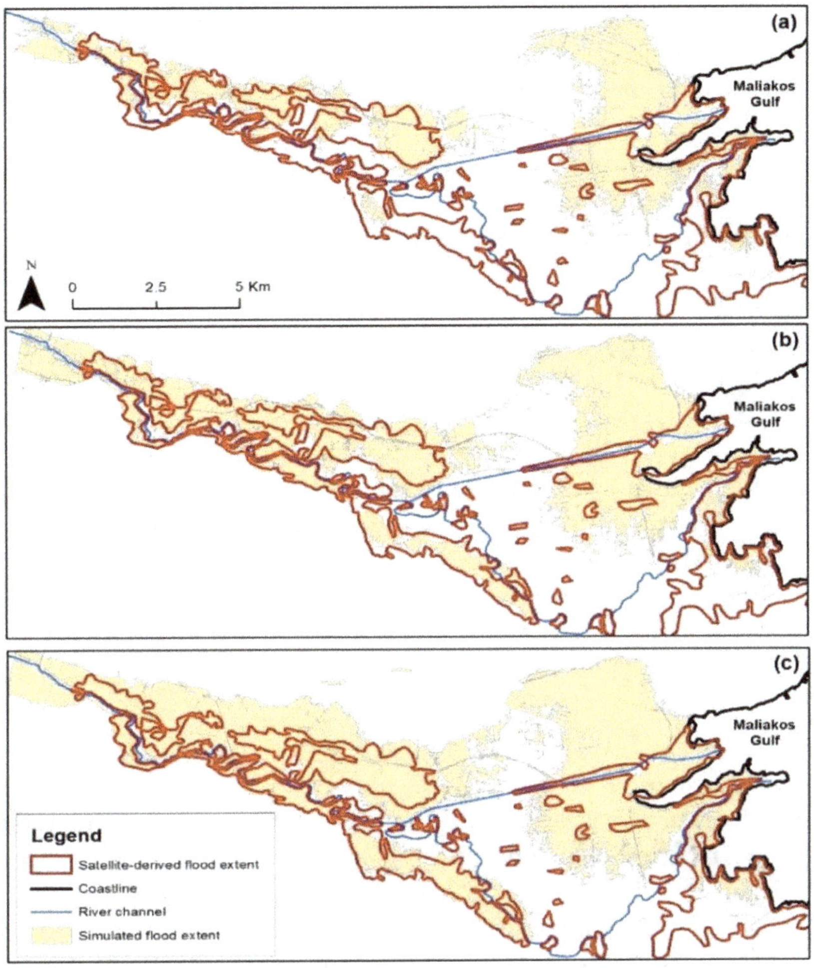

Figure 13. Maps of the study area comparing the flood extent in different discharge scenarios indicatively of (**a**) 950 m^3/s, (**b**) 1100 m^3/s and (**c**) 1250 m^3/s, as simulated by HEC-RAS and plotted against the satellite-derived flood extent. The red line represents the satellite-derived flood extent.

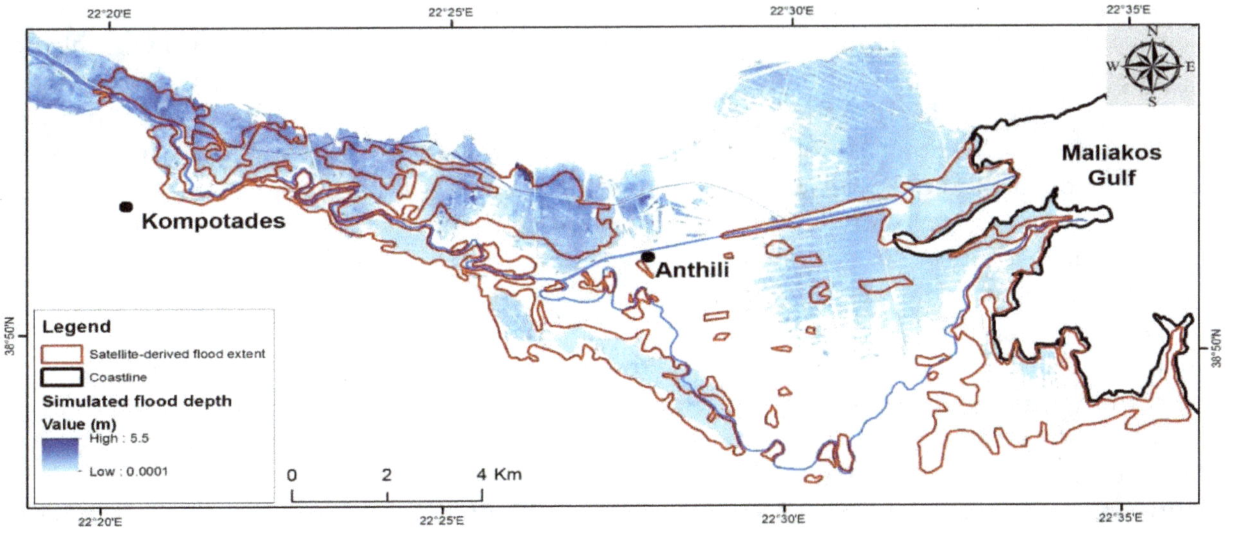

Figure 14. Map of the study area illustrating the extent of the simulated discharge scenario that agrees to the highest degree with the satellite-derived extent. The grading of blue colors indicates floodwater depth at this discharge value.

However, as can be seen in Figures 13 and 14, near the outlet of the river (eastern part of the maps), it was found that the model-derived extent was significantly more extended than the satellite-derived one in all scenarios tested. This can be attributed to the fact that these low-lying lands are artificially drained by a complicated system of culverts as they were naturally flooded in wet seasons and remained waterlogged for long time periods. Even if a very high-resolution DEM derived from the stereo pairs of very high accuracy photogrammetric aerial images was used in this study, difficulties can be still encountered in handling man-made hydraulic features such as ditches, canals, farm dams, constructed banks and stormwater infrastructure that can change natural drainage flow paths across the landscape [97]. Specially developed algorithms and manual editing of DEMs to consider these features have been proposed by many researchers [98,99]; however, artificial drainage flows through underground stormwater infrastructure still limit DEM-based hydrological analysis in situations that the topological relations among spatial characteristics are difficult to trace [97].

The huge effort required for the realistic representation of flow paths with hydraulic models in such situations highlights the importance of RS approaches in flood crisis management.

Additionally, the discrepancies that were observed in the southeastern part of the affected area, where the model underestimates the flood extent in comparison to RS results (Figures 13 and 14), could be the result of lateral flows contributed by small torrents in that area, which are not considered by the model.

5. Discussion

In this study, Sentinel-1 radar and Landsat images were utilized in conjunction with hydraulic modelling to obtain flood characteristics and land use/cover (LULC) to assess flood impact in agricultural areas. The findings demonstrate a process that allows accurate flood mapping and contributes to a better understanding of flood metrics, including flood depth, peak discharge, recurrence interval, inundation depth and duration, which are all crucial information in flood risk management in ungauged basins.

Natural disasters, and especially floods, can cause great damage to agriculture, which is one of the most crucial economic sectors and is also one of the most affected by extreme events [100]. Crop growth necessitates water; nevertheless, excess water occurring during flood events and long duration inundation is sometimes negative or even harmful. Several previous studies have described various adverse effects of flooding on crops, which vary with flood seasonality, crop type, growth stage and environmental conditions, when flooding occurs and according to the inundation duration and severity [100,101]. In the Sperchios plain area, the crop systems mainly include summer crops such as cotton, corn and rice, fewer winter crops such as cereals (wheat, oat etc.), fodders and vegetables, and a small number of olives, orchards and vineyards, and accordingly, flood impact was found to be limited.

The obtained results of this study revealed that the proper estimation of inundation duration is also critical for the evaluation of flood impacts on agriculture. A key issue that is yet to be determined is related to the expansion of RS observation frequency to an extent that would allow us to record and eventually to follow the evolution of accidental flood phenomena. For that purpose, dense acquisitions at a high spatial resolution (20, 10 m or better) are necessary. Although the EO capacity has been constantly increasing over the last two decades, at the moment, unless initiatives like the International Charter are activated, the chances of capturing flood events at their peak with high-resolution EO data are quite low. In this direction, current and near-future SAR, and optical satellite missions are expected to significantly enhance the efficiency by which floods are examined utilizing both systematic and on-demand EO data acquisitions. This would be accomplished through improved repeat cycles, finer spatial resolution, as well as polarimetric and interferometric capabilities.

Multi-seasonal optical data are unique for a detailed LULC classification. The LULC and especially crop type supervised classification deriving from multitemporal L8 data has been proved as a very accurate and reliable way in large cultivated areas from a lot of studies, such as those of Mtibaa and Irie [102], de Oliveira Santos et al. [103], Schultz et al. [104] and Murthy et al. [105]. The above findings are in agreement with the highly accurate LULC classification achieved in the present study. Besides, it must be noted that the present-day availability of the recent open-access Sentinel-2 satellites have already give prospects for the increase of the LULC classification accuracy, due to its highest spatial (10 m), spectral (four red edge bands) and temporal characteristics (5 days) [106,107].

The adaptation of a multi-sensor integration approach increases the likelihood of the imaging flood extent when cloud contamination is low [20]. Multi-source methods and multitemporal data are an encouraging research path worthy of evaluation. The synergistic use of optical and radar data can be crucial for flood mapping and inundation duration detection applications [108]. The MNDWI, which provided adequate results in this study, has also been proven sufficient in several studies in the past, such as those of Psomiadis et al. [13], Conde and Muñoz [23], Kyriou and Nikolakopoulos [109] for flood events occurring in Greece and Clement et al. [74], Singh et al. [110], Baig et al. [111], in other flood events and waterlogged areas. As well as water body detection, it is exceptionally suitable for the delineation and monitoring of water surfaces, since it highlights the strong absorption of SWIR radiation by water bodies. Moreover, the MNDWI is a better index than NDWI for extracting water features mixed with vegetation (depth of standing water varied from 60 to 75 cm) from the satellite image [110,112].

Radar systems are more valuable in providing data and observations during floods, which mostly occur on cloudy days. Cloud cover was found to be an important obstacle in the use of optical RS data for the delineation of the flooded area in this study. With the new opportunities in spatial resolution, polarimetry, wavelength, and other features provided by the new available SAR sensors, new aspects are emerging on space-based flood monitoring. From the evaluation of the results, it can be concluded that from the implementation of the two techniques (threshold and change detection), the flooded areas had nearly the same size and the differences among the two maps are quite insignificant, especially at the western riparian area. Nevertheless, it must be noted that the change detection method better highlighted the small, flooded areas near the river delta.

Two crucial factors for the accurate delineation of a flooded area are the presence of speckle and polarization. A significant drawback of SAR images is the existence of speckle, a signal-dependent granular noise, that visually degrades the image appearance and hinders its interpretation. A firmly increasing number of papers investigating the despeckling filter selection has appeared in the literature over the last two decades, e.g., Conde and Muñoz [23], Rana et al. [77], Argenti et al. [72], Medeiros et al. [113], Xiao et al. [114] and Kushwaha et al. [115]. The speckle filtering implementation does not change the spatial detail and it is ordinarily applied to attain smoothing of the limits of the different forms represented, avoiding the loss of image details. Sentinel-1 SAR data provide VV and VH polarizations in which VV backscatter offer better delineation of water as compared with VH backscatter, an outcome that has been observed from several previous studies, such as those of Clement et al. [74], Agnihotri et al. [116] and Tsyganskaya et al. [117]. As it has resulted from the present study, following the previous ones, the VH polarization overestimates the flooded areas, while the VV offers a more reasonable depiction, especially when Sentinel-1 data are used [74]. Additionally, it is notable that the C-band wavelength is too short to decompose the accurate scattering processes of highly vegetated zones where canopies are dense. Future efforts are required towards the investigation of the multitemporal synergistic studies between Sentinel-2 and X- or L-band SAR EO systems, to explore whether the issue of dense canopy penetration can be overcome [118].

The synergistic use of 1D HEC-RAS hydraulic modelling (based on high-resolution DEM), RS data and GIS analysis provided relevant and valuable results that can contribute to flood detection and to estimate flood metrics including peak discharge and return period, through reverse estimations. Indirect estimation of the flood recurrence interval can also be a valuable aspect for future flood management efforts, particularly in areas with a scarcity of instrumental hydrological data. In this study, we selected intentionally specific RS and hydraulic modelling tools that are available in a large part of the world, to increase the applicability of the approach applied. Numerous previous studies have also reached similar conclusions, adopting RS data and techniques, GIS and DEM to assess the flood characteristics, such as those of Liang and Liu [119], Huang et al. [120], Cian et al. [121] and Psomiadis et al. [13]

Besides, Khalfallah and Saidi [122] and Qi and Altinakar [123] used Geographical Information System (GIS)-based tools and HEC-RAS for integrated flood management, while Ezzine et al. [124], Gkouma et al. [125] and Quang et al. [126] utilized mainly Sentinel-1 and other SAR imagery and hydraulic modelling for flood mapping.

Topographic datasets and accurate DEM perform a significant role in hydraulic modelling and the correct prediction of flood inundation areas [127]. Therefore, particularly for local-scale events, the prospect of using in the future more systematically high-resolution elevation data derived from Unmanned Aerial Vehicles (UAV) and/or laser scanning (light detection and ranging (LIDAR)) aerial photogrammetry would provide more accurately DEM products and will improve the estimation of flood extent and depth [20,128]. However, it should be mentioned that artificial drainage flows through underground stormwater infrastructure can still limit DEM-based hydrological analysis in situations that the topological relationships among spatial features are difficult to trace as seems to be the case in specific places of the current study area [97]. It is also essential to be noted the fact that the accuracy of hydraulic modelling predictions can probably be improved by considering other detailed information offered by 2D flood modelling, such as flood velocity and duration, except flood depth.

Besides this, more accurate hydraulic modelling outcomes can be useful for the determination of an optimal threshold of the SAR images, which sometimes it is difficult to be approached.

The hydraulic model estimation of the approximately eight-year flood return period is in agreement with the frequent and continuous flood events at the plain area of the Sperchios basin [54]. The multitemporal and multi-seasonal RS data has proved that it can deliver an integrated monitoring approach of LULC and flood events, necessary for the post-flood inundation duration detection and its impact on crops. These detection results could lead stakeholders and decision-makers to locate the most vulnerable areas and to propose changes in the agricultural management of the area, by utilizing a more reasonable selection of crops and/or more drastic flood prevention structures. A synthesis of existing remote and proximal sensing technologies could provide routine, nationwide flood susceptibility under all conditions [20].

Finally, it is noteworthy that the rainfall amounts between 31/1/15 and 1/2/15 do not constitute a very rare amount of rainfall. Nevertheless, it has to be stressed that due to the size of the catchment, the rainfall measurements through rain gauges are not representative of the entire area of the watershed. This fact shows that a large amount of precipitation probably took place in the mountainous parts of the catchment that was not recorded by the meteorological stations of Makrakomi and Lamia located at lower altitudes. Moreover, due to the large size of the Sperchios drainage basin, the water deriving from precipitation or snow melting from the mountainous part needs a longer time to reach its outflow point [54].

6. Conclusions

The present study highlights the advantages of high-resolution multisource RS imagery and hydraulic modelling for flood extent and depth mapping. Furthermore, it evaluates the ability of multitemporal RS data to support the assessment of flood impact on agricultural areas through the estimation of the inundation duration.

The results confirm that the Sentinel-1 mission is capable of providing adequate spatial and temporal resolution data, from which high accuracy flood maps can be extracted. Thus, S1 data, along with optical data when it is possible according to the weather conditions, are able to enhance the capacity to map the extent and the temporal dynamics of each flood more accurately, supporting improved assessments of flood impact on agricultural areas. However, the current analysis indicated that the chances of capturing a flood event at its peak and to follow its evolution with high spatial and temporal resolution EO data are still quite low. The obtained results reveal that the proper estimation of inundation duration is critical for the evaluation of flood impacts on agriculture. Accordingly, improved repeat cycles and finer spatial resolution RS observations may further improve flood monitoring.

Both processing methods of radar data, thresholding and image differencing, implemented for the inundated area mapping appeared to have adequate and similar results. Furthermore, the multitemporal optical and radar data deliver unique information for permanent water cover, LULC mapping and inundation duration monitoring, which can delineate ideally the flood event impact on the crops of the area. Therefore, the obtained results can provide a better insight into future flood risk reduction planning in the study area and act as an example for similar studies in other similar regions.

HEC-RAS 1D hydraulic modelling was also proved as a useful tool when combined with RS for studying and understanding key flood metrics in a post-flood event environment. The results of this study indicated that the conjunction of hydraulic modelling with multitemporal RS data contributes to a better understanding of flood characteristics, such as flood extent and depth, recurrence interval, peak discharge and inundation duration. The important implication of this synergistic use of the two techniques is that it shows the unique value in poorly gauged basins. Besides, given the wide availability of the required data, the developed approach can be applied in diverse environments around the world where flood monitoring is poorly developed.

The availability of flood observations is of high importance, mainly during a flood disaster, when decision-makers and relief organizations require a quick overview of the condition and comprehensive insights into the affected area to distribute their resources with maximum efficiency.

Author Contributions: Conceptualization, E.P., M.D. and K.X.S.; methodology, E.P., M.D. and K.X.S.; remote sensing, E.P.; hydraulic modelling, M.D. and K.X.S.; validation, E.P., M.D. and K.X.S., resources, E.P.; data curation, E.P., M.D. and K.X.S., writing—original draft preparation, E.P., M.D. and K.X.S.; supervision, E.P.; All authors have read and agreed to the published version of the manuscript.

Acknowledgments: The authors would like to thank the Greek Agricultural Insurance Organization (ELGA) for providing them with statistical data for flood damage compensations. We would also like to acknowledge the European Space Agency and USGS-NASA for the availability of Sentinel and Landsat data.

References

1. Diakakis, M.; Boufidis, N.; Salanova Grau, J.M.; Andreadakis, E.; Stamos, I. A systematic assessment of the effects of extreme flash floods on transportation infrastructure and circulation: The example of the 2017 Mandra flood. *Int. J. Disaster Risk Reduct.* **2020**, *47*, 101542. [CrossRef]
2. Schlaffer, S.; Matgen, P.; Hollaus, M.; Wagner, W. Flood detection from multi-temporal SAR data using harmonic analysis and change detection. *Int. J. Appl. Earth Obs. Geoinf.* **2015**, *38*, 15–24. [CrossRef]
3. Tsakiris, G. Flood risk assessment: Concepts, modelling, applications. *Nat. Hazards Earth Syst. Sci.* **2014**, *14*, 1361–1369. [CrossRef]
4. Barredo, J.I. Major flood disasters in Europe: 1950-2005. *Nat. Hazards* **2007**, *42*, 125–148. [CrossRef]
5. Diakakis, M.; Priskos, G.; Skordoulis, M. Public perception of flood risk in flash flood prone areas of Eastern Mediterranean: The case of Attica Region in Greece. *Int. J. Disaster Risk Reduct.* **2018**, *28*, 404–413. [CrossRef]
6. Salvati, P.; Petrucci, O.; Rossi, M.; Bianchi, C.; Pasqua, A.A.; Guzzetti, F. Gender, age and circumstances analysis of flood and landslide fatalities in Italy. *Sci. Total Environ.* **2018**, *610–611*, 867–879. [CrossRef]
7. Jonkman, S.N.; Kelman, I. An analysis of the causes and circumstances of flood disaster deaths. *Disasters* **2005**, *29*, 75–97. [CrossRef]
8. Douben, K.-J. Characteristics of river floods and flooding: A global overview, 1985–2003. *Irrig. Drain.* **2006**, *55*, S9–S21. [CrossRef]
9. Halgamuge, M.N.; Nirmalathas, A. Analysis of large flood events: Based on flood data during 1985–2016 in Australia and India. *Int. J. Disaster Risk Reduct.* **2017**, *24*, 1–11. [CrossRef]
10. Diakakis, M.; Mavroulis, S.; Deligiannakis, G. Floods in Greece, a statistical and spatial approach. *Nat. Hazards* **2012**, *62*, 485–500. [CrossRef]
11. Psomiadis, E. Flash flood area mapping utilising Sentinel-1 radar data. In *Proceedings of the SPIE—The International Society for Optical Engineering*; SPIE-Intl Soc Optical Eng: Edinburgh, UK, 2016; Volume 10005.
12. Abdelkareem, M. Targeting flash flood potential areas using remotely sensed data and GIS techniques. *Nat. Hazards* **2017**, *85*, 19–37. [CrossRef]

13. Psomiadis, E.; Soulis, K.; Zoka, M.; Dercas, N. Synergistic Approach of Remote Sensing and GIS Techniques for Flash-Flood Monitoring and Damage Assessment in Thessaly Plain Area, Greece. *Water* **2019**, *11*, 448. [CrossRef]
14. Borga, M.; Anagnostou, E.N.; Blöschl, G.; Creutin, J.D. Flash floods: Observations and analysis of hydro-meteorological controls. *J. Hydrol.* **2010**, *394*, 1–3. [CrossRef]
15. Diakakis, M.; Andreadakis, E.; Nikolopoulos, E.I.; Spyrou, N.I.; Gogou, M.E.; Deligiannakis, G.; Katsetsiadou, N.K.; Antoniadis, Z.; Melaki, M.; Georgakopoulos, A.; et al.. An integrated approach of ground and aerial observations in flash flood disaster investigations. The case of the 2017 Mandra flash flood in Greece. *Int. J. Disaster Risk Reduct.* **2019**, *33*, 290–309. [CrossRef]
16. Klemas, V. Remote Sensing of Floods and Flood-Prone Areas: An Overview. *J. Coast. Res.* **2015**, *314*, 1005–1013. [CrossRef]
17. Brivio, P.A.; Colombo, R.; Maggi, M.; Tomasoni, R. Integration of remote sensing data and GIS for accurate mapping of flooded areas. *Int. J. Remote Sens.* **2002**, *23*, 429–441. [CrossRef]
18. Ramsey, E.; Rangoonwala, A.; Bannister, T. Coastal Flood Inundation Monitoring with Satellite C-band and L-band Synthetic Aperture Radar Data. *JAWRA J. Am. Water Resour. Assoc.* **2013**, *49*, 1239–1260. [CrossRef]
19. Cian, F.; Marconcini, M.; Ceccato, P. Normalized Difference Flood Index for rapid flood mapping: Taking advantage of EO big data. *Remote Sens. Environ.* **2018**, *209*, 712–730. [CrossRef]
20. O'Hara, R.; Green, S.; McCarthy, T. The agricultural impact of the 2015-2016 floods in Ireland as mapped through Sentinel 1 satellite imagery. *Irish J. Agric. Food Res.* **2019**, *58*, 44–65. [CrossRef]
21. Wagner, W.; Freeman, V.; Cao, S.; Matgen, P.; Chini, M.; Salamon, P.; McCormick, N.; Martinis, S.; Bauer-Marschallinger, B.; Navacchi, C.; et al.. Data processing architectures for monitoring floods using Sentinel-1. *ISPRS Ann. Photogramm. Remote Sens. Spat. Inf. Sci.* **2020**, *V-3*, 641–648. [CrossRef]
22. User Guides—Sentinel-1 SAR—Sentinel Online. Available online: https://sentinel.esa.int/web/sentinel/user-guides/sentinel-1-sar (accessed on 25 August 2020).
23. Carreño Conde, F.; De Mata Muñoz, M. Flood Monitoring Based on the Study of Sentinel-1 SAR Images: The Ebro River Case Study. *Water* **2019**, *11*, 2454. [CrossRef]
24. Baghdadi, N.; Bernier, M.; Gauthier, R.; Neeson, I. Evaluation of C-band SAR data for wetlands mapping. *Int. J. Remote Sens.* **2001**, *22*, 71–88. [CrossRef]
25. Manjusree, P.; Prasanna Kumar, L.; Bhatt, C.M.; Rao, G.S.; Bhanumurthy, V. Optimization of threshold ranges for rapid flood inundation mapping by evaluating backscatter profiles of high incidence angle SAR images. *Int. J. Disaster Risk Sci.* **2012**, *3*, 113–122. [CrossRef]
26. Sanyal, J.; Lu, X.X. Application of remote sensing in flood management with special reference to monsoon Asia: A review. *Nat. Hazards* **2004**, *33*, 283–301. [CrossRef]
27. Sun, Z.; Li, X.; Fu, W.; Li, Y.; Tang, D. Long-term effects of land use/land cover change on surface runoff in urban areas of Beijing, China. *J. Appl. Remote Sens.* **2013**, *8*, 084596. [CrossRef]
28. Chatziantoniou, A.; Petropoulos, G.P.; Psomiadis, E. Co-Orbital Sentinel 1 and 2 for LULC mapping with emphasis on wetlands in a mediterranean setting based on machine learning. *Remote Sens.* **2017**, *9*. [CrossRef]
29. Souza, C.M.; Shimbo, J.Z.; Rosa, M.R.; Parente, L.L.; Alencar, A.A.; Rudorff, B.F.T.; Hasenack, H.; Matsumoto, M.; Ferreira, L.G.; Souza-Filho, P.W.M.; et al.. Reconstructing Three Decades of Land Use and Land Cover Changes in Brazilian Biomes with Landsat Archive and Earth Engine. *Remote Sens.* **2020**, *12*, 2735. [CrossRef]
30. Mohamed, M.; Anders, J.; Schneider, C. Monitoring of Changes in Land Use/Land Cover in Syria from 2010 to 2018 Using Multitemporal Landsat Imagery and GIS. *Land* **2020**, *9*, 226. [CrossRef]
31. Xu, Y.; Zhang, L.M. Breaching Parameters for Earth and Rockfill Dams. *J. Geotech. Geoenvironmental Eng.* **2009**, *135*, 1957–1970. [CrossRef]
32. Psomiadis, E.; Soulis, K.X.; Efthimiou, N. Using SCS-CN and earth observation for the comparative assessment of the hydrological effect of gradual and abrupt spatiotemporal land cover changes. *Water (Switzerland)* **2020**, *12*. [CrossRef]
33. Qi, S.; Brown, D.G.; Tian, Q.; Jiang, L.; Zhao, T.; Bergen, K.M. Inundation extent and flood frequency mapping using LANDSAT imagery and digital elevation models. *GISci. Remote Sens.* **2009**, *46*, 101–127. [CrossRef]
34. Wang, Y.; Colby, J.D.; Mulcahy, K.A. An efficient method for mapping flood extent in a coastal floodplain using Landsat TM and DEM data. *Int. J. Remote Sens.* **2002**, *23*, 3681–3696. [CrossRef]

35. Yamagata, Y.; Akiyama, T. Flood damage analysis using multitemporal Landsat Thematic Mapper data. *Int. J. Remote Sens.* **1988**, *9*, 503–514. [CrossRef]
36. Pantaleoni, E.; Engel, B.A.; Johannsen, C.J. Identifying agricultural flood damage using Landsat imagery. *Precis. Agric.* **2007**, *8*, 27–36. [CrossRef]
37. Xu, H. Modification of normalised difference water index (NDWI) to enhance open water features in remotely sensed imagery. *Int. J. Remote Sens.* **2006**, *27*, 3025–3033. [CrossRef]
38. Campos, J.C.; Sillero, N.; Brito, J.C. Normalized difference water indexes have dissimilar performances in detecting seasonal and permanent water in the Sahara-Sahel transition zone. *J. Hydrol.* **2012**, *464–465*, 438–446. [CrossRef]
39. McFeeters, S.K. The use of the Normalized Difference Water Index (NDWI) in the delineation of open water features. *Int. J. Remote Sens.* **1996**, *17*, 1425–1432. [CrossRef]
40. Memon, A.A.; Muhammad, S.; Rahman, S.; Haq, M. Flood monitoring and damage assessment using water indices: A case study of Pakistan flood-2012. *Egypt. J. Remote Sens. Sp. Sci.* **2015**, *18*, 99–106. [CrossRef]
41. Rättich, M.; Martinis, S.; Wieland, M. Automatic Flood Duration Estimation Based on Multi-Sensor Satellite Data. *Remote Sens.* **2020**, *12*, 643. [CrossRef]
42. Bhatt, C.M.; Rao, G.S.; Farooq, M.; Manjusree, P.; Shukla, A.; Sharma, S.V.S.P.; Kulkarni, S.S.; Begum, A.; Bhanumurthy, V.; Diwakar, P.G.; et al.. Satellite-based assessment of the catastrophic Jhelum floods of September 2014, Jammu & Kashmir, India. *Geomat. Nat. Hazards Risk* **2017**, *8*, 309–327. [CrossRef]
43. Rahman, M.; Ningsheng, C.; Islam, M.M.; Dewan, A.; Iqbal, J.; Washakh, R.M.A.; Shufeng, T. Flood Susceptibility Assessment in Bangladesh Using Machine Learning and Multi-criteria Decision Analysis. *Earth Syst. Environ.* **2019**, *3*, 585–601. [CrossRef]
44. Anusha, N.; Bharathi, B. Flood detection and flood mapping using multi-temporal synthetic aperture radar and optical data. *Egypt. J. Remote Sens. Sp. Sci.* **2020**, *23*, 207–219. [CrossRef]
45. Diakakis, M.; Deligiannakis, G.; Katsetsiadou, K.; Antoniadis, Z.; Melaki, M. Mapping and classification of direct flood impacts in the complex conditions of an urban environment. The case study of the 2014 flood in Athens, Greece. *Urban Water J.* **2017**, *14*, 1065–1074. [CrossRef]
46. Saksena, S.; Merwade, V. Incorporating the effect of DEM resolution and accuracy for improved flood inundation mapping. *J. Hydrol.* **2015**, *530*, 180–194. [CrossRef]
47. Vesakoski, J.-M.; Alho, P.; Hyyppä, J.; Holopainen, M.; Flener, C.; Hyyppä, H. Nationwide Digital Terrain Models for Topographic Depression Modelling in Detection of Flood Detention Areas. *Water* **2014**, *6*, 271–300. [CrossRef]
48. Dasallas, L.; Kim, Y.; An, H. Case Study of HEC-RAS 1D–2D Coupling Simulation: 2002 Baeksan Flood Event in Korea. *Water* **2019**, *11*, 2048. [CrossRef]
49. Hicks, F.E.; Peacock, T. Suitability of HEC-RAS for Flood Forecasting. *Can. Water Resour. J.* **2005**, *30*, 159–174. [CrossRef]
50. Albu, L.-M.; Enea, A.; Iosub, M.; Breabăn, I.-G. Dam Breach Size Comparison for Flood Simulations. A HEC-RAS Based, GIS Approach for Drăcșani Lake, Sitna River, Romania. *Water* **2020**, *12*, 1090. [CrossRef]
51. Anees, M.T.; Abdullah, K.; Nordin, M.N.M.; Rahman, N.N.N.A.; Syakir, M.I.; Kadir, M.O.A. One- and Two-Dimensional Hydrological Modelling and Their Uncertainties. In *Flood Risk Management*; InTech: London, UK, 2017.
52. Zamani, A.; Maroukian, H. A morphological study of an old delta of the Sperchios River. In *Proceedings of the 6th Colloquium on the Geology of the Aegean Region*; Kallergis, G., Ed.; Institute of Geological and Mining Research: Athens, Greece, 1979; pp. 261–282.
53. Psomiadis, E. Research of Geomorphological and Environmental Changes in the Sperchios' River Basin Utilizing New Technologies. Ph.D. Thesis, Agricultural University of Athens, Athina, Greece, 2010.
54. Psomiadis, E.; Charizopoulos, N.; Soulis, K.X.; Efthimiou, N. Investigating the Correlation of Tectonic and Morphometric Characteristics with the Hydrological Response in a Greek River Catchment Using Earth Observation and Geospatial Analysis Techniques. *Geosciences* **2020**, *10*, 377. [CrossRef]
55. Ferrière, J. Recent Developments in the Maliakos Isopic Zone, Eastern Central Greece. In *Proceedings of the 6th Colloquium Geology of the Aegean Region*; Institute of Geological and Mining Research: Athens, Greece, 1977; pp. 197–210.

56. Psomiadis, E.; Migiros, G.; Parcharidis, I.; Poulos, S. Short Period Change Detection of Sperchios Lower Delta Area Using Space Radar Images. *Bull. Geol. Soc. Greece* **2004**, *36*, 919. [CrossRef]
57. European Space Agency (ESA) Copernicus Open Access Hub. Available online: https://sentinel.esa.int/web/sentinel (accessed on 23 September 2020).
58. Sentinel Online—ESA User Guides and Technical Guides of Sentinel-1 SAR. Available online: https://sentinel.esa.int/web/sentinel (accessed on 23 September 2020).
59. United States Geological Survey (USGS) EarthExplorer. Available online: https://earthexplorer.usgs.gov/ (accessed on 23 September 2020).
60. Stathopoulos, N.; Kalogeropoulos, K.; Louka, P.; Karymbalis, E.; Papadopoulos, A.G.; Chalkias, C. Flood Impact Assessment via SAR Data Analysis and GIS. Application in Sperchios River Basin, Greece (in Greek). Available online: https://www.researchgate.net/publication/317098438_FLOOD_IMPACT_ASSESSMENT_VIA_SAR_DATA_ANALYSIS_AND_GIS_APPLICATION_IN_SPERCHIOS_RIVER_BASIN_GREECE (accessed on 18 September 2020).
61. Tsakalias, G.; Koutsoyiannis, D. Hydrological Characteristics of the Sperchios Basin—ITIA. Available online: https://www.itia.ntua.gr/en/docinfo/98/ (accessed on 18 September 2020).
62. Bernstein, L.S.; Adler-Golden, S.M.; Jin, X.; Gregor, B.; Sundberg, R.L. Quick atmospheric correction (QUAC) code for VNIR-SWIR spectral imagery: Algorithm details. In Proceedings of the Workshop on Hyperspectral Image and Signal Processing, Evolution in Remote Sensing, Shanghai, China, 4–7 June 2012.
63. Psomiadis, E.; Charizopoulos, N.; Efthimiou, N.; Soulis, K.X.; Charalampopoulos, I. Earth Observation and GIS-Based Analysis for Landslide Susceptibility and Risk Assessment. *ISPRS Int. J. Geo-Inf.* **2020**, *9*, 552. [CrossRef]
64. Sonobe, R.; Yamaya, Y.; Tani, H.; Wang, X.; Kobayashi, N.; Mochizuki, K. Mapping crop cover using multi-temporal Landsat 8 OLI imagery. *Int. J. Remote Sens.* **2017**, *38*, 4348–4361. [CrossRef]
65. Asgarian, A.; Soffianian, A.; Pourmanafi, S. Crop type mapping in a highly fragmented and heterogeneous agricultural landscape: A case of central Iran using multi-temporal Landsat 8 imagery. *Comput. Electron. Agric.* **2016**, *127*, 531–540. [CrossRef]
66. Soulis, K.X.; Psomiadis, E.; Londra, P.; Skuras, D. A New Model-Based Approach for the Evaluation of the Net Contribution of the European Union Rural Development Program to the Reduction of Water Abstractions in Agriculture. *Sustainability* **2020**, *12*, 7137. [CrossRef]
67. Psomiadis, E.; Dercas, N.; Dalezios, N.R.; Spyropoulos, N.V. Evaluation and cross-comparison of vegetation indices for crop monitoring from sentinel-2 and worldview-2 images. In Proceedings of the SPIE—The International Society for Optical Engineering, Warsaw, Poland, 2 November 2017; Volume 10421.
68. Psomiadis, E.; Papazachariou, A.; Soulis, K.X.; Alexiou, D.S.; Charalampopoulos, I. Landslide mapping and susceptibility assessment using geospatial analysis and earth observation data. *Land* **2020**, *9*, 133. [CrossRef]
69. Jia, K.; Wei, X.; Gu, X.; Yao, Y.; Xie, X.; Li, B. Land cover classification using Landsat 8 Operational Land Imager data in Beijing, China. *Geocarto Int.* **2014**, *29*, 941–951. [CrossRef]
70. Worden, J.; de Beurs, K.M. Surface water detection in the Caucasus. *Int. J. Appl. Earth Obs. Geoinf.* **2020**, *91*, 102159. [CrossRef]
71. Ji, L.; Zhang, L.; Wylie, B. Analysis of dynamic thresholds for the normalized difference water index. *Photogramm. Eng. Remote Sens.* **2009**, *75*, 1307–1317. [CrossRef]
72. Argenti, F.; Lapini, A.; Alparone, L.; Bianchi, T. A tutorial on speckle reduction in synthetic aperture radar images. *IEEE Geosci. Remote Sens. Mag.* **2013**, *1*, 6–35. [CrossRef]
73. Gan, T.Y.; Zunic, F.; Kuo, C.C.; Strobl, T. Flood mapping of danube river at romania using single and multi-date ERS2-SAR images. *Int. J. Appl. Earth Obs. Geoinf.* **2012**, *18*, 69–81. [CrossRef]
74. Clement, M.A.; Kilsby, C.G.; Moore, P. Multi-temporal synthetic aperture radar flood mapping using change detection. *J. Flood Risk Manag.* **2018**, *11*, 152–168. [CrossRef]
75. Step by Step: Recommended Practice Flood Mapping | UN-SPIDER Knowledge Portal. Available online: http://www.un-spider.org/advisory-support/recommended-practices/recommended-practice-flood-mapping/step-by-step (accessed on 22 September 2020).
76. Filipponi, F. Sentinel-1 GRD Preprocessing Workflow. *Proceedings* **2019**, *18*, 11. [CrossRef]
77. Rana, V.K.; Suryanarayana, T.M.V. Evaluation of SAR speckle filter technique for inundation mapping. *Remote Sens. Appl. Soc. Environ.* **2019**, *16*, 100271. [CrossRef]

78. Lin, Y.N.; Yun, S.-H.; Bhardwaj, -A.; Hill, E.M. Urban Flood Detection with Sentinel-1 Multi-Temporal Synthetic Aperture Radar (SAR) Observations in a Bayesian Framework: A Case Study for Hurricane Matthew. *Remote Sens* **2019**, *11*, 1778. [CrossRef]
79. Henry, J.-B.; Chastanet, P.; Fellah, K.; Desnos, Y.-L. Envisat multi-polarized ASAR data for flood mapping. *Int. J. Remote Sens.* **2006**, *27*, 1921–1929. [CrossRef]
80. Tavus, B.; Kocaman, S.; Gokceoglu, C.; Nefeslioglu, H.A. Considerations on the use of Sentinel-1 data in flood mapping in urban areas: Ankara (Turkey) 2018 floods. *ISPRS Int. Arch. Photogramm. Remote Sens. Spat. Inf. Sci.* **2018**, *XLII-5*, 575–581. [CrossRef]
81. Horritt, M.S.; Bates, P.D. Evaluation of 1D and 2D numerical models for predicting river flood inundation. *J. Hydrol.* **2002**, *268*, 87–99. [CrossRef]
82. Kasper, K.E.; Thornton, C.I.; Abt, S.R.; Robeson, M.D.; Watson, C.C. *Accuracy of HEC-RAS to Calculate Flow Depths and Total Energy Loss with and without Bedway Weirs in a Meander Bend*; Colorado State University: Fort Collins, CO, USA, 2005.
83. US Army Corps of Engineers. *HEC-RAS River Analysis System Hydraulic Reference Manual*; USACE: Davis, CA, USA, 2016.
84. Bates, P.D.; Lane, S.N.; Ferguson, R.I. *Computational Fluid Dynamics*; Bates, P.D., Lane, S.N., Ferguson, R.I., Eds.; John Wiley & Sons, Ltd.: Chichester, UK, 2005; ISBN 9780470015193.
85. Vieux, B.E. *Distributed Hydrologic Modeling Using GIS*; Water Science and Technology Library, Ed.; Springer: Dordrecht, The Netherlands, 2001.
86. Candela, A.; Noto, L.V.; Aronica, G. Influence of surface roughness in hydrological response of semiarid catchments. *J. Hydrol.* **2005**, *313*, 119–131. [CrossRef]
87. Christofidis, A. Development of a GIS-Based Rainfall-Runoff Model. PhD Thesis, Faculty of Civil Engineering, National Technical University of Athens, Athens, Greece, 2008.
88. Diakakis, M. A method for flood hazard mapping based on basin morphometry: Application in two catchments in Greece. *Nat. Hazards* **2011**, *56*, 803–814. [CrossRef]
89. Congalton, R.G. A review of assessing the accuracy of classifications of remotely sensed data. *Remote Sens. Environ.* **1991**, *37*, 35–46. [CrossRef]
90. Cohen, J. A Coefficient of Agreement for Nominal Scales. *Educ. Psychol. Meas.* **1960**, *20*, 37–46. [CrossRef]
91. Rahman, M.R.; Thakur, P.K. Detecting, mapping and analysing of flood water propagation using synthetic aperture radar (SAR) satellite data and GIS: A case study from the Kendrapara District of Orissa State of India. *Egypt. J. Remote Sens. Sp. Sci.* **2018**, *21*, S37–S41. [CrossRef]
92. Tiwari, V.; Kumar, V.; Matin, M.A.; Thapa, A.; Ellenburg, W.L.; Gupta, N.; Thapa, S. Flood inundation mapping- Kerala 2018; Harnessing the power of SAR, automatic threshold detection method and Google Earth Engine. *PLoS ONE* **2020**, *15*, e0237324. [CrossRef]
93. Long, S.; Fatoyinbo, T.E.; Policelli, F. Flood extent mapping for Namibia using change detection and thresholding with SAR. *Environ. Res. Lett.* **2014**, *9*, 9. [CrossRef]
94. Mittra, M.K.; Stickler, F.C. Excess Water Effects on Different Crops. *Trans. Kansas Acad. Sci.* **1961**, *64*, 275. [CrossRef]
95. Salamin, P. *Relations entre les irrigations et l'evacuation de l'eau*; International Commission on Irrigation & Drainage (ICID): New Delhi, India, 1957.
96. Tsakalias, G.; Koutsoyiannis, D. Hydrological characteristics of the Sperchios basin. In Proceedings of the Sperchios 2000+; Sterea Hellas District, National Technical University of Athens: Athens, Greece, 1995; pp. 89–98.
97. Choi, Y. A new algorithm to calculate weighted flow-accumulation from a DEM by considering surface and underground stormwater infrastructure. *Environ. Model. Softw.* **2012**, *30*, 81–91. [CrossRef]
98. Choi, Y.; Yi, H.; Park, H.D. A new algorithm for grid-based hydrologic analysis by incorporating stormwater infrastructure. *Comput. Geosci.* **2011**, *37*, 1035–1044. [CrossRef]
99. Callow, J.N.; Smettem, K.R.J. The effect of farm dams and constructed banks on hydrologic connectivity and runoff estimation in agricultural landscapes. *Environ. Model. Softw.* **2009**, *24*, 959–968. [CrossRef]
100. Chen, H.; Liang, Q.; Liang, Z.; Liu, Y.; Xie, S. Remote-sensing disturbance detection index to identify spatio-temporal varying flood impact on crop production. *Agric. For. Meteorol.* **2019**, *269–270*, 180–191. [CrossRef]

101. Gould, I.J.; Wright, I.; Collison, M.; Ruto, E.; Bosworth, G.; Pearson, S. The impact of coastal flooding on agriculture: A case-study of Lincolnshire, United Kingdom. *L. Degrad. Dev.* **2020**, *31*, 1545–1559. [CrossRef]
102. Mtibaa, S.; Irie, M. Land cover mapping in cropland dominated area using information on vegetation phenology and multi-seasonal Landsat 8 images. *Euro-Mediterranean J. Environ. Integr.* **2016**, *1*, 1–16. [CrossRef]
103. de Oliveira Santos, C.L.M.; Lamparelli, R.A.C.; Figueiredo, G.; Dupuy, S.; dos Santos Luciano, A.C.; da Silva Torres, R.; le Maire, G.; Boury, J. Classification of Crops, Pastures, and Tree Plantations along the Season with Multi-Sensor Image Time Series in a Subtropical Agricultural Region. *Remote Sens.* **2019**, *11*, 334. [CrossRef]
104. Schultz, B.; Immitzer, M.; Formaggio, A.; Sanches, I.; Luiz, A.; Atzberger, C. Self-Guided Segmentation and Classification of Multi-Temporal Landsat 8 Images for Crop Type Mapping in Southeastern Brazil. *Remote Sens.* **2015**, *7*, 14482–14508. [CrossRef]
105. Murthy, C.S.; Raju, P.V.; Badrinath, K.V.S. Classification of wheat crop with multi-temporal images: Performance of maximum likelihood and artificial neural networks. *Int. J. Remote Sens.* **2003**, *24*, 4871–4890. [CrossRef]
106. Forkuor, G.; Dimobe, K.; Serme, I.; Tondoh, J.E. Landsat-8 vs. Sentinel-2: Examining the added value of sentinel-2's red-edge bands to land-use and land-cover mapping in Burkina Faso. *GISci. Remote Sens.* **2018**, *55*, 331–354. [CrossRef]
107. Csillik, O.; Belgiu, M.; Asner, G.P.; Kelly, M. Object-Based Time-Constrained Dynamic Time Warping Classification of Crops Using Sentinel-2. *Remote Sens.* **2019**, *11*, 1257. [CrossRef]
108. Volpi, M.; Petropoulos, G.P.; Kanevski, M. Flooding extent cartography with Landsat TM imagery and regularized kernel Fisher's discriminant analysis. *Comput. Geosci.* **2013**, *57*, 24–31. [CrossRef]
109. Kyriou, A.; Nikolakopoulos, K. Flood mapping from Sentinel-1 and Landsat-8 data: A case study from river Evros, Greece. In Proceedings of the Earth Resources and Environmental Remote Sensing/GIS Applications VI; Michel, U., Schulz, K., Ehlers, M., Nikolakopoulos, K.G., Civco, D., Eds.; SPIE: Toulouse, France, 2015; Volume 9644, p. 964405.
110. Singh, K.V.; Setia, R.; Sahoo, S.; Prasad, A.; Pateriya, B. Evaluation of NDWI and MNDWI for assessment of waterlogging by integrating digital elevation model and groundwater level. *Geocarto Int.* **2015**, *30*, 650–661. [CrossRef]
111. Baig, M.H.A.; Zhang, L.; Wang, S.; Jiang, G.; Lu, S.; Tong, Q. Comparison of MNDWI and DFI for water mapping in flooding season. In Proceedings of the International Geoscience and Remote Sensing Symposium (IGARSS), Melbourne, Australia, 21–26 July 2013; pp. 2876–2879.
112. Zhou, W.; Li, Z.; Ji, S.; Hua, C.; Fan, W. A new index model NDVI-MNDWI for water object extraction in hybrid area. In *International Conference on Geo-Informatics in Resource Management and Sustainable Ecosystem*; Springer: Berlin/Heidelberg, Germany, 2015; Volume 482, pp. 513–519.
113. Medeiros, F.N.S.; Mascarenhas, N.D.A.; Costa, L.F. Evaluation of speckle noise MAP filtering algorithms applied to SAR images. *Int. J. Remote Sens.* **2003**, *24*, 5197–5218. [CrossRef]
114. Xiao, J.; Li, J.; Moody, A. A detail-preserving and flexible adaptive filter for speckle suppression in SAR imagery. *Int. J. Remote Sens.* **2003**, *24*, 2451–2465. [CrossRef]
115. Kushwaha, S.P.S.; Dwivedi, R.S.; Rao, B.R.M. Evaluation of various digital image processing techniques for detection of coastal wetlands using ERS-1 SAR data. *Int. J. Remote Sens.* **2000**, *21*, 143–1161. [CrossRef]
116. Agnihotri, A.K.; Ohri, A.; Gaur, S.; Shivam; Das, N.; Mishra, S. Flood inundation mapping and monitoring using SAR data and its impact on Ramganga River in Ganga basin. *Environ. Monit. Assess.* **2019**, *191*, 1–16. [CrossRef] [PubMed]
117. Tsyganskaya, V.; Martinis, S.; Marzahn, P.; Ludwig, R. Detection of Temporary Flooded Vegetation Using Sentinel-1 Time Series Data. *Remote Sens.* **2018**, *10*, 1286. [CrossRef]
118. Whyte, A.; Ferentinos, K.P.; Petropoulos, G.P. A new synergistic approach for monitoring wetlands using Sentinels -1 and 2 data with object-based machine learning algorithms. *Environ. Model. Softw.* **2018**, *104*, 40–54. [CrossRef]
119. Liang, J.; Liu, D. Estimating daily inundation probability using remote sensing, riverine flood, and storm surge models: A case of hurricane harvey. *Remote Sens.* **2020**, *12*, 1495. [CrossRef]
120. Huang, C.; Chen, Y.; Wu, J.; Chen, Z.; Li, L.; Liu, R.; Yu, J. Integration of remotely sensed inundation extent and high-precision topographic data for mapping inundation depth. In Proceedings of the 2014 The 3rd

International Conference on Agro-Geoinformatics, Agro-Geoinformatics 2014, Beijing, China, 11–14 August 2014; Institute of Electrical and Electronics Engineers Inc.: Beijing, China, 2014.

121. Cian, F.; Marconcini, M.; Ceccato, P.; Giupponi, C. Flood depth estimation by means of high-resolution SAR images and lidar data. *Nat. Hazards Earth Syst. Sci.* **2018**, *18*, 3063–3084. [CrossRef]
122. Ben Khalfallah, C.; Saidi, S. Spatiotemporal floodplain mapping and prediction using HEC-RAS—GIS tools: Case of the Mejerda river, Tunisia. *J. African Earth Sci.* **2018**, *142*, 44–51. [CrossRef]
123. Qi, H.; Altinakar, M.S. A GIS-based decision support system for integrated flood management under uncertainty with two dimensional numerical simulations. *Environ. Model. Softw.* **2011**, *26*, 817–821. [CrossRef]
124. Ezzine, A.; Saidi, S.; Hermassi, T.; Kammessi, I.; Darragi, F.; Rajhi, H. Flood mapping using hydraulic modeling and Sentinel-1 image: Case study of Medjerda Basin, northern Tunisia. *Egypt. J. Remote Sens. Sp. Sci.* **2020**. [CrossRef]
125. Gkouma, A.; Zotou, I.; Bellos, V.; Karathanassi, V.; Tsihrintzis, V.A. Evaluating the performance of a hydrodynamic model using SAR images. In Proceedings of the 11th World Congress on Water Resources and Environment (EWRA 2019), Madrid, Spain, 25–29 June 2019; European Water Resources Association: Madrid, Spain, 2019; pp. 425–426.
126. Hong Quang, N.; Tuan, V.A.; Thi Thu Hang, L.; Manh Hung, N.; Thi The, D.; Thi Dieu, D.; Duc Anh, N.; Hackney, C.R. Hydrological/Hydraulic Modeling-Based Thresholding of Multi SAR Remote Sensing Data for Flood Monitoring in Regions of the Vietnamese Lower Mekong River Basin. *Water* **2019**, *12*, 71. [CrossRef]
127. Schumann, G.J.P.; Bates, P.D.; Neal, J.C.; Andreadis, K.M. Measuring and Mapping Flood Processes. In *Hydro-Meteorological Hazards, Risks, and Disasters*; John, F., Paron Giuliano, S.P., Baldassarre, D., Eds.; JAI-Elsevier Science Inc.: Boston, MA, USA, 2015; pp. 35–64, ISBN 9780123964700.
128. Coveney, S.; Roberts, K. Lightweight UAV digital elevation models and orthoimagery for environmental applications: Data accuracy evaluation and potential for river flood risk modelling. *Int. J. Remote Sens.* **2017**, *38*, 3159–3180. [CrossRef]

2

Earth Observation and GIS-Based Analysis for Landslide Susceptibility and Risk Assessment

Emmanouil Psomiadis [1,*], Nikos Charizopoulos [1], Nikolaos Efthimiou [2], Konstantinos X. Soulis [1] and Ioannis Charalampopoulos [3]

[1] Department of Natural Resources Management and Agricultural Engineering, Agricultural University of Athens, 75 Iera Odos st., 11855 Athens, Greece; nchariz@aua.gr (N.C.); soco@aua.gr (K.X.S.)
[2] Faculty of Environmental Sciences, Czech University of Life Sciences Prague, Kamýcká 129, 165 00 Prague–Suchdol, Czech Republic; efthimiou@fzp.czu.cz
[3] Department of Crop Science, Agricultural University of Athens, 75 Iera Odos st., 11855 Athens, Greece; icharalamp@aua.gr
* Correspondence: mpsomiadis@aua.gr

Abstract: Landslides can cause severe problems to the social and economic well-being. In order to effectively mitigate landslide hazards, the development of detailed susceptibility maps is required, towards implementing targeted risk management plans. This study aims to create detailed landslide susceptibility (LS) and landslide risk (LR) maps of the Sperchios River basin by applying an expert semi-quantitative approach that integrates the Geographic Information Systems (GIS)-based multicriteria analysis and Earth Observation (EO) data. Adopting the analytic hierarchy process (AHP) for a weighted linear combination (WLC) approach, eleven evaluation parameters were selected. The results were validated using a historic landslide database, enriched with new landslide locations mapped by satellite and aerial imagery interpretation and field surveys. Moreover, the landslide risk map of the area was also developed, based on the LS delineation, considering additionally the anthropogenic exposure and overall vulnerability of the area. The results showed that the most susceptible areas are located at the west and south-west regions of the basin. The synergistic use of GIS-based analysis and EO data can provide a useful tool for the design of natural hazards prevention policy at highly susceptible to risk landslide risk areas.

Keywords: landslide susceptibility; GIS-based analysis; Earth Observation; human risk mapping; Sperchios River basin

---

1. Introduction

Landslides are determined as the motion of debris, mass of rock, or earth down a slope due to gravity effect [1,2]. They are classified among the most dangerous and catastrophic natural hazards, being a significant threat to property and human life, moreover causing several indirect implications such as blocking of streams and aggregation of rivers, flash-flood occurrence, destruction of agricultural land, etc. [3–7]. In recent decades landslide risk has increased due to population growth, leading to the expansion of residential areas; establishment of new settlements towards steeper slopes and/or to areas of inappropriate foundation (bedrock, soil) in terms of structure and stability [2,8,9].

The definition of independent factors that reflect conditions prior to the landslide occurrence is of great significance. Data quality is particularly important in the field of landslide research, and highly related to the accuracy of the derivative results. The components responsible for landslide episodes variously and complexly interdepend. They can be distinguished into preparatory and triggering mechanisms [10,11]. The preparatory mechanisms comprise general meteorological conditions, geomorphological processes, lithology (ground conditions and rock distribution), tectonic activity (the distance from the rupture plane and the seismic factor play a significant role on determining the landslide magnitude), vegetation cover, etc. The triggering mechanisms include a variety of external factors, such as intensive rainfalls, earthquakes (seismic-triggered landslides are classified as coherent, disrupted, and lateral spreads), speedy river erosion, landscape processes, and anthropogenic activities (deforestation and road construction in steep mountainous areas, uncontrollable irrigation, etc.) [12–17]. The designation of landslide risk areas can be achieved in one (or more) of the following ways: (a) landslide inventory maps, that display (at least) the geographical distribution of past events, followed by associated databases of landslide and terrain properties; (b) landslide susceptibility (LS) maps that refer to the tendency of an area to landslide occurrence—they depict the possibility of occurrence of a landslide event of a specific type at a particular place (where); (c) landslide hazard maps, which define the likelihood of occurrence of a potentially damaging landslide that may take place within a given area and period of time (where, how often, and how large)—this concept contains both spatial and temporal dimensions [8,18,19]; and (d) landslide risk maps that show potential damage or losses to individuals, infrastructure and property [20].

A landslide hazard map demonstrates regions susceptible to landslides by taking into account the phenomenon's causative and triggering factors (e.g., geomorphological, geological, and meteorological) with data concerning the past distribution of slope failures [12,21]. Landslide hazard assessment is crucial for the evaluation and management of natural disasters. It is also an essential stage for natural and urban planning in government policies worldwide [22,23]. Due to the potential lack of past landslides records, hazard estimation is not always possible [18,24]. Hazard assessment remains the only way to overcome this problem, identifying the place of potential landslides in a zone based on a set of terrain features [25]. Landslide risk associates the probability information from a landslide hazard map, with an analysis of all probable effects, taking into consideration the vulnerability and exposure of the area, i.e., the level of population; property; economic activity, the state of people; infrastructure; accommodation; and other actual human assets situated in hazard-prone areas [26].

Hence, the primary step in evaluating landslide risk is defining LS [5,12,27,28]. The LS term is commonly used to describe the position of potential landslides at a given area, based on a set of landscape characteristics. LS analysis considers that future landslides might be triggered by the same conditioning parameters from which present and past landslides were triggered [25]. Although these approaches make available information on potentially unstable slopes, they do not provide direct information on landslide extent and frequency. There are three LS assessment methods, categorized as qualitative, quantitative and semi-quantitative [8,12,29–31]. Qualitative methods tend to be very subjective, since they depend on expert opinion, with the weight of the factors leading to vulnerability being evaluated based on expert knowledge and experience [30,32,33]. The elementary types of qualitative methods utilize landslide index to identify areas with analogous geological and geomorphologic features that are susceptible to landslides [12]. Quantitative methods produce numerical estimates among controlling factors and landslides. They are categorized into deterministic and statistical [34]. The former depend on slope stability studies, signified in terms of the safety factor [35]. The latter examine the historical connection concerning landslide-controlling factors and the allocation of landslides, involving analytic bivariate, logistics regression, fuzzy logic, neural network, etc. Throughout the last decades, quantitative procedures have been applied for LS zonation researches in different areas [2,7,8,12,23,25,33,34]. Additionally, qualitative methodologies that operate in weighting and rating procedures are called semi-quantitative techniques [7,24]. These methodologies are the analytic hierarchy process (AHP) [2,5,13,26,36–40] and the weighted linear combination (WLC) [24,41]. The WLC method includes a combination of several landslide factors. Each factor is divided into classes, the layers of which are thereafter weighted according to the significance of each feature relative to the remaining ones. Finally, all the derived maps are overlaid in order to construct the final LS map [11,12].

Numerous LS mapping methodologies have been presented over the last decades [2,8]. Their aim is to categorize the different parts of the land surface according to the grade of actual or potential landslide susceptibility [5]. The new perspectives offered by earth observation and geographic information systems (GIS) technologies, provide the opportunity of an easy and integrated way to collect and process a variety of spatial data, necessary for LS estimation. GIS-based and remote sensing techniques have proven to be excellent tools in the spatial analysis of terrain factors for LS zonation [2,27,31,40,42–45].

One of the most significant variables affecting landslide susceptibility, apart from geology, weather conditions, and land cover, is topography (i.e., slope angle, aspect, and curvature) [46]. Therefore, the selection of a proper digital elevation model (DEM) is essential in LS assessment [47]. Japan's Advanced Spaceborne Thermal Emission Radiometer (ASTER) of Terra satellite makes available valuable geomorphic for landslide modelling, through the potential development of accurate and detailed DEM [48]. Likewise, the innovative Sentinel 2 (S2) mission of the European Space Agency Copernicus program comprises two identical satellites in the same orbit, 180° apart. Together they cover all Earth's surface, providing imagery of high spatial and temporal resolution (10 m and 5 day cycle, at the equator, respectively). Given its Multi-Spectral Imager (MSI) sensor, it operates on 13 different spectral bands covering visible (VIS, 4 bands, 1-2-3-4), near infrared (NIR, 5 bands, 5-6-7-8-8A), and short wave infrared (SWIR, 4 bands, 9-10-11-12) part of electromagnetic spectrum, having a wide swath coverage (290 km width). This satellite imagery is bringing land monitoring to a completely new level as a unique tool in the monitoring of vegetation conditions and classification. Land cover type and changes perform a crucial role in landslide appearance as they are an essential factor for stabilizing slopes [11,49].

The study objective is to identify the impact of the several natural and anthropogenic factors on LS at the Sperchios River basin, Central Greece, an area highly vulnerable to such phenomena. By adopting the analytic hierarchy process (AHP) for a weighted linear combination (WLC) approach, eleven evaluation parameters including lithology, slope angle, slope curvature, slope aspect, relative relief, land use/cover, soil depth, rainfall distribution, proximity to faults, and distance to roads and streams were initially selected as evaluation factors. All layers were subsequently converted to a raster layout, and a geospatial database was developed. Finally, the potential landslide risk of the area was estimated, considering the anthropogenic exposure (presence and activities) in the area. The study innovates by employing the applicability and efficiency of two state-of-the-art technologies, such as GIS and EO to competently collect process and incorporate the various spatial data necessary, complemented by well-established methods (AHP–WLC), for landslide susceptibility assessment and risk delineation.

2. Study Area

The study area is the Basin of the Sperchios River in Central Greece. It is located between 38°44′ and 39°05′ N lat., and 21°50′ and 22°45′ E long., occupying an area of approximately 1830 km^2 (Figure 1). The River's western and southern flanks comprise high elevations and steep slopes (Mt. Tymphristos 2315 m and Mt. Oiti 2152 m), contrary to its northern flanks where lower elevations (Mt. Othrys 1720 m) and milder slopes are met [50–53]. Mean annual precipitation for the period 1980–2010 was measured as 788 mm, considering the records of eight pluviometric stations [52].

The climatic of the study area is characterized as Csa (Mediterranean with hot summer) according to Köppen [54] indicating a dry warm period and a wet period distributed from late autumn to early spring. The mountainous areas of the studied basin are anticipated to form colder and wet conditions in terms of precipitation. Moreover, the topographic relief of the area enhances the formation of orographic precipitation during the year period.

According to Ferrière (1977) [55], there are three distinct lithological units in Sperchios basin, i.e., (i) the western part, dominated of Paleocene-Eocene flysch of the East Pindus and Parnassos geotectonic zones, including alternating beds of argillite-siltstone-fine conglomerate and intercalations of shale; (ii) the southeastern part, consisting of Middle Triassic-Jurassic massive dolomites and limestones of the Pelagonian zone; Upper Cretaceous flysch consisting of coarse sandstone alternating with shale and sandy marl belonging to the Boeotia zone; and Upper Cretaceous thickly-bedded limestones and Eocene flysch, composed of sandstone, clay and marl of the Parnassos geotectonic zone; and (iii) the north and northeastern part (of utterly different lithological composition) with the presence of an ophiolitic complex in a shale-chert formation includes shale, chert, and limestone with ultra-mafic and mafic igneous rocks peridotites, dunites, pyroxenites, gabbros, serpentinites, diavases, dolerites, and metamorphosed green phyllite and schist; belonging to the Maliakos (Sub-Pelagonian) zone. The central and lower (elevations < 500 m) part of the watershed is occupied by Neogene and Quaternary unconsolidated deposits [50,56–58].

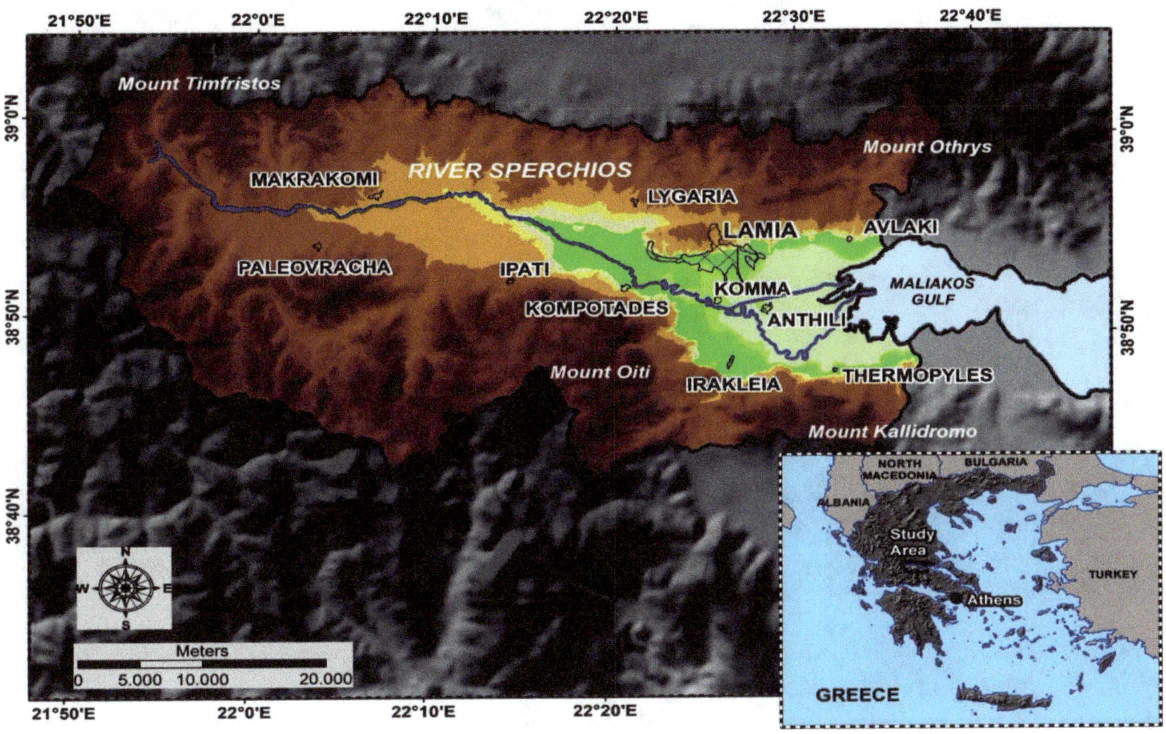

Figure 1. Location and geomorphological characteristics of the study area.

2.1. Landslides

In Greece, numerous landslide events have taken place over the years, especially at its western part, due to the effect of the presence of several preparatory factors such as the geological formations and the steep slopes (high relief energy). Moreover, the intense folding and joint tectonics, and the extreme weather conditions have favored the occurrence of significant landslide events [19,59–62].

According to the landslide hazard zonation map of Greece, the western regions of the Sperchios basin exhibit the highest frequency of landslides events (Figure 2a) [19,61]. The main preparatory factors of the landslide mechanism comprise the vulnerable geological formations (flysch) and the steep slopes, while the triggering factors are the extreme rainfall events and the ongoing tectonic activity [11,48]. Several landslide phenomena have occurred at the highly vulnerable western and southwestern parts of the catchment, mainly due to the extended presence of flysch formations (43% of the entire basin area). Such formations are related as much as 70–80% to the landslide's occurrence in Greece [62,63]. Apart from the bedrock characteristics, intense morphological relief with steep slopes, dense drainage network and deep valleys, and extensive human activities, such as relatively dense road network, and arable-irrigated areas are also responsible for landslide manifestations in the Sperchios watershed [12]. In the village Plaio Mikro Chorio, an area with similar geological and geomorphological characteristics, which is located approximately 30 km away from the basin's western edge, one of the deadliest landslides in the history of modern Greek state, occurred on the 13 January 1963. The landslide destroyed the largest part of the settlement with severe consequences, causing 13 casualties [64]. Contrary, at the south-central and eastern parts (Neogene and calcareous formations), the few events that have been taken place are mainly related mainly to rock falls [19,52,58,65–67] (Figure 2b).

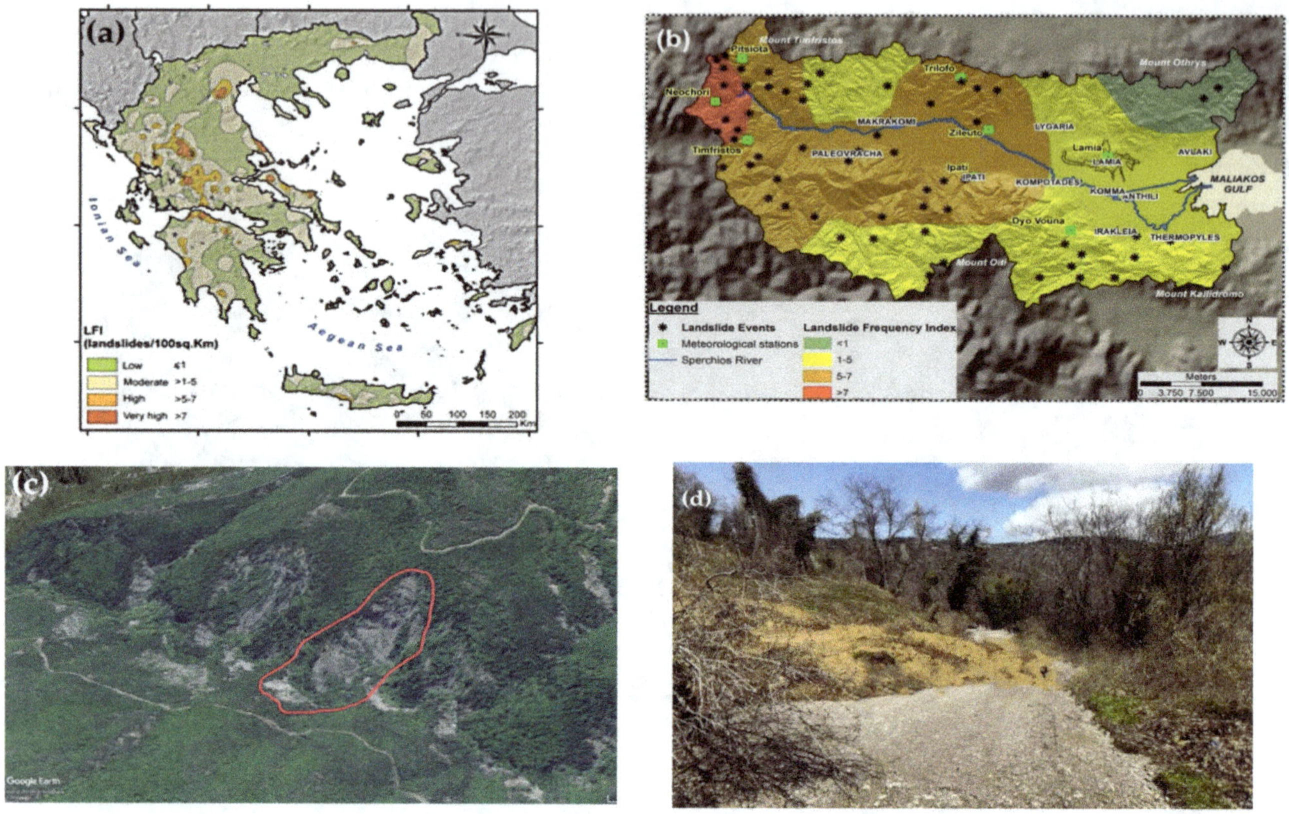

Figure 2. (**a**) The Landslide Frequency Index (LFI-landslides/100 km^2) map of Greece [16,56], (**b**) LFI of Sperchios river basin and the points of landslide events constitute the inventory map of the area (**c**) detection of past landslide event in Google Earth (red line), and (**d**) landslide detection during the fieldwork.

2.2. *Inventory Map*

A landslide inventory map provides valuable information on the spatial distribution of landslide events. The map is critical for examining the correlation among an event and its contributory factors.

The landslide inventory map of the study area was developed through the following stages: (a) collection of landslides historical information from previous studies (Figure 2a,b), (b) detection of landslides in very high resolution aerial photos and Google Earth satellite images (Figure 2c), and (c) fieldwork for the mapping of recent events (Figure 2d). Fifty-nine (59) landslide sites in total were gathered throughout the study area. Based on random sampling, the new landslide inventory map was separated into two datasets: 70% (40 landslide spots) for the triggering factors classification and weighting, and 30% (19 landslide spots) for the validation process, respectively [11,27].

3. Materials and Methods

3.1. *Dataset*

The different types of datasets utilized in the present study comprise (i) geological maps from the Institute of Geology and Mineral Exploration (IGME) at 1:50,000 scale [68]; (ii) topographic maps from the Hellenic Military Geographical Service at a scale of 1:50,000 [69]; (iii) satellite imagery (Sentinel 2 scenes); (iv) digital elevation model (DEM) of the area deriving from Terra/ASTER satellite data, level 1A, acquired on 30 July 2003 [70]; (v) precipitation records from eight (8) meteorological stations (Figure 2b), overseen by the Hellenic National Meteorological Service (1 station), the Ministry of Environment, Energy, and Climate Change (4 stations and the Public Power Cooperation (3 stations); (vi) soil maps from the Directorates of Forests (Ministry of Agriculture) [71]; and (vii) population demographic characteristics from the Hellenic Statistical Authority [72].

More specifically, the geological maps were used to depict the basin's lithological and structural units, and the S2 images to delineate and supplement the lineaments-possible faults (derived from the geological maps) and extract the land cover types. The ASTER DEM (15 m) along with the topographic maps (20-m contour interval) was used to develop a detailed DEM, and to describe local topography (slope angle, slope aspect, slope curvature, and relative relief) and stream network. Finally, the precipitation records served as means to estimate mean annual precipitation for the period of 1971–2010 [73]. The data were incorporated towards compiling the final LS map. The geospatial database manipulation and the S2 and Aster satellite data processing were accomplished by utilizing ArcGIS (Environmental Systems Research Institute, Redlands, CA, USA) and ENVI (Harris Geospatial Solutions, Boulder, CO, USA) software, respectively. Moreover, the data pre-processing and a part of workflow's automation was made via R-Language scripts [74].

3.2. Methodology

The accuracy of susceptibility mapping improves when all landslide-controlling factors are incorporated in the analytical procedure, which is often hard to achieve, given the difficulty to acquire detailed data. Thus, analyses mainly depend on physical and anthropogenic factors.

The study's first task involved the selection of the landslide controlling factors, based on expert knowledge, extended field observations, literature review, and collection of available landslide historical data. Eleven factors were eventually selected, i.e., geological characteristics, such as (i) lithology, (ii) proximity to major faults; geomorphological features, such as (iii) slope angle, (iv) slope aspect, (v) slope curvature, (vi) relative relief, (vii) drainage network, (viii) soil depth; and anthropogenic factors and weather information, such as (ix) road network, (x) land use/cover, and (xi) rainfall distribution [12,19]. The next step was to assign weights and rank values per factor (i.e., raster layer; pixel size 20 × 20 m) and factor class (within each layer), respectively. Two methods were used for the LS index calculation, the simple and the geometric mean (GM). Regarding the first method, the weight calculation was based on the analytic hierarchy process [36], providing an objective approach to such an assignment [13,27]. Regarding the second method, the rating of the classes considers the most stable regions of each parameter and classifies them with zero values. The landslide susceptibility indices (LSI) were developed utilizing the raster layers and their corresponding weights using the two methods [12–14]. Finally, the landslide risk map was produced, through the evaluation of landslide hazard map impact on anthropogenic exposure and vulnerability of the study area. The methodology applied is schematically presented in Figure 3 and is analyzed in the following sections.

3.2.1. Digital Elevation Model from ASTER Data

A strong advantage of the along-track mode of the ASTER data acquisition system is that it provides stereo pairs of images acquired a few seconds separately under the same conditions that are appropriate for DEM creation by stereo-correlation techniques [48,75].

The stereo pair of the near-infrared backward and nadir images (3N and 3B) was used for the DEM creation. The synchronized along-track stereo-data acquisition offers a robust benefit in terms of radiometric differences contrasted. Stereo-correlation is a computational and statistical method used to obtain a DEM from the registered images of stereo pairs. For the photogrammetric matching rational polynomial coefficients (RPC) of the images along with well-distributed ground control points were applied. As a result, a very precise DEM with a 15 m spatial resolution was produced.

The ASTER DEM was compared to a DEM developed from 1:50,000 topographic maps' digitized contours. The visual comparison of the DEMs led to the detection of errors and significant alterations in the elevation data of the DEM deriving from the digitized contours; thus, it required updating.

The vertical accuracy of the ASTER elevation data was checked using three hundred points of known elevation with the root mean square error (RMSE)-z being was ± 3.2 m. Therefore, the accuracy of the ASTER DEM is considered appropriate for updating the DEM deriving from the 1:50,000 topographic maps [75].

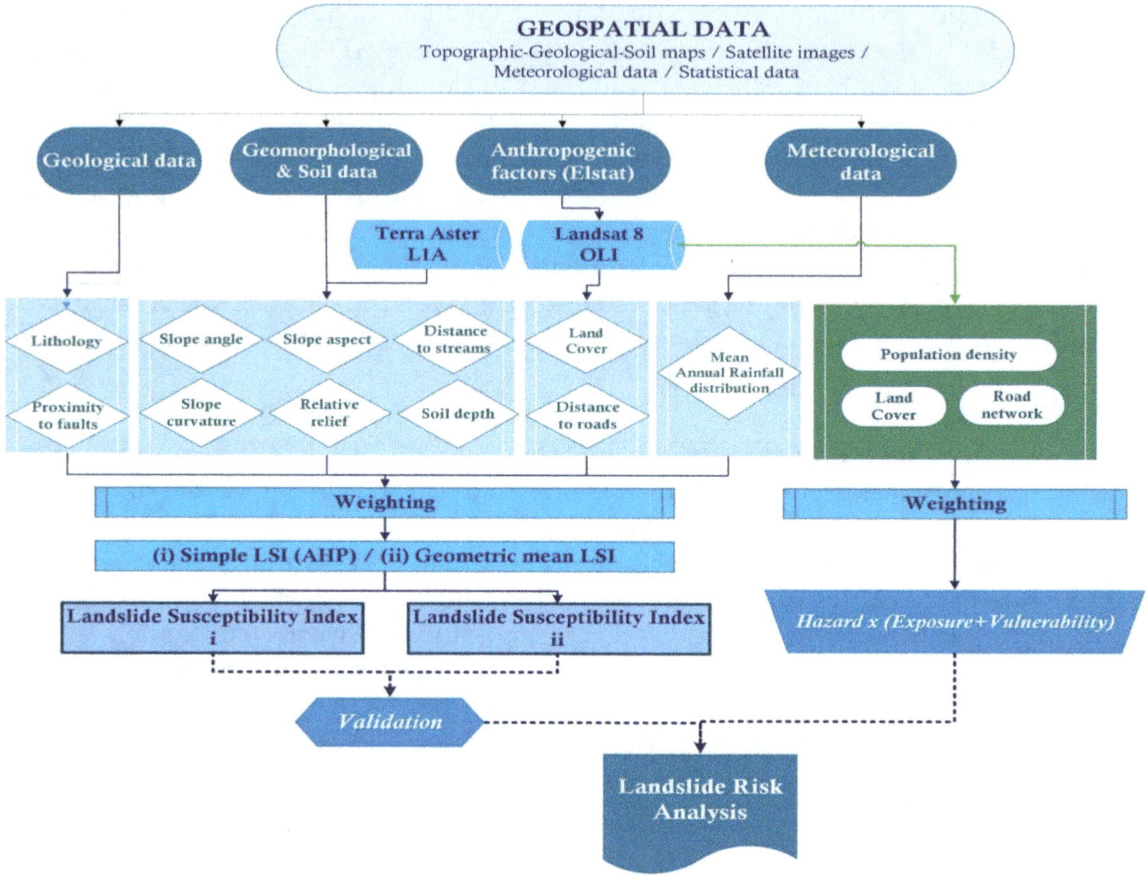

Figure 3. Methodology flowchart.

3.2.2. Land Cover Classification from Sentinel Data

Land cover classification based on S2 images has progressed over the last five decades, having been used for the identification of different land cover types through various classification methods applied on EO data [76]. S2 MSI data belongs to the new generation of satellites with improvements that are principally characterized by enhanced spectral, spatial, and radiometric resolution and were utilized for land cover mapping [77]. Four cloud-free, geometrically, and atmospherically corrected images, of the year 2019 (one for every season), were acquired, to distinguish the different land cover types (Figure 4). The images were acquired free of charge via the European Space Agency (ESA) portal (https://scihub.copernicus.eu/) [11,78].

The land cover map derived from the supervised classification, utilizing the maximum likelihood method with 440 ground control samples (collected through field survey and orthophoto maps). The data samples were divided into the training (70%) and the validation data set (30%). Using the first (training) set, and twelve classes, a two-step classification procedure was followed. The first step comprised a broad distinction between general land cover types (forest, water, built-up classes, etc.). In the second step, the general LC types were further classified into summer and winter arable crops, and different kinds of forest types (coniferous, broad-leaved, etc.). The accuracy assessment of the classification map was accomplished using the second validation set. The results were assessed using overall accuracy (OA), and the Kappa coefficient [79,80]. The classification OA and Kappa coefficient were estimated at 95.8% (295 correctly classified parcels divided to 308 total reference parcels) and 0.88, respectively, indicating a highly accurate result. Finally, post-classification filtering was performed to remove high-frequency deviations and misclassified pixels.

Figure 4. Sentinel-2 images acquired on (**a**) December 2018 (winter image), (**b**) April 2019, (**c**) July 2019, and (**d**) September 2019, and used for the natural vegetation and crop classification procedure.

3.2.3. Data Layers Formation

The factors chosen for the estimation of landslide susceptibility in the study area were incorporated into a geospatial database. All thematic layers were georeferenced to the Greek Geodetic Reference System (GGRS'87). The rating classes of each factor were determined based on the first dataset of the inventory map maintained for their classification and weighting, along with previous researches [12,19] and our experience on the local conditions of the study area.

Lithology: Lithology is one of the most critical controlling factors of landslides since each lithological unit displays different susceptibility rates. Several researchers [5,28,62,65] emphasized the role of geology on slope stability. The lithology layer derived by digitizing the IGME (1957–1991) geological maps. The bedrock formations were classified into five general categories concerning LS, i.e., (a) limestones and marbles, (b) Neogene sediments, (c) schists and ophiolites, (d) alluvial deposits and debris, and (e) flysch (Figure 5a—Table 1).

Land Cover: Vegetation is a potential driver of LS. Some land cover types, e.g., forest trees with strong and extensive rooting system, enhance slope stability through their effect on the soil's hydrological and mechanical attributes [12,81]. The land use/cover data deriving from L8 data was reclassified into 7 classes based on their effect on LS, i.e., (a) coniferous and broadleaves forest; (b) deciduous forest; (c) tree crops; (d) urban areas; (e) arable land; (f) vineyards and nude soils; and (g) pasture land, grassland, and burnt areas (Figure 5b—Table 1) [82].

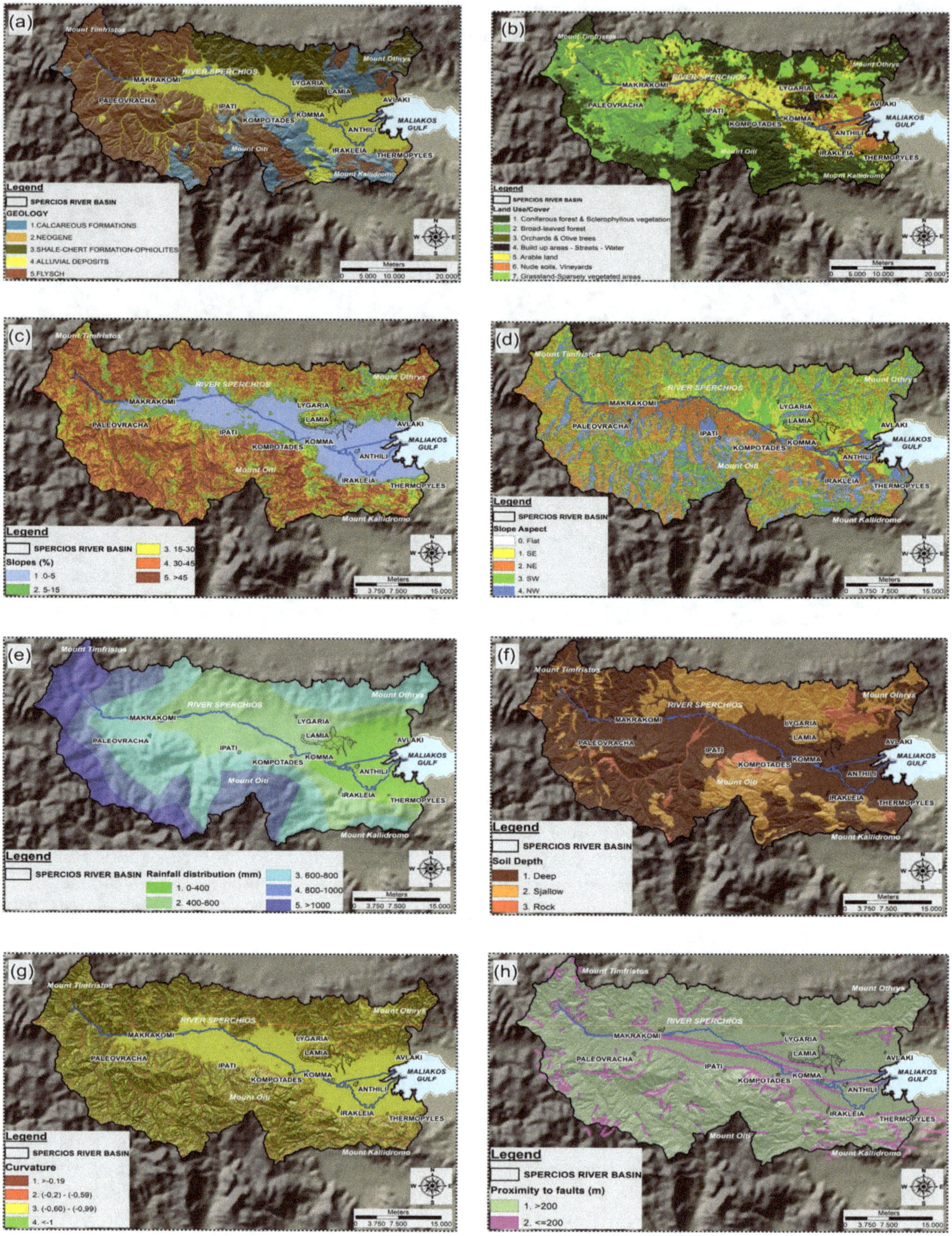

Figure 5. *Cont.*

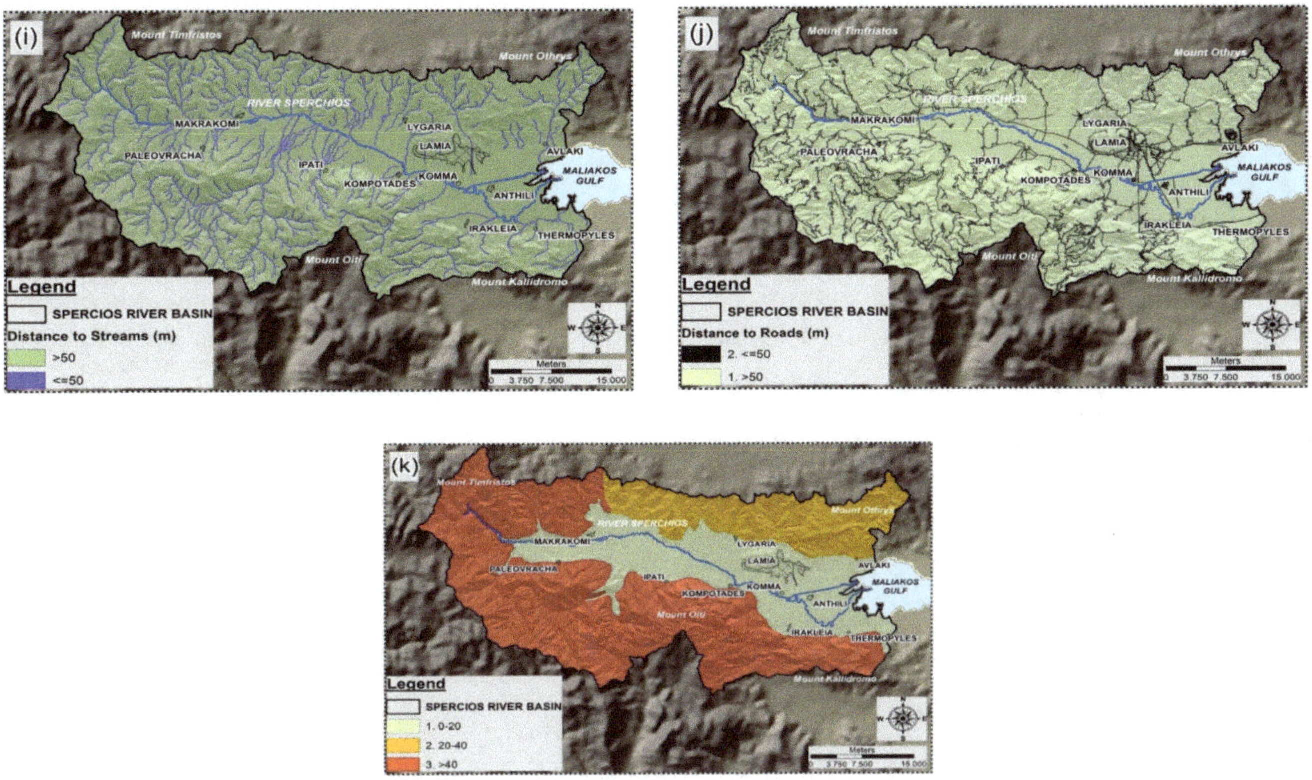

Figure 5. The thematic raster maps of the eleven (11) factors, used for the estimation of Landslide Susceptibility of Sperchios River basin: (**a**) Geological map, (**b**) Land use/cover map, (**c**) Slope map, (**d**) Slope aspect map, (**e**) Rainfall distribution map, (**f**) Soil depth map, (**g**) Curvature map, (**h**) Proximity to faults, (**i**) Distance to streams, (**j**) Distance to roads, and (**k**) Relative relief map.

Slope angle: Slope angle and geometry control slope stability and affect surface runoff, in conjunction to soil permeability and saturation capacity [5,7,12,33,40,83]. Gentle (low gradient) slopes are expected to display less LS than the steep ones, as in the case of the study area (although in some other cases surface landslides occur on low slopes in recent soils). Slope angle was modelled using a grid-based DEM, developed from topographic maps contour lines (20 m intervals) validated from the ASTER-derived DEM [28]. The slope angle was classified into five classes based on their effect on LS, as (a) very gentle slopes; 0–5%, (b) gentle slopes; 5–15%, (c) moderate steep slopes; 15–30%, (d) steep slopes; 30–45%, and (e) very steep slopes; >45% (Figure 5c—Table 1).

Slope aspect: The slope aspect defines the direction of slopes. Slope orientation indirectly influences landslides as it controls the exposure of slopes to various climatic features, such as rainfall, sun, and by extension, vegetation cover [84]. Slope aspect was grouped into five classes, as (a) 0–90° as NE, (b) 90–180° as SE, (c) 180–270° as SW, (d) 270–360° as NW, and (e) flat regions (Figure 5d—Table 1).

Mean Annual Rainfall distribution (MARd): Rainfall distribution affects overland water runoff volume and soil moisture. In the present study, the data of eight meteorological stations were used, covering 40 years (1971–2010). Precipitation was spatially distributed based on its mean annual values, utilizing the co-kriging interpolation method, using as covariate the basin elevation, as described in detail by Soulis et al. (2018) [85]. Thus, MARd was classified into five classes, as (a) 0–400 mm, (b) 400–600 mm, (c) 600–800 mm, (d) 800–1000 mm, and (e) >1000 mm (Figure 5e—Table 1) [11,48].

Soil depth: Soil depth affects soil percolation and the shape of the slope. Three classes were developed, based on the thickness of the top-soil layer, as (a) deep soils (>80 cm), (b) shallow soils (20–80 cm) and (c) rock (without soil cover or <20 cm) (Figure 5f—Table 1) [48].

Table 1. Classification of the landslide controlling factors and categorization into rating classes according to their significance on Landslide Susceptibility.

Factor	Class	Class Value Rating (i)	Class Value Ratting (ii)
Slope Angle (%)	0–5	1	0
	5–15	2	2
	15–30	3	3
	30–45	4	4
	>45	5	8
Lithology	Carbonate formations	1	0
	Neogene	2	1
	Schists-Ophiolites	3	2
	Alluvial dep.-Debris	4	3
	Flysch	5	6
Land Use/Cover	Coniferous forests	1	0
	Deciduous Forests	2	0
	Orchards-Olives	3	3
	Urban-Roads	4	4
	Arable land	5	5
	Nude soil and rocks	6	8
	Pastures	7	8
Slope Aspect	Flat areas	0	0
	SE	1	1
	NE	2	2
	SW	3	3
	NW	4	4
Relative Relief (m)	<20	1	0
	20–40	2	2
	>40	3	3
Slope Curvature (m^{-1})	>−0.19	1	0
	(−0.2)–(−0.59)	2	2
	(−0.6)–(−0.99)	3	3
	≤−1	4	4
Soil Depth (m)	Deep	1	0
	Shallow	2	2
	Rock	3	3
Distance to streams (m)	>50	1	0
	≤50	2	2
Proximity to faults (m)	>200	1	0
	≤200	2	2
Distance to roads (m)	>50	1	0
	≤50	2	2
Rainfall (mm)	0–400	1	0
	400–600	2	2
	600–800	3	3
	800–1000	4	4
	>1000	5	8

Slope Curvature: Slope curvature is a measure of a slope's morphology and topography. Positive values express upward curved surfaces, while negative values upward concave surfaces [30,86]. The lower negative values indicate greater LS. The classification of the slope curvature map comprises the following classes (a) ≥−0.19, (b) −0.2 to −0.59, (c) −0.60 to −0.99, and (d) ≤−1 (Figure 5g—Table 1).

Proximity to faults: Faulting earthquakes are considered as an important triggering factor of landslide events [16,17,87]. Moreover, contiguity to tectonic structures creates the possibility of a landslide phenomenon since erosion processes and water flow along a crack might be caused. For the fractures (both active and non-active faults), a buffer zone of 200 m width was developed, and the area was divided into two sections, i.e., inside and outside the zone (Figure 5h—Table 1) [11,48].

Distance to streams: Vicinity to streams is an important controlling factor of landslides since it can cause significant erosion processes (gully erosion, i.e., Gorgopotamos gorge) [5,88]. Third or higher-order streams (according to Strahler's classification) were selected and processed by creating a 50 m buffer zone. Thus, the basin area was divided into two sections, i.e., within and outside the zone (Figure 5i—Table 1).

Distance to roads: During the development of road networks, extensive excavations, and deforestation; removal of natural vegetation takes place quite often. Such actions, in conjunction with other natural factors (e.g., intense precipitation events), can cause significant landslide phenomena. A buffer zone of 50 m around the main road network of the study area was set, dividing it into two sections, i.e., within and outside the region (Figure 5j—Table 1) [7,11,48].

Relative relief: Relative relief portrays the absolute maximum difference in elevation at a specific point. Comparing slopes with identical geo-mechanical and geometrical parameters, the ones with the greater elevation deficit are more susceptible to landslides. In such landscapes, higher runoff and lower infiltration rates are expected [28,89]. Relative relief was created by using a unit area of 100 × 100 m and classified into 3 classes, as (a) 0–20, (b) 20–40, and (c) >40 (Figure 5k—Table 1).

3.2.4. Analytic Hierarchy Process

In AHP, all factors are compared pairwise in terms of the intensity of their importance using a continuous 1 to 9 point scale according to Saaty [90].

The nine-point scale of preference between two parameters in AHP is analyzed as follows 1: Equal importance i.e., two factors contribute equally to the objective, 3: Moderate prevalence of one over another i.e., experience and judgment slightly to moderately favor one factor over another, 5: Strong or essential prevalence i.e., experience and judgment strongly or essentially favor one factor over another, 7: Very strong or demonstrated prevalence i.e., a factor is strongly favored over another and its dominance is shown in practice, 9: Extremely high prevalence i.e., the evidence of favoring one factor over another is of the highest degree possible of an affirmation, 2; 4; 6; 8: Intermediate values i.e., reciprocals for inverse comparison, used to represent compromises between the preferences in weights 1, 3, 5, 7 and 9; Opposites i.e., used for inverse comparison.

Thus, the pairwise comparison matrix was created to calculate factor weights in AHP (Table 2). The diagonal boxes are assigned with a unit value, while the boxes in the upper and lower halves are symmetrical to one another and the corresponding values, and consequently reciprocal with each other. When the factor on the vertical axis is more important than the factor on the horizontal one, the values vary between 1 and 9. At the same time, and contrary to the previous notion, the value varies between the 1/2 and 1/9 reciprocals.

An essential feature of the AHP is that allows defining rating inconsistencies by the consistency index (CI), which is used defined by Equation (1) [36,91]:

$$CI = \frac{\lambda_{max} - N}{N - 1} \tag{1}$$

where λmax = the largest eigenvalue and N is the order of comparison matrix.

Table 2. Pair-wise comparison matrix and weights (normalized principal eigenvector) for landslide causative factors, derived from the analytic hierarchy process (AHP) method application.

	(i)	(ii)	(iii)	(iv)	(v)	(vi)	(vii)	(viii)	(ix)	(x)	(xi)	Weights
Geology (i)	1	2	2	4	4	4	2	5	6	7	8	0.227
Slope Gradient (ii)	1/2	1	1	2	3	3	3	3	6	7	8	0.155
Rainfall Distribution (iii)	1/2	1	1	1	3	3	3	3	5	6	7	0.140
Land Use/Cover (iv)	1/4	1/2	1	1	3	3	3	3	6	7	8	0.133
Slope Curvature (v)	1/4	1/3	1/3	1/3	1	3	3	3	6	7	8	0.10
Slope Aspect (vi)	1/4	1/3	1/3	1/3	1/3	1	3	3	6	7	8	0.081
Soil Depth (vii)	1/2	1/3	1/3	1/3	1/3	1/3	1	3	5	6	7	0.067
Relative Relief (viii)	1/5	1/3	1/3	1/3	1/3	1/3	1/3	1	3	5	6	0.044
Proximity to faults (ix)	1/6	1/6	1/5	1/6	1/6	1/6	1/5	1/4	1	4	3	0.024
Distance to Rivers (x)	1/7	1/7	1/6	1/7	1/7	1/7	1/6	1/5	1/4	1	2	0.015
Distance to Roads (xi)	1/8	1/8	1/7	1/8	1/8	1/8	1/7	1/6	1/3	1/2	1	0.012
CR = 0.072												

Saaty (1980) [36] developed an average random consistency index (RI) for different matrix orders and defined the consistency ratio (CR) as the ratio between the consistency index (*CI*) and the random consistency index (RI). The CR depends on the number of parameters. In case that CR is greater than 0.1, the comparison matrix is inconsistent and should be revised. In the present case, the CR value is less than 0.10 (0.08), which clarifies that the preferences used to create the comparison matrices are consistent [7,27,92].

3.2.5. Weighting of Parameters—Landslide Susceptibility Index

The various classes of the thematic layers were designated with the corresponding rating values inputted as attribute records in a GIS environment, and a raster map was developed for every data layer. Subsequently, the reclassified raster layers of the eleven produced maps were used as input data for the LSI estimation. After the weights' assignment, all factors were combined using the raster calculator procedure. The equations deriving from the implementation of the two different techniques, the Simple LSI and the GM-based LSI method [12,93] are analyzed as follows.

(a) Simple LSI (i):

Simple LSI (i) was calculated by multiplying the assigned rate values (Table 1) of the raster layers with the corresponding weights deriving from the AHP method (Equation (2)),

$$\text{LSI (i)} = \sum_{\text{i}=1}^{n} weight_{AHP} \times \text{class rate (i)} \tag{2}$$

where n is the total number of data layers.

(ii) Geometric mean LSI (ii):

The geometric mean is a multiplication-based method that ends when a null (0) value arises and is useful for concealing areas characterized by factors considered unconnected (the most stable areas) to the potential incident. By implementing the zero values the GM excludes the generally stable areas from the estimation as it depicts the areas which are predominantly vulnerable to landslides. The GM is assessed by the following Equation (3) [12,91,94],

$$\text{LSI (ii)} = \text{GM} = \left(\prod_{\text{i}=1}^{\text{n}} class\ rate\ \text{(ii)} \right)^{1/n} \tag{3}$$

where n is the total number of data layers; in order to serve the mathematical calculations, the zero values were replaced with the value 0.001.

The final step of the process involves the LSI maps validation alongside with landslide events and their distribution, having derived from the inventory database, and supplemented with potential landslide locations. The latter were mapped utilizing satellite images, very high-resolution orthophotos, and field observations serving as ground truth (Figure 2a,b) [11,19,62,65]. The final maps of the already applied methods were also compared and validated in order to select the most suitable for the study area [95,96].

3.2.6. Validation

The two LSI maps were validated based on two methodological approaches. The first method makes a comparison with the inventory map of the historic landslide events that occurred in the study area. Therefore, using the dataset that "preserved" for validation purposes, a GIS-based statistical analysis was made. The second method was formulated by creating a fishnet (3000 × 3000 m/point) that comprises 202 points located in the basin, which were intersected with the two LSI maps and the landslide frequency index (Figure 2b). In this case, the validation, in this case, was made using the root mean square error (RMSE) (Equation (4)). RMSE calculates the predicting errors of the two indices or in other words, measures the difference between the value predicted by each method and the LFI map,

$$\mathrm{RMSE} = \sqrt{\frac{1}{\mathrm{n}} \sum_{i=1}^{n} \left(\mathrm{V}_{\mathrm{predicted}} - \mathrm{V}_{\mathrm{actual}}\right)^2} \tag{4}$$

where n is the number of samples in the dataset used, $V_{predicted}$ is the value predicted from each method, and V_{actual} is the actual value in the LFI map.

3.2.7. Landslide Risk Analysis

In recent decades, increasing population and expansion of settlements joint by the subsequent intensification of anthropogenic activities (such as cultivations and transportation network) over vulnerable areas have essentially increased the impact of natural disasters. In the Sperchios basin the population increased approximately by 13.3% during the last four decades (with a small decrease in the last decade), while the extent of the urban fabric increased by almost 19.9%. Moreover, other intense human activities like land cultivation have increased by approximately 5.1%, while several public construction projects (some of noteworthy scale), have taken place in the region, involving the reconstruction of the main National Highway and Railway, the construction of the E65 highway, and enhancement and expansion of numerous secondary road network.

The LS map was managed in order to initially assess the landslide hazard zoning (LHZ) of the study area. LHZ assigns the estimated frequency (i.e., annual probability) to potential landslides and can be expressed as the specific landslide type of a given volume [97,98]. The determination of landslide hazard range requires the detection of the vulnerable areas, and the calculation of the probability of landslide occurrence within a pre-defined period. Since this period is often challenging to be demarcated, landslide hazard is frequently represented by LS [99].

Landslide risk (LR) utilizes the hazard mapping results to assess the landslide probability of occurrence and its severity to individuals (annual probability of human loss), property (the annual value of property loss), and environmental assets. Hence, LR depends on temporal and spatial probability (hazard), the vulnerability to such phenomena, and the human exposure to danger [12,84,100,101]. LR can be estimated based on Equation (5):

$$\mathrm{LR} = \mathrm{Hazard} \times \mathrm{Exposure} \times \mathrm{Vulnerability} \tag{5}$$

Exposure was assessed by creating a map of human activity in the area. The map was based on a combination of three factors, population density, land use/cover, and the road network extent. These factors were classified, with their classes being ranked based on the exposure to risk (Table 3). Vulnerability signifies the degree of loss to a given component within the area affected by the landslide, and it is expressed on a scale of 0 (no loss) to 1 (total loss). For human life and property, the loss is the value of the damage concerning the probability that peoples can be affected by the natural disaster (landslide) [12,101].

Table 3. Classification of risk factors and categorization into rating classes according to their significance on Landslide Risk.

Risk Factors	Class (Exposure)	Vulnerability	Class Rating	Importance
Population density	Low	Low to Moderate (0)	1	10
	Moderate		2	
	High	High to Very High (1)	3	
	Very High		4	
Land Use/Cover	Areas with low human intervention (Forests, pastures, etc.)	Low	1	8
	Cultivated areas	Moderate	2	
	Urban areas	High	3	
Road network	Distance to road >50 m	Low	1	6
	Distance to road ≤50 m	High	2	

Subsequently, the region's vulnerability and the weights of the three exposure factors were calculated (Table 3). The highest values were attributed to areas where human presence and activity are intense. Therefore, the impact of landslides in the corresponding risk assessment is high (Table 3). After that, the "human exposure" map was created by reclassified the map into three categories of Low to Moderate, Moderate to High, and Very High anthropogenic presence (Figure 6). The percentage of each category in the basin was 38.02%, 35.17%, and 26.81%, respectively.

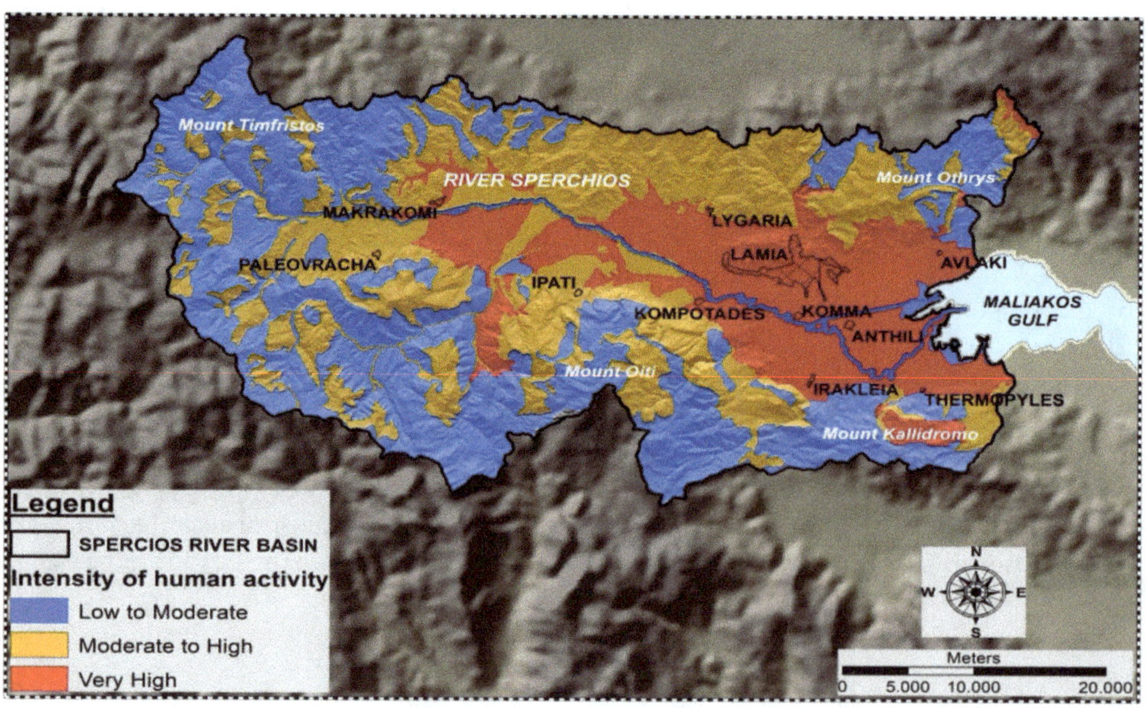

Figure 6. Representation of anthropogenic activity distribution in the Sperchios river basin.

4. Results and Discussion

From the statistical evaluation of the eleven selected factors related to landslide events, the distribution and correlation of past landslides to each factor was estimated. The analysis showed that in the geological map, 61.0% of past landslides are located on flysch formation (37 events), 13.6% on carbonate formations, and 13% on ophiolites (8 and 7 events, respectively). This high percentage of landslide events on flysch, which covers 42.8% of the basin, is due to its compound characteristics and especially to its composition, consisting of clayey or psammitic materials. This clayey, and usually thick soil layer, due to its high capacity to absorb water and then soften, or due to the existence of weathered basement rocks beneath the humus, creates potential slip surfaces (at the upper zone or between the weathered area and the bedrock), which are very susceptible to landslide occurrences. It was evident that the phenomenon is directly related to extreme weather events of high rainfall amounts that interact with the clayey flysch formation triggering numerous of landslide events. Likewise, very steep slopes lead to intensive instability conditions that contribute directly to the landslide phenomena. Slides that usually take place in the gentle slopes of the flysch mantle are typically quite shallow and take the form of a sheet of weathered zone sliding on a slip surface parallel to the ground [60,102,103]. Contrary, at the south-central and eastern parts where calcareous formations appear, the events are primarily related to rock falls.

60% of the study area is characterized by mean annual rainfall amounts higher than 600 mm. From these, 29% are greater than 800m, while the 44% falls on steep and very steep slopes, and 25% falls on moderate slopes. These characteristics and the fact that 39% of the basin's steep and very steep slopes, amounts are situated on flysch formation (where 38% of the mean annual rainfall amounts falls), generates large susceptible areas prone to fast weathering and gradual weakening, mostly triggered by prolonged or short intensive rainfall events, which result in many rotational slides and mudflows [52,104].

Furthermore, 74.5%, 43.1%, and 83.05% of the past landslide events occurred in forested areas, deep soils, and high relative relief, probably due to the fact that these characteristics mainly appears in the mountainous southern part which is also composed of flysch. This part of the basin exhibits intense tectonic activity (the basin is a graben which functions as a tectonic dipole, where the south part is lifted, and the north part is sunk), higher relief with very steep slopes (at a relatively small distance, an altitude difference of 300 m can be noted) and very high precipitation depths (>1000 mm) [52]. At the same time, the northern part displays milder topography with lower altitudes. Surprisingly, the number of landslides on slope aspects is almost equal to all directions, contrary to the fact that NW and SW aspects are more susceptible to landslide events in Greece [59]. The reason for this is probably the existence of all the above-mentioned characteristics in the SW part of the basin, which inevitably have a SE or NE orientation [52,60].

The correlation of landslide manifestations to faults, streams, and roads proximity, appears to be stronger to roads (27.1%), whereas the link is very weak (11.86%) to faults and streams (10.2%). Hence, these factors seem to have a meager contribution to landslides.

4.1. Landslide Susceptibility Indices

The LSI maps were classified into five categories, Very Low, Low, Moderate, High, and Very High susceptibility (Figure 7a,b). The generated maps were processed (dissolved and smoothed out using a 5 × 5 low-pass filter) to decrease the significance of possible misclassified cells. The percentage distribution of the two created LSI maps classes is demonstrated in Table 4. Higher LSI values indicate areas where the combination of factors is more likely to result in landslide events.

The GIS-based statistical analysis revealed that the most susceptible areas to landside manifestation (high and very high values) are placed at (a) slope angles with very high and high values approximately 66.17% for LSI (i) and 31.11% for LSI (ii), (b) flysch formation of about 72.33% for LSI (i) and 49.77% for LSI (ii), (c) mean annual precipitation > 1000 mm around 69.69% for LSI (i) and 34.41%, for LSI (ii), and (d) forest land cover approximately 59.21% and 48.62% for LSI (i) and LSI (ii), respectively.

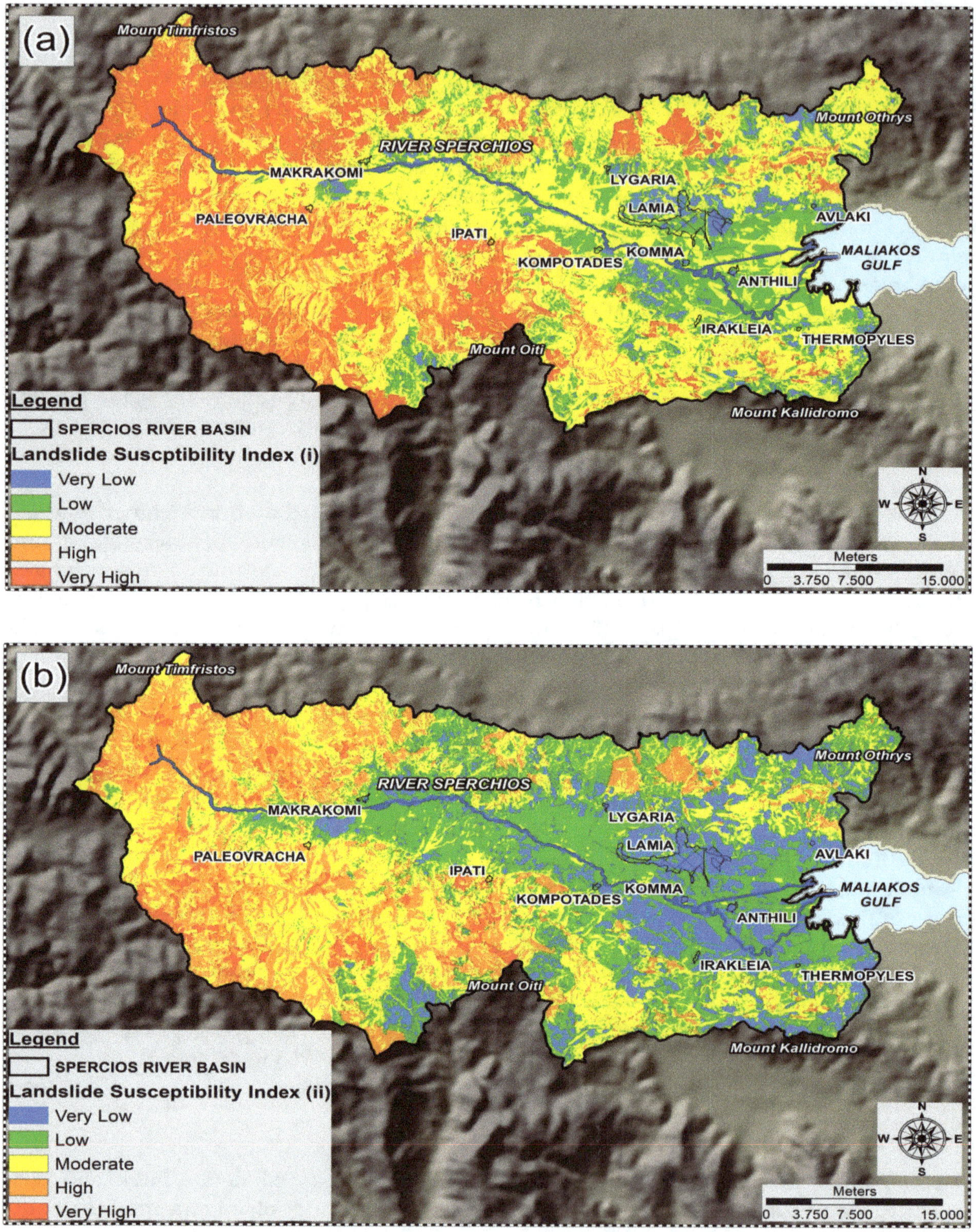

Figure 7. (**a**,**b**). The Landslide Susceptibility Indices (i) and (ii), after reclassifying the calculated values into five classes of potential susceptibility.

Table 4. Percentage distribution of the five classes of the two created landslide susceptibility (LSI) (i) and LSI (ii) maps, using the two different methods.

Susceptibility Classes	Very Low (%)	Low (%)	Moderate (%)	High (%)	Very High (%)
LSI (i)	4.52	16.03	43.09	21.00	15.35
LSI (ii)	14.34	30.39	29.79	23.18	2.31

It is apparent that the most sensitive areas (high and very high susceptibility) to landslides are mainly located in the western and southwestern parts of the basin, areas where flysch formations appear on steep slopes in conjunction with the highest rainfall amounts, and less in the lowland and coastal areas. These regions are dominated by flysch (capable of absorbing and retaining large amounts of water due to the formed clay-rich soils); moreover, they are characterized by intense and prolonged rainfalls, very steep slopes, and convex slope curvatures. Consequently, when the moisture amount in these soils exceeds a "moisture limit", these soils become extremely susceptible to landslides.

4.2. Accuracy Assessment

Using the first method, the validation analysis showed satisfactory results for LSI (i) since the overall matching acquires a value of 89.5% (Table 5). Contrary, the LSI (ii) showed a much lesser accuracy with a value of approximately 68.4% [11]. The RMSE indicated that the LSI (i) and LSI (ii) maps display a value of 0.21 and 0.43, respectively. The results of the method are perfect when RMSE values are equal to 0 [105,106]. Apparently, from the two validation approaches, can be determined that the LSI (i) method provides a more accurate assessment of landslide susceptibility in the study area.

Table 5. Confusion matrix of LS map validation.

Validation Sample		Target Class (Observed)	
		Susceptible Areas (Moderate, High, Very High Classes)	No Susceptible Areas (Low, Very Low Classes)
LSI (i)	Landslide areas	17	2
	Landslide-free areas	1	18
LSI (ii)	Landslide areas	13	6
	Landslide-free areas	3	16

4.3. Landslide Risk

Utilizing the classified LSI (i) map which appears to have better accuracy for the landslide susceptibility assessment of the area, the landslide risk analysis was designed. The risk map was developed by taking as a fact that the areas occupied by intense human presence (settlements, cultivated fields, and proximity to the road network) are ranked as moderate to high-risk zones, with continuous human presence, and are related to constant economic activities.

Based on the data inputs of hazard, exposure, vulnerability, the adopted weighting-ranking system, and the reclassification of the original values, the LR map was created (Figure 8). The LR values were grouped into five relative risk classes, which along with their percentage distribution into the study area, are demonstrated in Table 6.

Table 6. Percentage distribution of the five LR classes.

Landslide Risk Classes	Very Low (%)	Low (%)	Moderate (%)	High (%)	Very High (%)
Area (percentage)	16.60	26.60	12.10	39.72	4.98

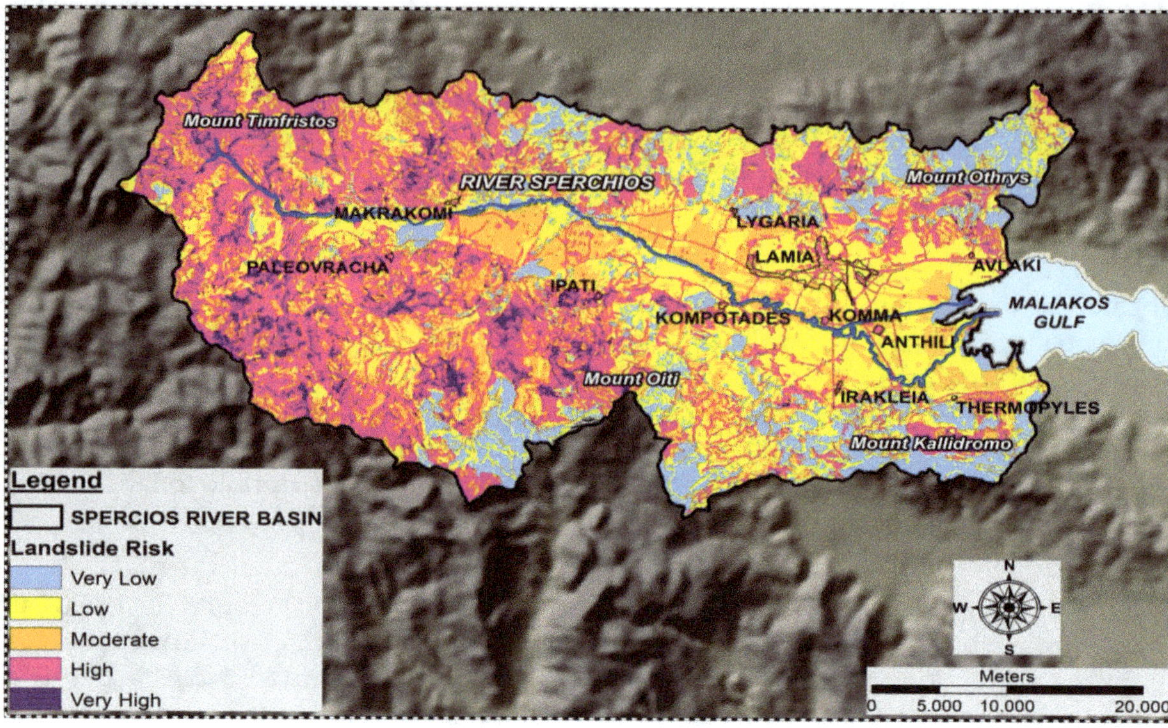

Figure 8. Landslide Risk map of the study area.

The statistical analysis of the LR map depicts that 44.7% of the area appears to have high and very high risk values, principally due to the increase of human activities in hilly areas nowadays. On the other hand, 43.2% of the region presents very low and low risk conditions to landslide hazard, which can be effortlessly explicated considering that the development of the main residential constructions and human activities are concentrated at the eastern coastal zone of the basin where landslides are more limited than on the southern and western mountainous regions.

4.4. Discussion

Landslides can trigger massive casualties and severe problems to the social and economic well-being [107,108]. The causes for landslides are numerous, complex, and every so regularly unidentified. For most morphometrical factors is challenging to be attempt to quantitatively measured them even in the field. Therefore, it is rather complicated to understand their contributions to the landslide occurrence mechanism [5,11,59,60,63]. Hence, in order to determine whether a factor affects the occurrence of landslides, several parameters were incorporated into a GIS-based analysis procedure [88,109].

The preparation of LS and LR maps is of great interest to planning organizations (at least) for introductory hazard studies, specifically when policy making is designed. Small-scale regional surveys are low-cost techniques by which larger areas can be covered in a relatively short time permitting an economical and rapid hazard assessment. Moreover, the LS map is essential for the delineation of the land use zones and the plan of future construction projects. Several methodologies have been applied for LS mapping. Eleven causative factors were considered, i.e., lithology, land use, slope angle, slope aspect, rainfall distribution, slope curvature, relative relief, proximity to faults, distance to roads, distance to streams, and soil depth. Their selection was based on the inventory map and several other supplementary data (aerial photos, satellite images, and fieldwork) [5,11,12,14,43].

Very high-resolution EO data and techniques along with Geographical Information System analysis provide powerful tools for the extraction of detailed geospatial layers, necessary for the landslide susceptibility assessment. In recent decades, many efforts have been made for integrated and synergistic use of these innovative technologies in order to achieve more accurate timely and cost-effective results, for large geographic areas [11,28,110–114].

From the two methods applied the simple landslide index provided more accurate results. The most hazardous to landslide occurrence regions are mainly concentrated in the southwest and west area of the basin, where the most significant factors such as geology (flysch), slope, rainfall distribution, (and much less land use, relative relief, curvature, and proximity to faults) are implicated in the major landslide events [11,19,61,63,65,115]. In addition, large historically observed landslides are clearly marked in the areas of high and extremely high susceptible zones. These results show that the predicted susceptibility levels are found to be in good agreement with the past landslides. Future research efforts mandate the examination of the seismic factor influence to landslide phenomena [17,87,116], since the study area belongs to a moderate-to-high seismic risk region of Greece (according to the Seismic Risk Map created by the Greek Earthquake Planning and Protection Organization) [117]. In the present study, this venture was limited by the lack of accurate data.

5. Conclusions

In this study, an attempt was made to create LS and LR maps by matching AHP and WLC methodologies and evaluate their results. The semi-quantitative method (AHP–WLC) is applicable and moreover accurate for LS mapping because of the pair-wise relative comparisons of the factors without discrepancies in the decision process. Moreover, the rating values reached in this study may be used in areas of similar geological, geomorphological, and hydro-climatic conditions.

The Sperchios River basin was the object of an investigation aiming to evaluate the risk of landslide disaster by using GIS and RS technology. It is confirmed that the integration of GIS techniques and earth observation (Sentinel-2) data is decisive in productively supporting researches regarding LS and LR of local and regional areas. The land cover map creation and the comparison of Aster-derived DEM showed suitable results, confirming the quality of the subsequent analysis, as LC and many of the factors produced from the DEM, influence LS and LR.

The landslide-prone areas, delineated by the LS and LR maps, represent an essential basis for the assessment of landslide hazard and risk over the study area. Therefore, the produced maps can be particularly useful to decision-makers for choosing suitable locations for future planning in large-scale regions and similarly in disaster management planning to prepare rescue routes, service centers, and shelters.

Author Contributions: Conceptualization, Emmanouil Psomiadis and Nikos Charizopoulos; methodology, Emmanouil Psomiadis, Nikos Charizopoulos, Nikolaos Efthimiou., and Konstantinos X. Soulis; software, Emmanouil Psomiadis, Nikos Charizopoulos, and Ioannis Charalampopoulos; data analysis, Emmanouil Psomiadis, Ioannis Charalampopoulos, Nikos Charizopoulos, Konstantinos X. Soulis, and Nikolaos Efthimiou; resources, Emmanouil Psomiadis, Konstantinos X. Soulis, and Nikos Charizopoulos; writing—original draft preparation, Emmanouil Psomiadis, Nikos Charizopoulos, Nikolaos Efthimiou, Konstantinos X. Soulis, and Ioannis Charalampopoulos; supervision, Emmanouil Psomiadis; and field work, Emmanouil Psomiadis and Nikos Charizopoulos. All authors have read and agreed to the published version of the manuscript.

Acknowledgments: The authors acknowledge USGS–NASA for the availability of Terra/Aster satellite data.

References

1. Cruden, D.M. A simple definition of a landslide. *Bull. Int. Assoc. Eng. Geol. Bull. l'Assoc. Int. Géol. l'Ing.* **1991**, *43*, 27–29. [CrossRef]
2. Yalcin, A.; Reis, S.; Aydinoglu, A.C.; Yomralioglu, T. A GIS-based comparative study of frequency ratio, analytical hierarchy process, bivariate statistics and logistics regression methods for landslide susceptibility mapping in Trabzon, NE Turkey. *Catena* **2011**, *85*, 274–287. [CrossRef]
3. Wang, J.; Peng, X. GIS-based landslide hazard zonation model and its application. *Procedia Earth Planet. Sci.* **2009**, *1*, 1198–1204. [CrossRef]

4. Martha, T.R.; Kerle, N.; Jetten, V.; van Westen, C.J.; Kumar, K.V. Characterising spectral, spatial and morphometric properties of landslides for semi-automatic detection using object-oriented methods. *Geomorphology* **2010**, *116*, 24–36. [CrossRef]
5. Rozos, D.; Bathrellos, G.D.; Skillodimou, H.D. Comparison of the implementation of rock engineering system and analytic hierarchy process methods, upon landslide susceptibility mapping, using GIS: A case study from the Eastern Achaia County of Peloponnesus, GREECE. *Environ. Earth Sci.* **2011**, *63*, 49–63. [CrossRef]
6. Chen, W.; Pourghasemi, H.R.; Naghibi, S.A. A comparative study of landslide susceptibility maps produced using support vector machine with different kernel functions and entropy data mining models in China. *Bull. Eng. Geol. Environ.* **2018**, *77*, 647–664. [CrossRef]
7. Achour, Y.; Boumezbeur, A.; Hadji, R.; Chouabbi, A.; Cavaleiro, V.; Bendaoud, E.A. Landslide susceptibility mapping using analytic hierarchy process and information value methods along a highway road section in Constantine, Algeria. *Arab. J. Geosci.* **2017**, *10*, 1–16. [CrossRef]
8. Guzzetti, F.; Carrara, A.; Cardinali, M.; Reichenbach, P. Landslide hazard evaluation: A review of current techniques and their application in a multi-scale study, Central Italy. *Geomorphology* **1999**, *31*, 181–216. [CrossRef]
9. Zhong, C.; Liu, Y.; Gao, P.; Chen, W.; Li, H.; Hou, Y.; Nuremanguli, T.; Ma, H. Landslide mapping with remote sensing: Challenges and opportunities. *Int. J. Remote Sens.* **2020**, *41*, 1555–1581. [CrossRef]
10. Van Westen, C.J.; Rengers, N.; Soeters, R. Use of geomorphological information in indirect landslide susceptibility assessment. *Nat. Hazards* **2003**, *30*, 399–419. [CrossRef]
11. Psomiadis, E.; Papazachariou, A.; Soulis, K.X.; Alexiou, D.S.; Charalampopoulos, I. Landslide mapping and susceptibility assessment using geospatial analysis and earth observation data. *Land* **2020**, *9*, 133. [CrossRef]
12. Kouli, M.; Loupasakis, C.; Soupios, P.; Vallianatos, F. Landslide hazard zonation in high risk areas of Rethymno Prefecture, Crete Island, Greece. *Nat. Hazards* **2010**, *52*, 599–621. [CrossRef]
13. Ladas, I.; Fountoulis, I.; Mariolakos, I. Using GIS & Multicriteria Decision analysis in landslide susceptibility mapping—A case study in Messinia prefecture area (SW Peloponnesus, Greece). *Bull. Geol. Soc. Greece* **2007**, *40*, 1973. [CrossRef]
14. Skilodimou, H.; Bathrellos, G.; Koskeridou, E.; Soukis, K.; Rozos, D. Physical and Anthropogenic Factors Related to Landslide Activity in the Northern Peloponnese, Greece. *Land* **2018**, *7*, 85. [CrossRef]
15. Serey, A.; Piñero-Feliciangeli, L.; Sepúlveda, S.A.; Poblete, F.; Petley, D.N.; Murphy, W. Landslides induced by the 2010 Chile megathrust earthquake: A comprehensive inventory and correlations with geological and seismic factors. *Landslides* **2019**, *16*, 1153–1165. [CrossRef]
16. Chunga, K.; Livio, F.A.; Martillo, C.; Lara-Saavedra, H.; Ferrario, M.F.; Zevallos, I.; Michetti, A.M. Landslides Triggered by the 2016 Mw 7.8 Pedernales, Ecuador Earthquake: Correlations with ESI-07 Intensity, Lithology, Slope and PGA-h. *Geosciences* **2019**, *9*, 371. [CrossRef]
17. Ferrario, M.F. Landslides triggered by multiple earthquakes: Insights from the 2018 Lombok (Indonesia) events. *Nat. Hazards* **2019**, *98*, 575–592. [CrossRef]
18. Guinau, M.; Vilajosana, I.; Vilaplana, J.M. GIS-based debris flow source and runout susceptibility assessment from DEM data? A case study in NW Nicaragua. *Nat. Hazards Earth Syst. Sci.* **2007**, *7*, 703–716. [CrossRef]
19. Sabatakakis, N.; Koukis, G.; Vassiliades, E.; Lainas, S. Landslide susceptibility zonation in Greece. *Nat. Hazards* **2013**, *65*, 523–543. [CrossRef]
20. Hervás, J.; Montanarella, L. Main issues on landslide mapping harmonization in EU member states in the framework of European Commission soil policy. In *Guidelines for Mapping Areas at Risk of Landslides in Europe*; Hervás, J., Ed.; EUR 23093 EN; Office for Official Publications of the European Communities: Luxembourg, 2007; pp. 7–10.
21. Fall, M.; Azzam, R.; Noubactep, C. A multi-method approach to study the stability of natural slopes and landslide susceptibility mapping. *Eng. Geol.* **2006**, *82*, 241–263. [CrossRef]
22. Carrara, A.; Crosta, G.; Frattini, P. Geomorphological and historical data in assessing landslide hazard. *Earth Surf. Process. Landf.* **2003**, *28*, 1125–1142. [CrossRef]
23. Hervás, J.; Bobrowsky, P. Mapping: Inventories, susceptibility, hazard and risk. In *Landslides—Disaster Risk Reduction*; Sassa, K., Canuti, P., Eds.; Springer Science and Business Media: Berlin/Heidelberg, Germany, 2009; pp. 321–349.
24. Ayalew, L.; Yamagishi, H.; Marui, H.; Kanno, T. Landslides in Sado Island of Japan: Part II. GIS-based

susceptibility mapping with comparisons of results from two methods and verifications. *Eng. Geol.* **2005**, *81*, 432–445. [CrossRef]

25. Carrara, A.; Guzzetti, F.; Cardinali, M.; Reichenbach, P. Use of GIS technology in the prediction and monitoring of landslide hazard. *Nat. Hazards* **1999**, *20*, 117–135. [CrossRef]
26. United Nations—SPIDER Knowledge Portal. Disaster Risk Management. Available online: http://www.un-spider.org/risks-and-disasters/disaster-risk-management (accessed on 7 July 2020).
27. Kayastha, P.; Dhital, M.R.; De Smedt, F. Application of the analytical hierarchy process (AHP) for landslide susceptibility mapping: A case study from the Tinau watershed, west Nepal. *Comput. Geosci.* **2013**, *52*, 398–408. [CrossRef]
28. Foumelis, M.; Lekkas, E.; Parcharidis, I. Landslide susceptibility mapping by GIS-based qualitative weighting procedure in Corinth area. *Bull. Geol. Soc. Greece* **2018**, *36*, 904. [CrossRef]
29. Carrara, A.; Cardinali, M.; Guzzetti, F. Uncertainty in assessing landslide hazard risk. *ITC J.* **1992**, *2*, 172–183.
30. Clerici, A.; Perego, S.; Tellini, C.; Vescovi, P. Landslide failure and runout susceptibility in the upper T. Ceno valley (Northern Apennines, Italy). *Nat. Hazards* **2010**, *52*, 1–29. [CrossRef]
31. Tsangaratos, P.; Rozos, D. Producing landslide susceptibility maps by applying expert knowledge in a GIS - based environment. *Bull. Geol. Soc. Greece* **2016**, *47*, 1539. [CrossRef]
32. Chalkias, C.; Kalogirou, S.; Ferentinou, M. Landslide susceptibility, Peloponnese Peninsula in South Greece. *J. Maps* **2014**, *10*, 211–222. [CrossRef]
33. Tsangaratos, P.; Ilia, I.; Hong, H.; Chen, W.; Xu, C. Applying Information Theory and GIS-based quantitative methods to produce landslide susceptibility maps in Nancheng County, China. *Landslides* **2017**, *14*, 1091–1111. [CrossRef]
34. Aleotti, P.; Chowdhury, R. Landslide hazard assessment: Summary review and new perspectives. *Bull. Eng. Geol. Environ.* **1999**, *58*, 21–44. [CrossRef]
35. Refice, A.; Capolongo, D. Probabilistic modeling of uncertainties in earthquake-induced landslide hazard assessment. *Comput. Geosci.* **2002**, *28*, 735–749. [CrossRef]
36. Saaty, T.L. *The Analytic Hierarchy Process*; McGraw-Hill, Ed.; Scientific Research Publishing: New York, NY, USA, 1980.
37. Saaty, T.L.; Vargas, L.G. *Prediction, Projection and Forecasting*; Kluwer Academic Publishers: Dordrecht, The Netherlands, 1991.
38. Mondal, S.; Maiti, R. Landslide Susceptibility Analysis of Shiv-Khola Watershed, Darjiling: A Remote Sensing & GIS Based Analytical Hierarchy Process (AHP). *J. Indian Soc. Remote Sens.* **2012**, *40*, 483–496. [CrossRef]
39. Peloponnese, N.; Papadakis, M.; Karimalis, A. Producing a Landslide Susceptibility Map through the Use of Analytic Hierarchical Process in Finikas watershed, North Peloponnese, Greece. *Am. J. GIS* **2017**, *6*, 14–22. [CrossRef]
40. Yalcin, A. GIS-based landslide susceptibility mapping using analytical hierarchy process and bivariate statistics in Ardesen (Turkey): Comparisons of results and confirmations. *Catena* **2008**, *72*, 1–12. [CrossRef]
41. Abay, A.; Barbieri, G.; Woldearegay, K. GIS-based Landslide Susceptibility Evaluation Using Analytical Hierarchy Process (AHP) Approach: The Case of Tarmaber District, Ethiopia. *Momona Ethiop. J. Sci.* **2019**, *11*, 14. [CrossRef]
42. Huabin, W.; Gangjun, L.; Weiya, X.; Gonghui, W. GIS-based landslide hazard assessment: An overview. *Prog. Phys. Geogr. Earth Environ.* **2005**, *29*, 548–567. [CrossRef]
43. Bathrellos, G.D.; Kalivas, D.P.; Skilodimou, H.D. GIS-based landslide susceptibility mapping models applied to natural and urban planning in Trikala, Central Greece. *Estud. Geológicos* **2009**, *65*, 49–65. [CrossRef]
44. Erener, A.; Sarp, G.; Duzgun, S.H. Use of GIS and Remote Sensing for Landslide Susceptibility Mapping. In *Advanced Methodologies and Technologies in Engineering and Environmental Science*; IGI Global: Hershey, PA, USA, 2019; pp. 384–398. [CrossRef]
45. Plank, S.; Twele, A.; Martinis, S. Landslide Mapping in Vegetated Areas Using Change Detection Based on Optical and Polarimetric SAR Data. *Remote Sens.* **2016**, *8*, 307. [CrossRef]
46. Fuchs, M.; Torizin, J.; Kühn, F. The effect of DEM resolution on the computation of the factor of safety using an infinite slope model. *Geomorphology* **2014**, *224*, 16–26. [CrossRef]
47. Schlögel, R.; Marchesini, I.; Alvioli, M.; Reichenbach, P.; Rossi, M.; Malet, J.P. Optimizing landslide susceptibility zonation: Effects of DEM spatial resolution and slope unit delineation on logistic regression models. *Geomorphology* **2018**, *301*, 10–20. [CrossRef]

48. Kawabata, D.; Bandibas, J. Landslide susceptibility mapping using geological data, a DEM from ASTER images and an Artificial Neural Network (ANN). *Geomorphology* **2009**, *113*, 97–109. [CrossRef]
49. Efthimiou, N.; Psomiadis, E.; Panagos, P. Fire severity and soil erosion susceptibility mapping using multi-temporal Earth Observation data: The case of Mati fatal wildfire in Eastern Attica, Greece. *Catena* **2020**, *187*. [CrossRef] [PubMed]
50. Zamani, A.; Maroukian, H. A morphological study of an old delta of the Sperchios River. In *Proceedings of the 6th Colloquium on the Geology of the Aegean Region*; Kallergis, G., Ed.; Institute of Geological and Mining Research: Athens, Greece, 1979; pp. 261–282.
51. Psomiadis, E.; Parcharidis, I.; Poulos, S.; Stamatis, G.; Migiros, G.; Pavlopoulos, A. Earth observation data in seasonal and long term coastline changes monitoring the case of Sperchios river delta (central Greece). *Z. Geomorphol. Suppl.* **2005**, *137*, 159–175.
52. Psomiadis, E. Research of Geomorphological and Environmental Changes in the Sperchios' River Basin Utilizing New Technologies. Ph.D. Thesis, Agricultural Univeristy of Athens, Athens, Greece, 2010. (In Greek).
53. Psomiadis, E. Flash flood area mapping utilising SENTINEL-1 radar data. In Proceedings of the Earth Resources and Environmental Remote Sensing/GIS Applications VII, Edinburgh, UK, 26–29 September 2016; SPIE: Bellingham, WA, USA, 2016; Volume 10005, p. 100051G.
54. Markonis, Y.; Batelis, S.C.; Dimakos, Y.; Moschou, E.; Koutsoyiannis, D. Temporal and spatial variability of rainfall over Greece. *Theor. Appl. Climatol.* **2017**, *130*, 217–232. [CrossRef]
55. Ferrière, J. Recent Developments in the Maliakos Isopic Zone, Eastern Central Greece. In *Proceedings of the 6th Colloquium Geology of the Aegean Region*; Institute of Geological and Mining Research: Athens, Greece, 1977; pp. 197–210.
56. Maroukian, H.; Lagios, E. Neotectonic movements in the Sperkhios River basin, Central Greece. *Z. Geomorphol. Suppl.* **1987**, *63*, 133–140.
57. Psomiadis, E.; Parcharidis, I.; Stamatis, G.; Foumelis, M. Remotely sensing data and thematic mapping for sustainable developing in Sperchios river basin (Central Greece). In Proceedings of the SPIE—The International Society for Optical Engineering, Bellingham, WA, USA, 21–26 January 2005; Volume 5983.
58. Psomiadis, E.; Migiros, G.; Antoniou, V. Geomorphological quantitative analysis of Sperchios River Basin area (Central Greece) utilizing geographical information systems. *Bull. Geol. Soc. Greece* **2013**, *47*, 325. [CrossRef]
59. Koukis, G.; Sabatakakis, N.; Ferentinou, M.; Lainas, S.; Alexiadou, X.; Panagopoulos, A. Landslide phenomena related to major fault tectonics: Rift zone of Corinth Gulf, Greece. *Bull. Eng. Geol. Environ.* **2009**, *68*, 215–229. [CrossRef]
60. Sabatakakis, N.; Koukis, G.; Mourtas, D. Composite landslides induced by heavy rainfalls in suburban areas: City of Patras and surrounding area, western Greece. *Landslides* **2005**, *2*, 202–211. [CrossRef]
61. Koukis, G.; Tsiambaos, G.; Sabatakakis, N. Landslides in Greece: Research evolution and quantitative analysis. In Proceedings of the 7th International Symposium on Landslides, Trondheim, Norway, 17–21 June 1996; pp. 1935–1940.
62. Koukis, G.; Sabatakakis, N.; Nikolaou, N.; Loupasakis, C. Landslide hazard zonation in Greece. In *Proceedings of the Open Symposium on Landslide Risk Analysis and Sustainable Disaster Management in the First General Assembly of International Consortium on Landslides*; Sassa, K., Fukuoka, H., Wang, F.W.G., Eds.; Springer: Berlin, Germany, 2005; pp. 291–296.
63. Koukis, G.; Sabatakakis, N.; Tsiambaos, G. Slope movements in the Greek territory: A statistical approach. In Proceedings of the 7th International Congress of International Association of Engineering Geology, Lisbon, Portugal, 5–9 September 1994; pp. 4621–4628.
64. Apostolidis, E. *Palaio Mikro Chorio Landslide, U-Geohaz & EOEG Field Trip*; Faculty of Geology and Geoenvironment: Athens, Greece, 2019.
65. Koukis, G.; Ziourkas, C. Slope instability phenomena in Greece: A statistical analysis. *Bull. Int. Assoc. Eng. Geol. Bull. l'Assoc. Int. Géol. l'Ing.* **1991**, *43*, 47–60. [CrossRef]
66. Mpliona, M. *Landslide Database Development in Greece*; University of Patras: Patras, Greece, 2008.
67. Rozos, D.; Apostolidis, E. Landslide phenomena in Kanalia village, Fthiotida Prefecture and handling their impacts in the residential development of the area. *Bull. Geol. Soc. Greece* **2004**, *36*, 1816. [CrossRef]

68. Institute of Geological and Mineral Exploration (IGME). *Geological Maps of Greece Series, Scale 1:50,000, Sheets: Lamia, Stylis, Sperkhias, Karpenision, Anavra, Domokos, Fourna, Leontarion, Lidorikion, Amfiklia, Amfissa*; IGME: Athens, Greece, 1991.
69. Hellenik Military Geographical Service (HMGS). *Topographic Maps of Greece Series, Scale 1:50,000, Sheets: Lamia, Stylis, Sperkhias, Karpenision, Efxinoupolis, Domokos, Fourna, Leontarion, Lidorikion, Amfiklia, Amfissa*; HMGS: Athens, Greece, 1990.
70. Lee, J.-D.; Han, S.-H.; Lee, S.-S.; Parkd, J.-S. Correcting DEM extracted from ASTER stereo images by combining cartographic DEM. In Proceedings of the International Society for Photogrammetry and Remote Sensing Congress, Beijing, China, 3–11 July 2008; pp. 829–834.
71. Directorates of Forests/Ministry of Agriculture. *Soil Maps of Greece Series, Scale 1:50.000, Sheets Lamia, Stylis, Sperkhias, Karpenision, Efxinoupolis, Domokos, Fourna, Leontarion, Lidorikion, Amfiklia, Amfissa*; Directorates of Forests: Athens, Greece, 1981.
72. Hellenic Statistical Authority/ELSTAT Population Demographic Data 1981. 1991. Available online: https://www.statistics.gr/en/statistics/-/publication/SAM03/2011 (accessed on 7 July 2020).
73. Soulis, K.X.; Manolakos, D.; Anagnostopoulos, J.; Papantonis, D. Development of a geo-information system embedding a spatially distributed hydrological model for the preliminary assessment of the hydropower potential of historical hydro sites in poorly gauged areas. *Renew. Energy* **2016**, *92*, 222–232. [CrossRef]
74. R Core Team. European Environment Agency. 2019. Available online: https://www.eea.europa.eu/data-and-maps/indicators/oxygen-consuming-substances-in-rivers/r-development-core-team-2006 (accessed on 7 July 2020).
75. Nikolakopoulos, K.G.; Chrysoulakis, N. Updating the 1:50.000 topographic maps using ASTER and SRTM DEM: The case of Athens, Greece. In Proceedings of the Remote Sensing for Environmental Monitoring, GIS Applications, and Geology VI, Stockholm, Sweden, 11–14 September 2006; Ehlers, M., Michel, U., Eds.; SPIE: Bellingham, WA, USA, 2006; Volume 6366, p. 636606.
76. Vadrevu, K.P. *Introduction to Remote Sensing*, 5th ed.; Campbell, J.B., Wynne, R.H., Eds.; Guilford Press: New York, NY, USA, 2011; ISBN 9781609181765. [CrossRef]
77. Csillik, O.; Belgiu, M.; Asner, G.P.; Kelly, M. Object-Based Time-Constrained Dynamic Time Warping Classification of Crops Using Sentinel-2. *Remote Sens.* **2019**, *11*, 1257. [CrossRef]
78. Psomiadis, E.; Soulis, K.X.; Efthimiou, N. Using SCS-CN and earth observation for the comparative assessment of the hydrological effect of gradual and abrupt spatiotemporal land cover changes. *Water* **2020**, *12*, 1386. [CrossRef]
79. Cohen, J. A Coefficient of Agreement for Nominal Scales. *Educ. Psychol. Meas.* **1960**, *20*, 37–46. [CrossRef]
80. Congalton, R.G. A review of assessing the accuracy of classifications of remotely sensed data. *Remote Sens. Environ.* **1991**, *37*, 35–46. [CrossRef]
81. Montgomery, D.R.; Schmidt, K.M.; Greenberg, H.M.; Dietrich, W.E. Forest clearing and regional landsliding. *Geology* **2000**, *28*, 311–314. [CrossRef]
82. Belgiu, M.; Csillik, O. Sentinel-2 cropland mapping using pixel-based and object-based time-weighted dynamic time warping analysis. *Remote Sens. Environ.* **2018**, *204*, 509–523. [CrossRef]
83. Clerici, A.; Perego, S.; Tellini, C.; Vescovi, P. A procedure for landslide susceptibility zonation by the conditional analysis method. *Geomorphology* **2002**, *48*, 349–364. [CrossRef]
84. Dai, F.C.; Lee, C.F.; Ngai, Y.Y. Landslide risk assessment and management: An overview. *Eng. Geol.* **2002**, *64*, 65–87. [CrossRef]
85. Soulis, K.; Kalivas, D.; Apostolopoulos, C. Delimitation of Agricultural Areas with Natural Constraints in Greece: Assessment of the Dryness Climatic Criterion Using Geostatistics. *Agronomy* **2018**, *8*, 161. [CrossRef]
86. Ohlmacher, G.C. Plan curvature and landslide probability in regions dominated by earth flows and earth slides. *Eng. Geol.* **2007**, *91*, 117–134. [CrossRef]
87. Livio, F.; Ferrario, M.F. Assessment of attenuation regressions for earthquake-triggered landslides in the Italian Apennines: Insights from recent and historical events. *Landslides* **2020**, 1–12. [CrossRef]
88. Chen, W.; Xie, X.; Peng, J.; Shahabi, H.; Hong, H.; Bui, D.T.; Duan, Z.; Li, S.; Zhu, A.X. GIS-based landslide susceptibility evaluation using a novel hybrid integration approach of bivariate statistical based random forest method. *Catena* **2018**, *164*, 135–149. [CrossRef]
89. Singh, V.P. *Hydrologic Systems: Watershed Modeling*; Prentice-Hall, Ed.; Prentice-Hall: Upper Saddle River, NJ, USA, 1989; Volume 2.

90. Saaty, T.L. A scaling method for priorities in hierarchical structures. *J. Math. Psychol.* **1977**, *15*, 234–281. [CrossRef]
91. Saaty, T.L. *Fundamentals of Decision Making and Priority Theory with the Analytic Hierarchy Process*; RWS Publications, Ed.; RWS Publications: Pittsburgh, PA, USA, 2000.
92. Malczewski, J. *GIS and Multicriteria Decision Analysis*; Wiley, Ed.; Wiley: New York, NY, USA, 1999.
93. Fourniadis, I.G.; Liu, J.G.; Mason, P.J. Landslide hazard assessment in the Three Gorges area, China, using ASTER imagery: Wushan-Badong. *Geomorphology* **2007**, *84*, 126–144. [CrossRef]
94. Liu, J.G.; Mason, P.J.; Clerici, N.; Chen, S.; Davis, A.; Miao, F.; Deng, H.; Liang, L. Landslide hazard assessment in the Three Gorges area of the Yangtze river using ASTER imagery: Zigui-Badong. *Geomorphology* **2004**, *61*, 171–187. [CrossRef]
95. Pourghasemi, H.; Gayen, A.; Park, S.; Lee, C.-W.; Lee, S. Assessment of Landslide-Prone Areas and Their Zonation Using Logistic Regression, LogitBoost, and NaïveBayes Machine-Learning Algorithms. *Sustainability* **2018**, *10*, 3697. [CrossRef]
96. Nguyen, V.-T.; Tran, T.H.; Ha, N.A.; Ngo, V.L.; Nadhir, A.-A.; Tran, V.P.; Duy Nguyen, H.; MA, M.; Amini, A.; Prakash, I.; et al. GIS Based Novel Hybrid Computational Intelligence Models for Mapping Landslide Susceptibility: A Case Study at Da Lat City, Vietnam. *Sustainability* **2019**, *11*, 7118. [CrossRef]
97. Varnes, D.J. *Slope Movement Types and Processes/Special Report 176: Landslides: Analysis and Control*; Transportation Research Board: Washington, DC, USA, 1978.
98. Hutchinson, J.N. General Report: Morphological and geotechnical parameters of landslides in relation to geology and hydrogeology. In Proceedings of the 5th International Conference on Landslides, Lausanne, Switzerland, 10–15 July 1988; Bonnard, C., Ed.; TRB, National Research Council: Lausanne, Switzerland, 1988; Volume 1, pp. 3–35.
99. Brabb, E.E. Innovative approaches to landslide hazard and risk mapping. In *Proceedings of the 4th International Symposium on Landslides*; Canadian Geotechnical Society, Ed.; Canadian Geotechnical Society: Toronto, ON, Canada, 1984; Volume 1, pp. 307–324.
100. Van Westen, C.J.; van Asch, T.W.J.; Soeters, R. Landslide hazard and risk zonation—Why is it still so difficult? *Bull. Eng. Geol. Environ.* **2006**, *65*, 167–184. [CrossRef]
101. Fell, R.; Corominas, J.; Bonnard, C.; Cascini, L.; Leroi, E.; Savage, W.Z. Guidelines for landslide susceptibility, hazard and risk zoning for land use planning. *Eng. Geol.* **2008**, *102*, 85–98. [CrossRef]
102. Christoulas, S.; Kalteziotis, N.; Gassios, E.; Sabatakakis, N.; Tsiambaos, G. Instability phenomena in weathered flysch in Greece. In Proceedings of the 5th International Symposium on Landslides, Lausanne, Switzerland, 10–15 July 1988; Balkema, Ed.; Balkema: Rotterdam, The Netherlands, 1988; pp. 103–108.
103. Petkovšek, A.; Fazarinc, R.; Kočevar, M.; Maček, M.; Majes, B.; Mikoš, M. The Stogovce landslide in SW Slovenia triggered during the September 2010 extreme rainfall event. *Landslides* **2011**, *8*, 499–506. [CrossRef]
104. Peternel, T.; Mikoš, M.; Đomlija, P.; Dugonjić-Jovančević, S.; Arbanas, Ž. Geological conditions of landslides in flysch deposits in Slovenia and Croatia. In Proceedings of the 2nd Regional Symposium on Landslides in the Adriatic-Balkan Region Belgrade, Belgrade, Serbia, 14–16 May 2015.
105. Faraji Sabokbar, H.; Shadman Roodposhti, M.; Tazik, E. Landslide susceptibility mapping using geographically-weighted principal component analysis. *Geomorphology* **2014**, *226*, 15–24. [CrossRef]
106. Nguyen, V.; Pham, B.; Vu, B.; Prakash, I.; Jha, S.; Shahabi, H.; Shirzadi, A.; Ba, D.; Kumar, R.; Chatterjee, J.; et al. Hybrid Machine Learning Approaches for Landslide Susceptibility Modeling. *Forests* **2019**, *10*, 157. [CrossRef]
107. Haque, U.; Blum, P.; da Silva, P.F.; Andersen, P.; Pilz, J.; Chalov, S.R.; Malet, J.P.; Auflič, M.J.; Andres, N.; Poyiadji, E.; et al. Fatal landslides in Europe. *Landslides* **2016**, *13*, 1545–1554. [CrossRef]
108. Schuster, R.L.; Fleming, R.W. Economic Losses and Fatalities Due to Landslides. *Environ. Eng. Geosci.* **1986**, *xxiii*, 11–28. [CrossRef]
109. Myronidis, D.; Papageorgiou, C.; Theophanous, S. Landslide susceptibility mapping based on landslide history and analytic hierarchy process (AHP). *Nat. Hazards* **2016**, *81*, 245–263. [CrossRef]
110. Pardeshi, S.D.; Autade, S.E.; Pardeshi, S.S. Landslide hazard assessment: Recent trends and techniques. *Springerplus* **2013**, *2*, 523. [CrossRef]
111. Negi, H.S.; Kumar, A.; Rao, N.N.; Thakur, N.K.; Shekhar, M.S. Snehmani Susceptibility assessment of rainfall induced debris flow zones in Ladakh–Nubra region, Indian Himalaya. *J. Earth Syst. Sci.* **2020**, *129*, 1–20. [CrossRef]

112. Tsangaratos, P.; Loupasakis, C.; Nikolakopoulos, K.; Angelitsa, V.; Ilia, I. Developing a landslide susceptibility map based on remote sensing, fuzzy logic and expert knowledge of the Island of Lefkada, Greece. *Environ. Earth Sci.* **2018**, *77*, 363. [CrossRef]
113. Lee, S.; Choi, J.; Min, K. Probabilistic landslide hazard mapping using GIS and remote sensing data at Boun, Korea. *Int. J. Remote Sens.* **2004**, *25*, 2037–2052. [CrossRef]
114. Sarkar, S.; Kanungo, D.P. An integrated approach for landslide susceptibility mapping using remote sensing and GIS. *Photogramm. Eng. Remote Sens.* **2004**, *70*, 617–625. [CrossRef]
115. Fernandez-Steeger, T.M.; Maessen, T.; Grenzdörfer, M.; Schneiderwind, S.; Papanikolaou, I.; Deligiannakis, G.; Migiros, G.; Psomiadis, E.; Kairis, O.; Pallikarakis, A. How geology influences the type and magnitude of postfire effects like landslides. In Proceedings of the EGU General Assembly 2016, Vienna, Austria, 17–22 April 2016; Volume 18, p. 10994-3.
116. Wartman, J.; Dunham, L.; Tiwari, B.; Pradel, D. Landslides in eastern Honshu induced by the 2011 Off the Pacific Coast of Tohoku earthquake. *Bull. Seismol. Soc. Am.* **2013**, *103*, 1503–1521. [CrossRef]
117. Earthquake Planning and Protection Organization, Greece. Seismic Risk Map of Greece. Available online: https://www.oasp.gr/node/87 (accessed on 9 July 2020).

3

Influence of Land Use and Land Cover Change on the Formation of Local Lightning

Sanjib Kumar Kar and Yuei-An Liou *

Center for Space and Remote Sensing Research, National Central University, No. 300, Jhongda Rd., Jhongli Dist., Taoyuan City 32001, Taiwan; sanjib_kar@hotmail.com

* Correspondence: yueian@csrsr.ncu.edu.tw

Abstract: Land use and land cover (LULC) play a crucial role in the interaction between the land and atmosphere, influencing climate at local, regional, and global scales. LULC change due to urbanization has significant impacts on local weather and climate. Land-cover changes associated with urbanization create higher air temperatures compared to the surrounding rural area, known as the "urban heat island (UHI)" effect. Urban landscapes also affect formation of convective storms. In recent years, the effect of urbanization on local convections and lightning has been studied very extensively. In this paper a long-term study has been carried out taking cloud-to-ground (CG) lightning data (1998–2012) from Tai-Power Company, and particulate matter (PM10), sulfur dioxide (SO_2) data (2003–2012) from the Environmental Protection Administration (EPA) of Taiwan, in order to investigate the influence of LULC change through urbanization on CG lightning activity over Taipei taking into account in situ data of population growth, land use change and mean surface temperature (1965–2010). The thermal band of the Land-Sat 7 satellite was used to generate the apparent surface temperature of New Taipei City. It was observed that an enhancement of 60–70% in the flash density over the urban areas compared to their surroundings. The spatial distribution of the CG lightning flashes follows closely the shape of the Taipei city heat island, thereby supporting the thermal hypothesis. The PM10 and SO_2 concentrations showed a positive linear correlation with the number of cloud-to-ground flashes, supporting the aerosol hypothesis. These results indicate that both hypotheses should be considered to explain the CG lightning enhancements over the urban areas. The results obtained are significant and interesting and have been explained from the thermodynamic point of view.

Keywords: Land use and land cover (LULC); cloud-to-ground (CG) lightning; particulate matter (PM10); sulfur dioxide (SO_2)

---

1. Introduction

The influence of urban environment on the lightning activity remains a debatable topic, beginning with first publication of Westcott [1], followed by several studies in US, Europe, Asia, Brazil and Australia [2–7]. The change of LULC in urban areas leads to a change of surface characteristics. Such change affects not only the dynamic and physical processes in the atmospheric boundary layer, but also affects the surface heat budget to form an urban heat island (UHI). The UHI effect and its associated impact on environment and regional climate have been receiving wide attention in recent years. It has already been established that the intensity of the UHI effect is strongly associated with the urban size, urban surface characteristics, anthropogenic heat release, topography, and meteorological conditions [8–13]. It has been hypothesized that convection is enhanced by a UHI-induced mesoscale circulation and as a result of such enhanced convection; increased thunderstorm and lightning activity accompanied with enhanced precipitation are observed over or downwind of major urban areas. During the Metropolitan Meteorological Experiment [14,15], this hypothesis was extensively investigated and substantiated. Modern urbanization, which caused dramatic change in LULC, is basically the demand of increased population and human activity. Contributions of manmade aerosols are also enhanced because of urbanization having significant impacts on local thunderstorm formation leading to an enhanced CG lightning activity particularly over urban area [16–18]. However, the physical mechanisms responsible for these effects are not fully understood due to the complex correlations.

The climate of coastal cities like Taipei is influenced by the development of cumulus convection caused by boundaries that occur when moist air is transported by the sea breeze from ocean to land. Taipei City is surrounded by mountains in all directions, except the two river valleys. The two river valleys, Tanshui and Keelung, respectively, in the northwest and northeast of Taipei City carry surface airflow to the open sea. Sea breezes funnel toward the southern part of the Taipei city through two river valleys. These low-level flows are expected to interact not only with the mountains situated south of Taipei City, but also with the UHI of Taipei City. Blending of UHI effect and sea breezes with the local synoptic and orographic structure causes the summer thunderstorm and lightning activity within the New Taipei City and Taipei City to develop a unique climate system. Exploring the impact of such triple interaction on the local atmospheric circulation Kitada et al. [19] indicated that urbanization has nominal effect on the diurnal airflow if urbanization is built up 100–200 km away from mountains. In contrast, a shift of the highest temperature zone from the center of the city to inland suburbs is evident in the case of extensive urbanized coastal areas. The development of Taipei City was started since 1960s, and with time it has become one of the largest trade centers in Eastern and Southeast Asia. Due to such rapid development in the past several decades, Taipei City has been well urbanized. The urban population has increased by a factor of 3.5. The land use for construction has also increased by a factor of three in the last few decades [20].

The effect of UHI leads to many serious environmental problems, such as regional climate change, air pollution, and visibility deterioration. For example, the diurnal temperature change has been decreasing since the 1970s in Taiwan [21]. The occurrence of fog in hours has decreased significantly over the last 40 years [21]. Temperature has been increasing at a rate of approximately 1.1–1.6 °C/century as recorded at eight lowland meteorology stations in Taiwan since 1900 to 2009. This rate of increase is consistent with the warming conditions of other East Asia countries, but significantly higher than the global rate of 0.74 °C/century [22,23]. During the last 30 years, the warming rate in Taiwan has almost doubled and reached 0.23–0.40 °C/decade. LULC caused by rapid urbanization has been attributed as one of the leading cause of such increasing trend of warming by researchers [24,25].

Taiwan, a mountainous island located at western Pacific coast region, is bisected by the Tropic of Cancer and has 300-km Central Mountain Range (CMR) stretching along north to south with more than 100 peaks above 3,000 m. The hot and humid weather condition provides plentiful lightning strikes over this island during premonsoon and monsoon seasons. Considerable research on thunderstorms and lightning has been conducted over tropical areas and in the United States. Compared to those studies the number of studies relating to Taiwan is limited [5,26,27]. The purpose of this study is to investigate the impact of LULC change caused by rapid urbanization on the formation of local lightning during the warm seasons (May–October) from 1998 to 2012 over Taipei, the capital of Taiwan. Possible influence of urban particulate matter on the enhancement of CG lightning activity is another focus of this study. More precisely we have tried to investigate in this study whether the local climate of Taipei is affected by this urbanization through an increase in lightning activity particularly during summer.

2. Data and Methodology

For the present study, we have collected lightning data from Tai-Power Company of Taiwan for the years 1998–2012 to determine the urban effect on lightning over Taipei and its surrounding places. The Lightning Location System (LLS) was built in 1989 with one APA (Advanced Position Analyzer), and six Direction Finders (DFs) installed at sites covering the entire area of Taiwan. The LLS was upgraded to a Total Lightning Detection System (TLDS) in 2002. The TLDS consist of seven lightning detection sensors (SAFIR 3000), which are located at the top of Ying-tsu-ling microwave tower, Wu-shih-pi microwave tower, Ji-shan microwave tower, Nan-Ke extra voltage substation, Feng-lin microwave tower, building roof of Ming-tan power plant, and building roof of Xiao-liou-chiou. The location of these seven sensors, distributed throughout Taiwan, is shown in Figure 1. The VHF interferometric technique is the main basis for the localization principles of SAFIR network [28,29]. The seven lightning detection sensors, formed a lightning detection network, could detect cloud-to-ground (CG) lightning discharges, intra cloud (IC) lightning discharges, and breakdown events. The lightning discharges detection is accomplished through the use of multiple, remote sensors that detect signals emitted by lightning discharges, and by filtering out the signals from non-lightning sources. The long rang localization of all lightning discharges (CG and CC lightning flashes) is governed by triangulation performed on GPS time synchronized direction of arrival provided by interferometric sensor of two different detection station in a SAFIR network. Each sensor detecting a lightning event sends data about the event to a central processor (SCM) that triangulates the results from each sensor creating an optimal estimate of location of the lightning event. The lightning detection network average efficiency is greater than 90%, and the lightning detection localization accuracy is less than 1 km. However, especially near the edges of the network the assumption of more than 90% uniform flash detection efficiency may not be realistic, but because of comparatively higher average detection efficiency and localization accuracy no attempt was taken to correct the detection efficiency because previous studies (e.g., [30] for a Lightning Position and Tracking System (LPATS); [31] for the National Lightning Detection Network in the United States; [32] for a LPATS in Germany, and [33] for a LPATS in Brazil) reported an overall detection efficiency of 90% for several lightning detection networks. For the present study we have only considered only CG lightning discharges and have ignored the IC discharges. CG lightning flash density was computed from the aggregated CG lightning flash data for the period 1998–2012.

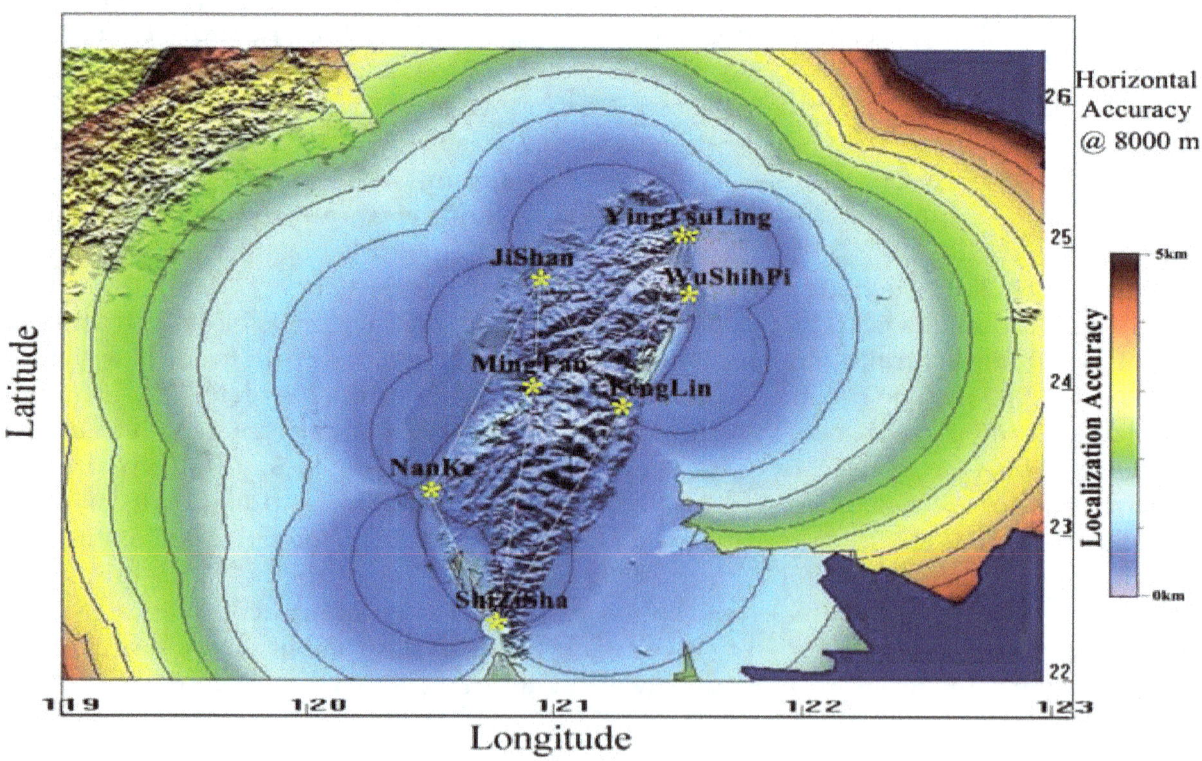

Figure 1. Location of the sensors used in the Total Lightning Detection System (TLDS) of Tai-Power Company, Taiwan.

Surface temperature data over Taipei are collected from Central Weather Bureau (CWB) for the years 1965–2010. With more than 400 stations, consisting of conventional surface station and Automatic

Rainfall and Meteorological Telemetry System (ARMTS), established by CWB, an extremely dense surface observation network was developed over Taiwan. Hourly observations at these stations were used to obtain the daily mean surface temperature. The census record and land use data are collected from various Internet sites of the Taiwan government agencies. Landsat satellite images are employed to demonstrate the urbanization of the city of Taipei and its surrounding areas and are downloaded from United State Geological Survey. Errors for these images refer especially to acquisition loss due to clouds covering the region of interest. Thermal band of the Land-Sat 7 satellite is used to generate apparent surface temperature of Taipei and New Taipei city. The sensors of Land-sat 7 acquire temperature data and store the information as a digital number (DN) with a range between 0 and 255. These DNs are first converted to radiance values using the bias and gain values specific to the individual scene. Thereafter an atmospheric correction using appropriate local values for several parameters are performed to generate more accurate surface temperature map. Air pollutants data were collected from a well-organized air quality-monitoring network operated by Taiwan Environmental Protection Administration (EPA) of Taiwan for the period 2003–2012. Taiwan's air quality monitoring network measures PM10 concentrations by the automatic Wedding β-gauge monitors, which is one of the US EPA-designated equivalent methods (no. EQPM-0391-081). The PM10 inlet is a cyclone operated at 18.9 min^{-1}. Particles are detected once every hour from its continuous collection on the filter tape and the daily average is computed for at least 16 effective hours every day. A comparison experiment had been made between Wedding β-gauge monitors and the manual samplers because of the frequent abundance of high humidity in the ambient air of Taipei. The results obtained from automatic Wedding β-gauge monitor and the manual samplers were very close. For the present study PM10 and SO_2 over Taipei City and New Taipei City are considered.

3. Results

Like the other countries, the population of Taiwan has been steadily increasing for decades. The population of New Taipei City and Taipei City, was 2.7 million in 1965 according to Taiwan government's census record while it reached 6.6 million in 2010. The proportion of people who live in the twin cities has almost increased by a factor of 2.5 since 1965. This increasing trend has been shown in Figure 2a. Since the population increase over a particular area generally demands a large change in land use and housing development, we have plotted the land use data of the Taipei City and New Taipei City also in Figure 2b for easy comparison. An increasing trend is clearly evident between population and land use change, which is again in phase to a great extent with that of daily mean surface temperature at Taipei, which has been shown in Figure 2c. This result corroborates well the findings of Pinto et al. [3] who found an increase in the surface temperature well correlated to the increased thunderstorm activity and population growth of the São Paulo and Campinas cities.

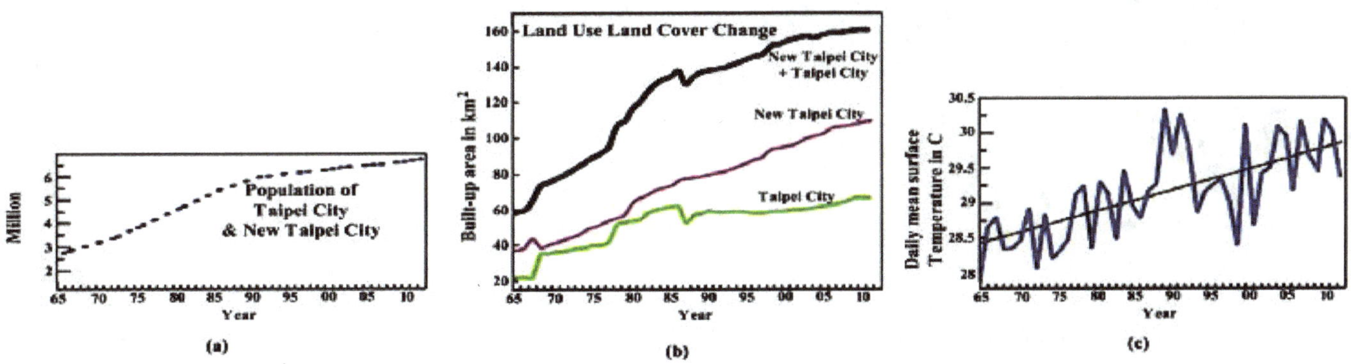

Figure 2. Long-term variation of (**a**) the population of New Taipei City and Taipei City; (**b**) built-up area in New Taipei City (pink line), Taipei City (green line), and the twin cities combined (black); and (**c**) the summer daily mean surface temperature of Taipei along with linear trend for the period 1965–2010 (according to the twin city governments).

A least squares–fit line for the past four decades indicates an increase of mean surface temperature approximately by 1.6 °C. This long-term increasing trend of daily surface temperature, despite a short-term cooling during 1992–97, strongly insists us to conclude that the increasing trend of surface temperature in Taipei City is likely a result of land use change due to urbanization and the effect of global warming. Increased land use in the New Taipei City has been further confirmed by images of inhabited region (cyan color) taken by Landsat-4 on 27th October, 1982 and by Landsat-8 on 29th January, 2014 (Figure 3). Based on our rough estimate using the satellite images shown in Figure 3, the built-up areas are 160.2 km^2 and 285 km^2 in 1982 and 2014, respectively. Thus, an expansion of over 78% during the past 32 years is clearly evident.

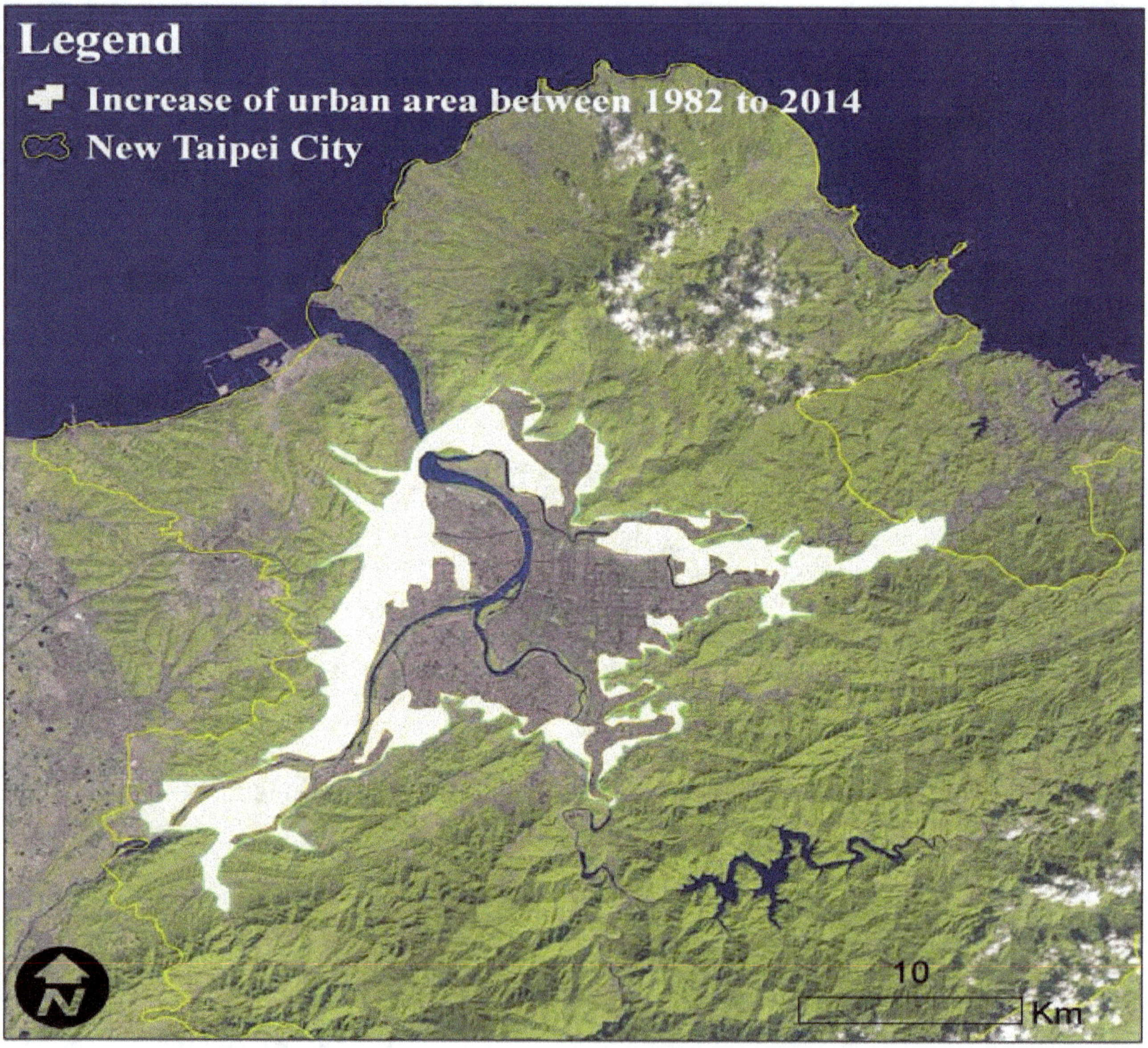

Figure 3. Landsat imageshowing the expansion of urbanization over New Taipei City and Taipei Cityand theirsurrounding areas. Areas in green are uninhabited land covered by vegetation or forest, and areas in light cyan are inhabited regions. Deep blue indicates water bodies. Off white areas are the expansions of urbanization. Bright green line and Yellow line indicates the boundary of Taipei City and New Taipei City, respectively.

Figures 3–5 compare the CG flash density in the northern Taiwan with the geographic position of Taipei City and its apparent surface temperature. Figure 3 presents the urban area of New Taipei City while the CG flash density and the apparent surface temperature over the same region are displayed in Figures 4 and 5, respectively. Figure 3 presents the region where the Taipei City is located (represented by the gray area) within its geopolitical limit. Figure 4 presents the CG flash density for the northern Taiwan computed from the aggregated CG lightning flash data for the period 1998–2012. Spatial distribution of lightning flash density shows clearly an enhancement of the CG flash density over Taipei.

The increase of the CG lightning activity for Taipei related to their surroundings was about 60–70%. It is worth mentioning in this context that Taipei has grown vertically over the past 15 years through construction of very tall buildings. It is possible that taller buildings shift the CG:IC ratio on a local scale. That is, a taller vertical profile might be responsible for enhanced CG flashes. Thermal band of the Land-Sat 7 satellite is used to generate Figure 5, which shows the heat island effect. A clear agreement among the three maps is evident, indicating that the CG flashes tend to roughly concentrate over the urbanized area, which also corresponds to the region of highest surface temperature. This result is consistent with the results obtained by Naccarato et al. [30] over large urban areas of Southern Brazil.

Furthermore, the CG lightning activity tends to concentrate where the human activities are more intense as is evident from the strong relationship between Figures 3 and 4. This anthropogenic influence results an increase in both the aerosol emissions due to the traffic and industrial activity and the local temperature due to the manmade structures and the lack of vegetation.

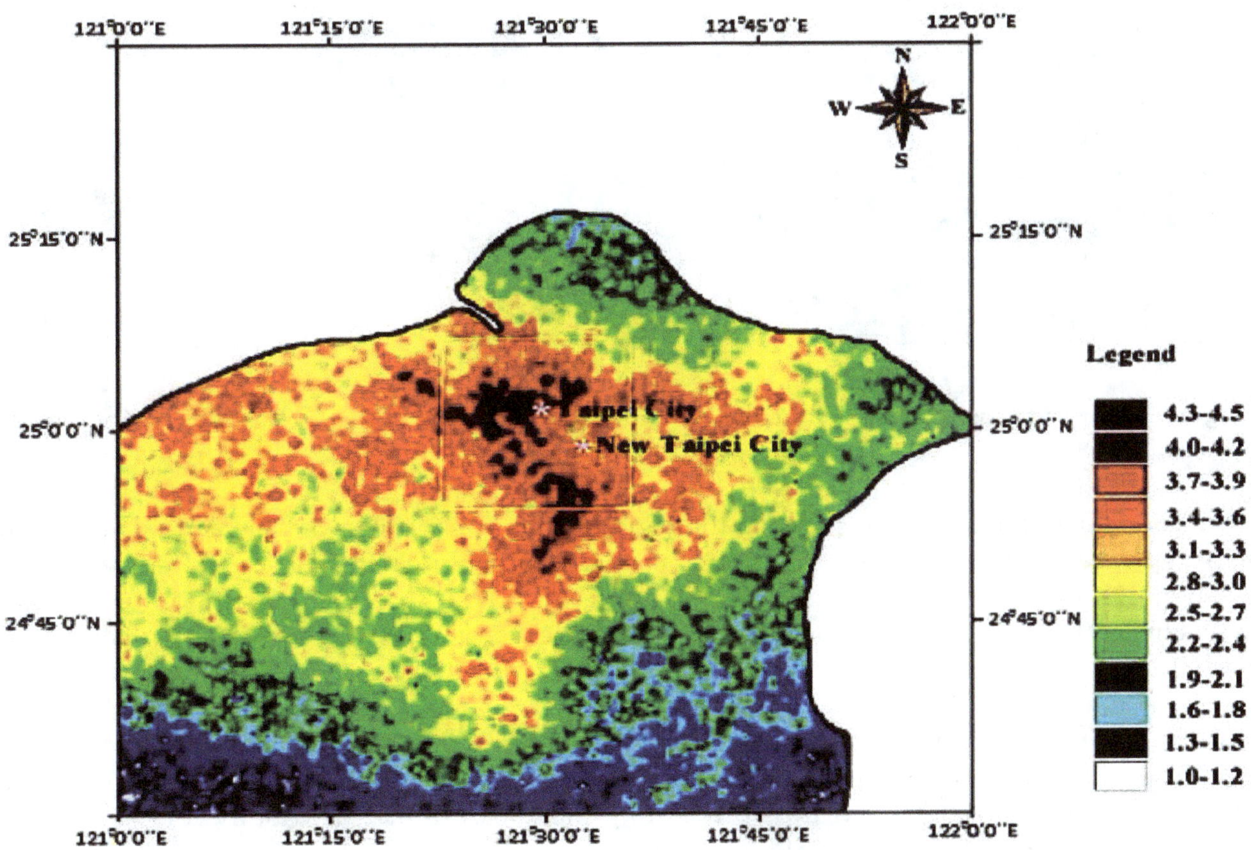

Figure 4. Lightning flash density in flashes.km^{-2} for Taipei City and New Taipei City with 1 km resolution.

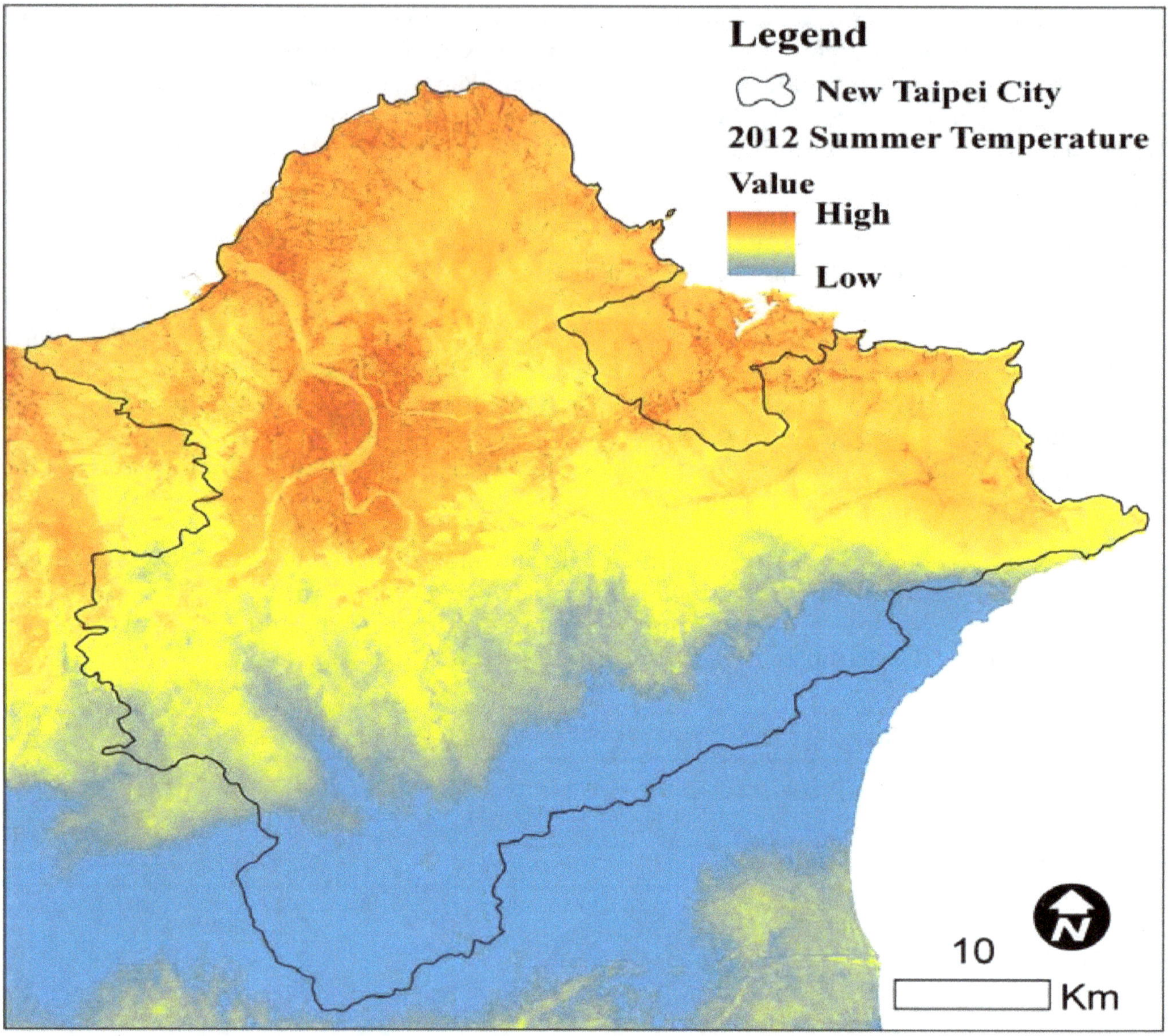

Figure 5. Apparent surface temperature of Taipei City and New Taipei City computed using LandSat-7 thermal data.

The good spatial relationship between the regions of higher number of CG flashes and the higher apparent surface temperature (Figures 4 and 5) might corroborate the thermal effect proposed by Williams and Stanfill [34]. Considering islands to discuss the influence of a portion of land on the cloud electrification and lightning they computed a critical area to assess how large an island should be to guarantee continental behavior. Thermal hypothesis indicated that the critical area required was 110 km^2 while that of was about 20,000–30,000 km^2 according to aerosol hypothesis. Moreover, the suppression of coalescence over the continent was discussed by them considering the traditional thermal hypothesis according to which the larger updrafts over land would permit less time for droplets to interact for coalescence, thus preventing the warm rain. They finally concluded that the thermal hypothesis could better explain the appreciable difference in the lightning activity between land and sea than the aerosol theory. In the case of our present study the thermal hypothesis could explain well the enhancement of CG flashes over these warmer regions particularly for the Taipei City, which is similar to islands in sea and where the heat island has area about 1600 km^2 much larger than the critical area. A further support to the thermal hypothesis showed that the land-sea breeze convergence over Houston combined with the UHI might intensify the thunderstorm generation over the city [35].

The total CG lightning flashes measured during warm seasons for 1998–2012 are presented along with best fit line in Figure 6. It is seen from Figure 6 that lightning increases systematically during this period. This increasing trend in lightning activity suggests that either cloud top heights or amount of deep clouds during monsoon have increased in last few years. This increasing trend of CG lightning is also in phase with population growth and land use change as shown in Figure 2 during this period.

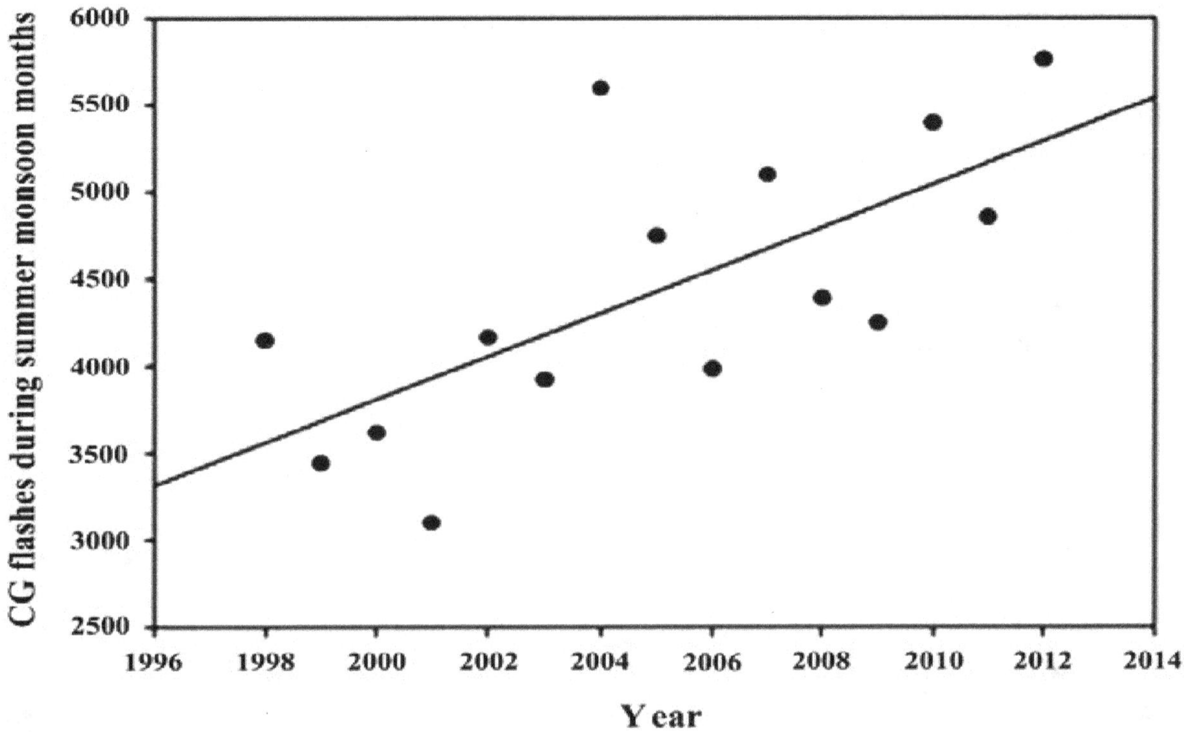

Figure 6. Variation of total CG lightning during warm seasons (May–October) over Taipei.

4. Discussion

Many studies in the recent years have shown that lightning activity in a region can be affected by changes in thermo-dynamical properties as well as increase in aerosol concentration [5,36,37]. The annual averages of PM10 and SO_2 concentrations for Taipei City have been considered in relation to the number of CG flashes to ascertain the possible effect of urban particulate matter on CG lightning activity. Figures 7 and 8 show the scatter plots of CG lightning flashes with PM10 and SO_2 concentrations, respectively, along with best fit line. Both the figures exhibit a positive correlation between the two plotted parameters with correlation coefficients 0.63 and 0.71 for the PM10 and SO_2 concentrations, respectively. The correlations in Figures 7 and 8 are statistically significant at the 1% significance level. A slight higher correlation coefficient is found between SO_2 concentrations and CG lightning flashes compared to that of PM10 and CG lightning flashes. Since the sulphate particles are usually more active in the formation of cloud droplets compared to PM10, a slight higher contribution from SO_2 concentration is expected to enhance the CG lightning compared with the PM10 concentration [38]. Our results partially correspond to the results of Soriano et al. [39], but corroborate well the reports of Westcott [1]. These results indicate a possible influence of aerosol concentrations on the number of CG lightning flashes. The results are consistent with previous studies [40–43], suggesting the enhancement of lightning activity in polluted atmospheric conditions when compared to aerosol clear ones. However, the production of CG lightning is enhanced by the increase cloud water in the mixed phase region and is paralleled by an increase in the electrical charge separation [35].

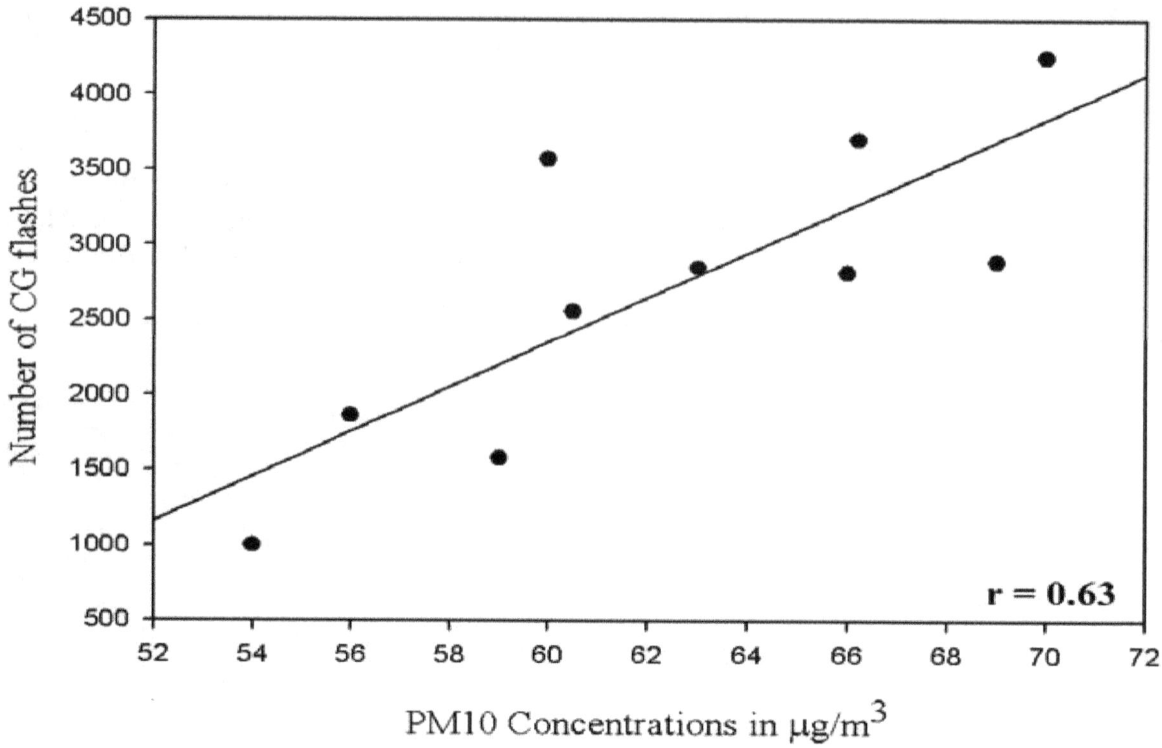

Figure 7. Scatter plot of the number of CG flashes and annual averages of PM10 concentrations.

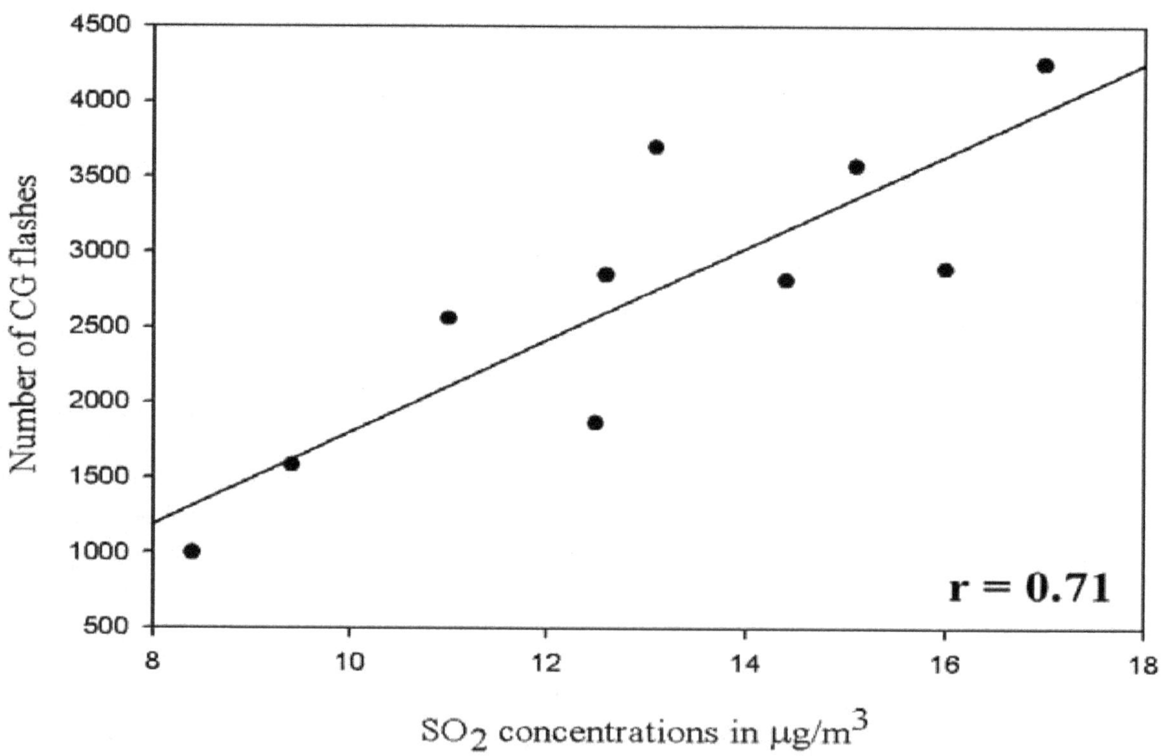

Figure 8. Scatter plot of the number of CG flashes and annual averages of SO_2 concentrations.

The Taipei City and its suburbs are surrounded by mountains with two major openings through River valleys. During the warm season, the comparatively cold moist air is transported by sea breezes into the Taipei City through these two river valleys. This moist air is then warmed by urban heat of Taipei City and converges towards the mountains situated south of the city. Convective overturning of moist air is expected to take place because of the orographic forcing facilitating the process of cumulus convection. Thus a zone of higher lightning frequency over Taipei City and its suburbs is formed which is likely enhanced by the warm moist sea breezes and the urban heat [27]. From our results, it is evident that for Taipei the lightning activity showed a significant increase with respect to their surrounding areas during the period of study with a simultaneously increase of surface temperature well correlated with the population growth of the city in terms of population. The results did not reflect anything expected from natural climate cycles and provide clear observational evidence for the anthropogenic influence related to the urban influence on enhanced lightning activity. It is also evident that during the past four decades the daily mean surface temperature of Taipei City has increased 1.6 °C. Therefore, it is highly expected that this urban environment change must have a significant effect on the increase in CG lightning frequency, as observed in Figure 4, over Taipei and its suburbs. Moreover, during the past four decades, the population of Taipei and its built-up area, both, have increased by a factor of 2.5 and 1.8, respectively. Most of the LULC change in Taipei and its suburbs is resulted either by building construction or by various urban surface developments. Penetration of sea breezes through two river valleys is expected to be obstructed by the increased surface roughness over the northwest slopes of the mountains south of Taipei City as indicated by Chen et al. [18], since most of the constructions are concentrated over these regions. This retardation of surface airflow may also lead to an enhancement of low-level convergence over the downwind of major constructed lands and subsequent increase in CG lightning flashes.

5. Conclusions

It has been found from the present study that over Taipei City the enhancement of CG lightning activity follows closely the increase of the apparent surface temperature, which indicates the heat island, thus corroborate the thermal hypothesis. In addition to population growth and land use increase, increased pollution and anthropogenic aerosols are also good indicators of urbanization of any city. Therefore, an assessment of the possible impact of aerosol microphysical forcing on the formation of urban CG lightning is equally important. Apart from the evidence in support of thermal hypothesis, a positive linear correlation between the number of CG flashes and a higher concentration of both SO_2 and PM10 has also been found, indicating that the aerosol hypothesis is equally important to support the increase of the number of CG flashes. Positive correlations of 0.63 and 0.71 are found for the PM10 and SO_2 concentrations, respectively, when compared separately with the number of CG flashes. This higher positive correlation strongly supports that aerosols play a key role in the enhancement of lightning activity over Taipei City. Williams et al. [44] first proposed that under continental and dirty boundary layer conditions, the available liquid water in the storm updraft is shared amongst an innumerable number of small droplets, thereby suppressing the mean droplet size and thwarting the coalescence process. As a result, the cloud water reaches the mixed phase region to participate in creating excess cloud buoyancy, in precipitation formation, and in electric charge separation and increasing the lightning activity. Therefore, based on possible increases of pollution and aerosols, the contribution of aerosol–microphysical forcing to CG lightning enhancement over Taipei city and its suburbs seems reasonable. It is worth mentioning in this context that both hypotheses act on the same way, probably with combined interaction, at least for the enhancement of CG lightning activity.

It is evident from the perspective of local atmospheric dynamics that the UHI, sea breeze and orographic interaction plays a significant role in the formation of urban CG lightning within the New Taipei City. The spatial variability of CG lightning activity over Taipei is also influenced by the aerosol microphysical forcing. This observation corroborates well the findings of Sequera et al. [45] that combined impact of aerosol and UHI plays crucial role in enhancing the lightning activity over Taipei. It is worth mentioning in this context that the aerosol effects or urban heat island may not only be responsible for the observed trends of lightning over Taipei; the synoptic variability can also affect these trends with equal probability. However, it is a challenging task to quantify the individual contributions of these two physical processes based on present observations because of the complexity of urban modifications of both dynamic and physical processes over this area. Therefore, extensive research is highly needed to fully understand this complex physical mechanism. The present study not only helps to understand the influence of urbanization on the local microclimate system, but also provide valuable environmental information for future urban planning and environmental management.

Author Contributions: "Conceptualization, S.K.K.; Y.-A.L.; Methodology, S.K.K.; Formal Analysis, S.K.K.; Investigation, S.K.K.; Data Curation, S.K.K.; Y.-A.L.; Writing-Original Draft Preparation, S.K.K.; Writing-Review & Editing, S.K.K.; Y.-A.L.; Visualization, Y.-A.L.; Supervision, Y.-A.L.; Project Administration, Y.-A.L.; Funding Acquisition, Y.-A.L.

Acknowledgments: The authors are thankful to Tai-Power Company of Taiwan for providing lightning data. We are grateful to United States Geological Survey (USGS) for providing Landsat images. Thanks are also given to the Ministry of Science and Technology (MOST) of Taiwan for financial support through grant 105-2221-E-008-056-MY3 and 107-2111-M-008-036. We are also extremely grateful to the anonymous reviewers for their valuable critical comments, which helped us a lot to improve this manuscript.

References

1. Westcott, N.E. Summertime cloud-to-ground lightning activity around major Midwestern urban areas. *J. Appl. Meteorol.* **1995**, *34*, 1633–1642. [CrossRef]
2. Steiger, S.M.; Orville, R.E.; Huffines, G. Cloud-to-ground lightning characteristics over Houston, Texas: 1989–2000. *J. Geophys. Res.* **2002**, *107*, 4117. [CrossRef]
3. Pinto, O., Jr.; Pinto, I.R.C.A.; Ferro, M.A.S. A study of the long-term variability of thunderstorm days in southeast Brazil. *J. Geophys. Res. Atmos.* **2013**, *118*, 5231–5246. [CrossRef]
4. Farias, W.R.G.; Pinto, O., Jr.; Pinto, I.R.C.A.; Naccarato, K.P. The influence of urban effect on lightning activity: Evidence of weekly cycle. *Atmos. Res.* **2014**, *135–136*, 370–373. [CrossRef]
5. Kar, S.K.; Liou, Y.A. Enhancement of cloud-to-ground lightning activity over Taipei, Taiwan in relation to urbanization. *Atmos. Rese.* **2014**, *147–148*, 11–120. [CrossRef]
6. Bourscheidt, V.; Pinto, O.; Naccarato, K.P. The effects of Sao Paulo urban heat island on lightning activity: Decadal analysis (1999–2009). *J. Geophys. Res. Atmos.* **2016**, *121*, 4429–4442. [CrossRef]
7. Allen, J.T.; Allen, E.R. A review of severe thunderstorms in Australia. *Atmos. Res.* **2016**, *178–179*, 347–366. [CrossRef]
8. Atkinson, B.W. Numerical modeling of urban heat island intensity. *Bound. Layer Meteorol.* **2003**, *109*, 285–310. [CrossRef]
9. Kim, Y.-H.; Baik, J.J. Daily maximum urban heat island intensity in large cities of Korea. *Theor. Appl. Climatol.* **2004**, *79*, 151–164. [CrossRef]
10. Galanaki, E.; Kotroni, V.; Lagouvardos, K.; Argiriou, A.A. ten-year analysis of cloud-to-ground lightning activity over the Eastern Mediterranean region. *Atmos. Res.* **2015**, *166*, 213–222.
11. Kingfield, D.M.; Calhoun, K.M.; De Beurs, K.M.; Henebry, G.M. Effects of City Size on Thunderstorm Evolution Revealed through a Multiradar Climatology of the Central United States. *J. Appl. Meteorol. Climatol.* **2018**, *57*, 295–317. [CrossRef]
12. Thanh Hoan, N.; Liou, Y.A.; Nguyen, A.K.; Sharma, R.C.; Phien, T.D.; Liou, C.L.; Cham, D.D. Assessing the effects of land-use types in surface urban heat islands for developing comfortable living in Hanoi City. *Remote Sens.* **2018**, *10*, 1965. [CrossRef]
13. Chang, T.-Y.; Liou, Y.A.; Lin, C.Y.; Liu, C.S.; Wang, Y.C. Evaluation of surface heat fluxes in Chiayi plain of Taiwan by remotely sensed data. *Int.l J. Remote Sens.* **2010**, *31*, 3885–3898. [CrossRef]
14. Changnon, S.A.; Braham, R.R.; Semonin, R.G.; Auer, A.H.; Hales, J. METROMEX: A Review and Summary. *Meteor. Monogr.* **1981**, *40*, 181.
15. Huff, F.A. Urban hydrometeorology review. *Bull. Am. Meteorol. Soc.* **1986**, *67*, 703–712. [CrossRef]
16. Rosenfeld, D. Suppression of rain and snow by urban air pollution. *Science* **2000**, *287*, 1793–1796. [CrossRef]
17. Ramanathan, V.; Crutzen, P.J.; Kiehl, J.T.; Rosenfeld, D. Aerosols, climate, and the hydrological cycle. *Science* **2001**, *294*, 2119–2124. [CrossRef]
18. Borys, R.D.; Lowenthal, D.H.; Cohn, S.A.; Brown, W.O.J. Mountain and radar measurements of anthropogenic aerosol effects on snow growth and snowfall rate. *Geophys. Res. Lett.* **2003**, *30*, 1538. [CrossRef]

19. Kitada, T.; Okamura, K.; Tanaka, S. Effects of topography and urbanization on local winds and thermal environment in the Nohbi Plain, coastal region of central Japan: A numerical analysis by mesoscale meteorological model with a k-ε turbulence model. *J. Appl. Meteorol.* **1998**, *37*, 1026–1046. [CrossRef]
20. Chen, T.-C.; Wang, S.-Y.; Yen, M.-C. Enhancement of afternoon thunderstorm activity by urbanization in a valley: Taipei. *J. Appl. Meteorol. Climatol.* **2007**, *46*, 1324–1340. [CrossRef]
21. Lin, C.-Y.; Chen, F.; Huang, J.C.; Chen, W.-C.; Liou, Y.-A.; Chen, W.-N.; Liu, S.-C. Urban heat island effect and its impact on boundary layer development and land–sea circulation over northern Taiwan. *Atmos. Environ.* **2008**, *42*, 5635–5649. [CrossRef]
22. Chung, Y.-S.; Yoon, M.-B.; Kim, H.-S. On climate variations and changes observed in South Korea. *Clim. Chang.* **2004**, *66*, 151–161. [CrossRef]
23. Trenberth, K.E.; Jones, P.D.; Ambenje, P.; Bojariu, R.; Easterling, D.; Tank, A.K.; Parker, D.; Rahimzadeh, F.; Renwick, J.A.; Rusticucci, M.; et al. Observations: Surface and atmospheric climate change. In *Climate Change 2007: The Physical Science Basis. Contribution of Working Group I to the Fourth Assessment Report of the Intergovernmental Panel on Climate Change*; Solomon, S., Qin, D., Manning, M., Chen, Z., Marquis, M., Averyt, K.B., Tignor, M., Miller, H.L., Eds.; Cambridge University Press: Cambridge, MA, USA, 2007.
24. Lin, W.-Z.; Tsai, H.-C.; Wang, C.-H.; Teng, W.-H. Urbanization-induced regional climate change on the western plain of Taiwan for the period 1964~1999. In Proceedings of the WSEAS International Conference on Environment, Ecosystems and Development, Venice, Italy, 2–4 November 2005; pp. 7–13.
25. Fujibe, F. Detection of urban warming in recent temperature trends in Japan. *Int. J. Climatol.* **2009**, *29*, 1811–1822. [CrossRef]
26. Kar, S.K.; Liou, Y.-A.; Ha, K.-J. Characteristics of cloud-to-ground lightning activity over Seoul, South Korea in relation to an urban effect. *Ann. Geophys.* **2007**, *25*, 2113–2118. [CrossRef]
27. Liou, Y.A.; Kar, S.K. Study of cloud-to-ground lightning and precipitation and their seasonal and geographical characteristics over Taiwan. *Atmos. Res.* **2010**, *95*, 115–122. [CrossRef]
28. Richard, P.; Auffray, G. VHF-UHF interferometric measurements, applications to lightning discharge mapping. *Radio Sci.* **1985**, *20*, 171–192. [CrossRef]
29. Richard, P.; Delannoy, A.; Labaune, G.; Laroche, P. Results of spatial and temporal characterization of the VHF–UHF radiation of lightning. *J. Geophys. Res.* **1986**, *91*, 1248–1260. [CrossRef]
30. Naccarato, K.P.; Pinto, O., Jr.; Pinto, I.R.C.A. Evidence of thermal and aerosol effects on the cloud-to-ground lightning density and polarity over large urban areas of Southeastern Brazil. *J. Geophys. Res. Lett.* **2003**, *30*, 1674–1677. [CrossRef]
31. Orville, R.E. Cloud-to-ground lightning flash characteristics in the contiguous United States: 1989–1991. *J. Geophys. Res.* **1994**, *99*, 10833–10841. [CrossRef]
32. Finke, U.; Hauf, T. The characteristics of lightning occurrence in Southern Germany. *Beitr. Phys. Atmos.* **1996**, *69*, 361–374.
33. Pinto, O., Jr.; Pinto, I.R.C.A.; Gomes, M.A.S.S.; Vitorello, I.; Padilha, A.L.; Diniz, J.H.; Carvalho, A.M.; CazettaFilho, A. Cloud-to-ground lightning in southeastern Brazil in 1993. 1: Geographical distribution. *J. Geophys. Res.* **1999**, *104*, 31369–31379. [CrossRef]
34. Williams, E.R.; Stanfill, S. The physical origin of the land-ocean contrast in lightning activity. *Comp. Rendus Phys.* **2002**, *3*, 1277–1292. [CrossRef]
35. Orville, R.E.; Huffines, G.R.; Nielsen-Gammon, J.; Zhang, R.; Ely, B.; Steiger, S.; Phillips, S.; Allen, S.; Read, W. Enhancement of cloud-to-ground lightning over Houston, Texas. *Geophys. Res. Lett.* **2001**, *28*, 2597–2600. [CrossRef]
36. Khain, A.; Lynn, B.; Dudhia, J. Aerosol effects on intensity of landfalling hurricanes as seen from simulations with the WRF model with spectral binmicrophysics. *J. Atmos. Sci.* **2010**, *67*, 365–384. [CrossRef]
37. Kar, S.K.; Liou, Y.-A.; Ha, K.-J. Aerosol effects on the enhancement of cloud-to-ground lightning over major urban areas of South Korea. *Atmos. Res.* **2009**, *92*, 80–87. [CrossRef]
38. Seinfeld, J.M. *Air Pollution: Physical and Chemical Fundamentals*; McGraw-Hill Co.: New York, NY, USA, 1975; p. 558.
39. Soriano, L.R.; Pablo, F. Effect of small urban areas in central Spain on the enhancement of cloud-to-ground lightning activity. *Atmos. Environ.* **2002**, *36*, 2809–2816. [CrossRef]
40. Farias, W.R.G.; Pinto, O., Jr.; Naccarato, K.P.; Pinto, I.R.C.A. Anomalous lightning activity over the Metropolitan Region of São Paulo due to urban effects. *Atmos. Res.* **2009**, *91*, 485–490. [CrossRef]

41. Kucienska, B.; Raga, G.B.; Koren, I.; Altaratz, O. Do aerosols affect lightning? A global study of a relation between aerosol optical depth and cloud to ground lightning. In Proceedings of the XV International Conference on Atmospheric Electricity, Norman, OK, USA, 15–20 June 2014.
42. Yang, X.; Li, Z. Increases in thunderstorm activity and relationships with air pollution in southeast China. *J. Geophys. Res. Atmos.* **2014**, *119*, 1835–1844. [CrossRef]
43. Zhao, P.; Yin, Y.; Xiao, H. The effects of aerosol on development of thunderstorm electrification: A numerical study. *Atmos. Res.* **2015**, *153*, 376–391. [CrossRef]
44. Williams, E.R.; Rosenfeld, D.; Madden, N.; Labrada, C.; Gerlach, J.; Atkinson, L. The role of boundary layer aerosol in the vertical development of precipitation and electrification: Another look at the contrast between lightning over land and over ocean. Reprints. In *Eleventh International Conference on Atmospheric Electricity*; American Meteorological Society: Boston, MA, USA, 1999; pp. 754–757.
45. Sequera, P.; González, J.E.; McDonald, K.; Bornstein, R.; Comarazamy, D. Combined impacts of land cover changes and large-scale forcing on southern California summer daily maximum temperatures. *J. Geophys. Res. Atmos.* **2015**, *120*, 9208–9219. [CrossRef]

4

Diurnal Cycle in Atmospheric Water over Switzerland

Klemens Hocke [1,2,*,†], Francisco Navas-Guzmán [1], Lorena Moreira [1], Leonie Bernet [1,2] and Christian Mätzler [1,2]

1 Institute of Applied Physics, University of Bern, CH-3012 Bern, Switzerland; francisco.navas@iap.unibe.ch (F.N.-G.); lorena.moreira@iap.unibe.ch (L.M.); Leonie.bernet@iap.unibe.ch (L.B.); christian.matzler@iap.unibe.ch (C.M.)

2 Oeschger Centre for Climate Change Research, University of Bern, CH-3012 Bern, Switzerland

* Correspondence: klemens.hocke@iap.unibe.ch

† Current address: K. Hocke, IAP, University of Bern, Sidlerstr. 5, CH-3012 Bern, Germany

Academic Editors: Yuei-An Liou, Jean-Pierre Barriot, Chung-Ru Ho, Yuriy Kuleshov, Chyi-Tyi Lee, Richard Müller and Prasad S. Thenkabail

Abstract: The TROpospheric WAter RAdiometer (TROWARA) is a ground-based microwave radiometer with an additional infrared channel observing atmospheric water parameters in Bern, Switzerland. TROWARA measures with nearly all-weather capability during day- and nighttime with a high temporal resolution (about 10 s). Using the almost complete data set from 2004 to 2016, we derive and discuss the diurnal cycles in cloud fraction (CF), integrated liquid water (ILW) and integrated water vapour (IWV) for different seasons and the annual mean. The amplitude of the mean diurnal cycle in IWV is 0.41 kg/m^2. The sub-daily minimum of IWV is at 10:00 LT while the maximum of IWV occurs at 19:00 LT. The relative amplitudes of the diurnal cycle in ILW are up to 25% in October, November and January, which is possibly related to a breaking up of the cloud layer at 10:00 LT. The minimum of ILW occurs at 12:00 LT, which is due to cloud solar absorption. In case of cloud fraction of liquid water clouds, maximal values of +10% are reached at 07:00 LT and then a decrease starts towards the minimum of −10%, which is reached at 16:00 LT in autumn. This breakup of cloud layers in the late morning and early afternoon hours seems to be typical for the weather in Bern in autumn. Finally, the diurnal cycle in rain fraction is analysed, which shows an increase of a few percent in the late afternoon hours during summer.

Keywords: cloud fraction; integrated liquid water; integrated water vapour; diurnal cycle; microwave radiometer

---

1. Introduction

Clouds and aerosols continue to contribute the largest uncertainty to estimates and interpretations of the Earth's changing energy budget [1]. There is a consensus in the climate research community that man-made global warming is amplified by an increase of water vapour in a warmer world [2]. The Earth's radiation budget between the incoming short-wave radiation of the Sun and the outgoing longwave radiation of the Earth is strongly influenced by the spatio-temporal distribution of clouds. Differing assumptions about how the Earth's cloud distribution is maintained for doubling of the CO_2 concentration in the atmosphere lead to estimated increases of global mean surface temperature in the range from 1.5 to 4.5 °C [2].

The physical description of clouds involves microphysics and nonlinear radiative-dynamic processes over temporal scales from about 1 s to 1 decade and spatial scales from about 1 m to 1000 km. Observing and modelling of the Earth's cloud distribution is still a challenge, even though large efforts have been undertaken in meteorology, atmospheric research, and climate sciences [3,4]. Diurnal variations in atmospheric water parameters are of high interest for measurement and atmospheric modelling since one day is the shortest regular period by which the Earth's surface and atmosphere absorb energy from the Sun. Cloud formation and rainfall depend on orography and land-sea contrast [5]. The phase of the diurnal variation can change significantly within a small horizontal distance. Parameterisation of cloud formation and rainfall is difficult since these phenomena depend for example on the variable surface flux of moisture, orography, convective processes, turbulence, eddies and the growth of the planetary boundary layer during the daytime

In the following, we focus on the diurnal cycle in integrated liquid water (or liquid water path). Two research communities are active in investigating the diurnal variation of cloud liquid water (ILW). The first group are scientists who generate world maps of cloud coverage, outgoing radiation, and rainfall data from polar-orbiting satellites imaging the Earth in the ultraviolet, visible, infrared, and microwave range [5–9]. Since the diurnal variations of cloud parameters and rainfall have strong amplitudes of about 10% or larger, construction of world maps and climatic trend analyses of cloud parameters from satellite observations have to correct for offsets caused by a varying local solar time at the observation places. The second group are atmospheric modellers who use the periodic signal of the diurnal cycle in atmospheric water to test the quality of weather and climate models. The models often provide incorrect phases and amplitudes of the diurnal variations of rainfall, cloud cover and ILW reflecting errors in the parameterization of clouds, convective processes, surface moisture flux, and microphysics [10,11].

There is no doubt that weather forecast and regional climate projections will not be correct if a fundamental process such as cloud formation and its diurnal variation is incorrectly simulated by the numerical model. Bergman, J.W. et al. [12] analysed that the cloud diurnal contribution to time-average surface energetics is as much as 20 W m^{-2} in a regional climate (Amazon basin) after deforestation, which caused a shift from a high cloud distribution to a low cloud distribution. Cross-validation studies fostered the progress in remote sensing, data analysis, and modelling. Observed and simulated diurnal variations were compared [9,13]. Numerous cross-validations were organized within the Global Energy and Water Experiment (GEWEX) and the International Satellite Cloud Climatology project (ISCCP) [14].

There are only a few studies investigating the diurnal variation in ILW by means of ground-based microwave radiometers. This is rather surprising since ground-based microwave radiometers are well suited for the continuous measurement of ILW during the daytime and nighttime in all seasons. Thus, data sets of microwave radiometers are convenient for intercomparisons with results from polar-orbiting satellites, other ground-based instruments (e.g., lidar), and models. A review of meteorological applications of ground-based microwave and millimeter wavelength radiometry was given by [15]. Roebeling, R.A. et al. [16] compared the diurnal variation in liquid water path derived from the Spinning Enhanced Visible and Infrared Imager (SEVIRI) on board Meteosat-8 with those observed by two ground-based microwave radiometers of the CloudNET in northern Europe and found a good agreement. Snider, J.B. et al. [17] presented diurnal variations in IWV and ILW observed with ground-based microwave radiometers operating near 20, 23, 31, and 90 GHz. Cross-validation with coincident measurements of infrared brightness temperature of the sky confirmed the diurnal variations in ILW obtained by microwave radiometers. In addition, the observed diurnal variations in ILW are in a qualitative agreement with expectations from theory and satellite observations that a maximum of cloud liquid water occurs before sunrise in oceanic areas during summer [9]. Model simulations by [18] showed quite similar shapes of the diurnal cycles in CF and ILW for marine boundary layer clouds with a maximum at 06:00 LT and a minimum at 18:00 LT.

Ground-based microwave radiometers are well suited for measurement of the diurnal cycle in atmospheric water parameters. We suggest that modellers and observers should take more advantage of ground-based microwave radiometers for research and validation studies of the diurnal cycle in ILW. Here, we analyse the long-term time series of integrated liquid water, integrated water vapour, and cloud fraction observed by the TROWARA radiometer, which is located in Bern on the Swiss Plateau. Section 2 describes the instrument and the measurement technique as well as the data analysis. Section 3 shows the climatology of IWV, ILW and CF averaged over the time interval 2004–2016. Furthermore, we discuss the results from the spectral data analysis of the time series. Section 3 also presents the diurnal cycles in IWV, ILW and CF. Conclusions are given in Section 4.

2. Instrument, Data and Analysis

2.1. The Microwave Radiometer TROWARA

Observations of the TROpospheric WAter RAdiometer (TROWARA) are central to our study. Peter, R. et al. [19,20] described the design and the construction of the TROWARA instrument, which is a dual-channel microwave radiometer. Two ferrite circulator switches at each frequency switch between the antenna and noise diodes where the latter are taken as hot and cold reference loads. A radiometer model was developed based on measurements of the reflection and transmission coefficients of all radiometer components up to and including the ferrite switches [20]. Tipping calibrations are carried out with an external mirror in order to correct the sky brightness temperature calculated by the model. The instrument is sufficiently stable that a tipping calibration is only necessary once every few weeks.

TROWARA provides the vertically-integrated water vapour (IWV) and vertically-integrated cloud liquid water (ILW), also known as liquid water path (LWP). TROWARA is operated inside a temperature-controlled room on the roof of the EXWI building of the University of Bern (46.95°N, 7.44°E, 575 m a.s.l.). The indoor operation of TROWARA permits the measurement of IWV even during rainy periods. TROWARA's antenna receives the atmospheric radiation through a microwave transparent window and is pointing the sky at a zenith angle of 50° towards the southeast.

The two microwave channels are at 21.4 GHz (bandwidth 100 MHz) and 31.5 GHz (bandwidth 200 MHz). The 21.4 GHz frequency is more sensitive to microwaves from water vapour, and the 31.5 GHz frequency is more sensitive to microwaves from atmospheric liquid water.

The radiative transfer equation of a non-scattering atmosphere can be expressed as

$$T_{B,i} = T_c e^{-\tau_i} + T_{mean,i}(1 - e^{-\tau_i}), \tag{1}$$

where $T_{B,i}$ is the observed brightness temperature of the i-th frequency channel (e.g., 21 GHz). τ_i is the opacity along the line of sight of the radiometer, and T_c is the contribution of the cosmic microwave background. The effective mean temperature of the troposphere is given by $T_{mean,i}$ [21,22].

Equation (1) leads to the opacities

$$\tau_i = -ln\left(\frac{T_{B,i} - T_{mean,i}}{T_c - T_{mean,i}}\right), \tag{2}$$

where the radiances $T_{B,i}$ are observed by TROWARA.

For a plane-parallel atmosphere, the opacity is linearly related to IWV and ILW

$$\tau_i = a_i'' + b_i'' IWV + c_i'' ILW, \tag{3}$$

where the coefficients a'' and b'' are not really constant since they can partly depend on air pressure. As shown by [22], the coefficients can be statistically derived by means of coincident measurements of radiosondes and fine-tuned at times of periods with a clear atmosphere. The coefficient c'' is the mass absorption coefficient of cloud water. It depends on temperature (and frequency), but not on pressure. It is derived from the physical expression of Rayleigh absorption by clouds [22]. Once the coefficients are determined, combined opacity measurements at 21 and 31 GHz permit the retrieval of IWV and ILW from Equation (3). Thus, a dual channel microwave radiometer can monitor IWV and ILW with a time resolution of 6–11 s and nearly all-weather capability during daytime and nighttime.

The physical temperature at the cloud base is derived for optically thick clouds (ILW > 30 g/m^2) from measurements of an infrared radiometer channel at a wavelength of $\lambda = 9.5$–11.5 μm. The narrowband infrared radiometer is a Heitronics KT15.85D pyrometer of type A, which was calibrated by the manufactor for the signal temperature range from −100 to +100 °C [23]. Temperature resolution, filter curve, view of field, radiometer model at low temperatures and other characteristics of the Heitronics KT15.85D are described in [23]. The temperature resolution is less than 2 °C for target temperatures from −100 to +100 °C and a response time of 1 s.

The antenna coil of TROWARA has a full width at half power of 4° and is pointing the sky at a zenith angle of 50 ° towards the southeast. The view direction is constant, and the microwave and infrared channels of TROWARA observe the short-term temporal variations of the brightness temperature in the same volume of the atmosphere. This contributes to the high sensitivity of TROWARA for cloud detection. Cossu, F., Mätzler, C. and Morland, J. [22,24] give further details of the sensors and the retrieval technique.

TROWARA delivered an almost uninterrupted time series of ILW since 2004, with a time resolution of 11 s until end of 2009 and 6 s afterwards. Clouds are detected in the line of sight of TROWARA with the time resolution of the ILW series. Cossu, F. [24] determined the instrumental noise $\sigma_{noise} = 0.77$ g/m^2 of TROWARA from the noise of ILW during 245 days in which the sky was free of clouds. We emphasize that this is a remarkable sensitivity for a microwave radiometer. If an ILW value exceeds the $3\sigma_{noise}$ level, then we are confident by 99.7% that the ILW value was generated by a cloud and not by instrumental noise. Thus, ILW $> 3\sigma_{noise} = 2.3$ g/m^2 is the criterion for the existence of a cloud. In contrast to the ILW series, the time series of IWV have been used since 1994 for trend analyses [25,26].

CF (cloud fraction) is easily determined in the time domain—for example, CF is the quotient of the time intervals when ILW >2.3 g/m^2 and the total observation time. The high spatio-temporal variability of clouds floating through the fixed line-of-sight of TROWARA requires a high temporal resolution of about 10 s for the cloud flag. CF for different categories of liquid water clouds were derived by [27] using the TROWARA measurements. TROWARA's coincident ILW and infrared brightness temperature measurements allow separation of the liquid water clouds into four categories:

1. CF1: thin liquid water clouds (2.3 g/m^2 < ILW < 30 g/m^2),
2. CF2: thick supercooled liquid water clouds (ILW > 30 g/m^2 and $T_{infrared} < 273.15$ K),
3. CF3: thick warm liquid water clouds (ILW > 30 g/m^2 and $T_{infrared} > 273.15$ K),
4. CF4: all liquid water clouds (ILW > 2.3 g/m^2).

Quite similar criteria for the separation of supercooled liquid water clouds were described by [28]. The critical point is that the derived cloud distributions are possibly biased towards the low level clouds since the infrared channel mainly sees the cloud base of thick clouds. Mätzler, C. et al. [29] avoided this bias by using additional satellite data for the cloud-top temperature.

Hirsch, E. et al. [30] determined the microphysical and optical properties of thin liquid water clouds and emphasized that these clouds should be considered in climate studies since they are frequent and they change the radiative forcing of the climate system. Measurements indicated that the downwelling infrared radiance of a thin liquid water cloud is about 60% greater than that of clear sky. Thin liquid water cloud areas often occur at the edges of and in the inter-region between clouds (*twilight* zone of clouds).

Since TROWARA is not sensitive to ice clouds, CF of TROWARA is in general smaller than that of synoptic observations. Cossu, F. [24] found a CF difference of about 17% between TROWARA and synoptic observations in the same region over a period of six years. In addition, TROWARA may not see some of the very thin and tenuous clouds that are still visible by the naked eye.

The present study is not a cloud type study that would require the evaluation of coincident observations by ceilometer, lidar, radiosonde and hemispherical sky camera. In our study, the terms *thin* and *thick* refer to the magnitude of the optical depth at microwave frequencies that are proportional to the liquid water path. The terms should not be misunderstood by the geometrical thickness of the clouds, which is not measured by the microwave radiometer. A statistical cloud type study was performed by [31] for Payerne. Payerne is representative for Bern since Payerne is located just 40 km west of Bern in the Swiss plateau. About 38% of the clouds classified by a sky camera have the type Sc (stratocumulus), while 17% are cirrus-cirrostratus (Cr-Cs) and 12% are cirrocumulus-altocumulus (Cc-Ac). Bernet, L. et al. [31] also found that the relationship between the ILW value and the cloud type is ambiguous. Cumulus and cirriform clouds generally have a small ILW, but no tendency for stratiform clouds was found.

2.2. Data Analysis

Hourly means of IWV, ILW, CF1, CF2, CF3 and CF4 were obtained by averaging of the 10 s sampling data. An upper threshold of 400 g/m^2 is used for ILW. This means that, in the presence of rain droplets, we take the value 400 g/m^2 as an estimate of the ILW of the cloud droplets. During precipitation, TROWARA overestimates ILW of the cloud droplets because of the strong microwave emission from the rain droplets ($d > 0.2$ mm). This is the reason why we take an upper threshold of 400 g/m^2 for vertically integrated cloud liquid water path during rainy periods. Generally, the results are not sensitive to the choice of the threshold as it was investigated by [32]. Furthermore, the rain periods are a small fraction of about 7% of the whole measurement time. Monthly means of IWV are well defined because of the continuous monitoring of IWV by TROWARA.

The arithmetic mean is removed from the time series of IWV, ILW or CF. Then, the Fast Fourier Transform (FFT) power spectra are obtained by folding these time series with a Hamming window and by applying zero padding at the beginning and end of the time series. The FFT power spectra are normalized by the power of the strongest spectral component, which is either the annual or the semi-annual oscillation.

Next, we derive amplitude spectra by means of bandpass filtering. The time series are filtered with a digital non-recursive, finite impulse response (FIR) bandpass filter performing zero-phase filtering by processing the time series in forward and reverse directions. The number of filter coefficients corresponds to a time window of three times the central period, and a Hamming window has been selected for the filter. Thus, the bandpass filter has a fast response time to temporal changes in the data series. The variable choice of the filter order permits the analysis of wave trains with a resolution that matches their scale. The bandpass cutoff frequencies are at $f_c = f_p \pm 10\% f_p$, where f_p is the central frequency. More details about the bandpass filtering are given by [33].

Climatologies of the time series are obtained by sorting the data for the month and taking the mean and the standard deviation. The mean diurnal cycles are obtained by sorting the data for the month and the hour of the day (in local time). Again, the arithmetic means of the sorted ensembles are taken. In order to intercompare the seasonal curves, we subtract the monthly mean values.

3. Results

3.1. Climatologies

We evaluated in total about 48 million samples of TROWARA measurements from 2004 to 2016 (sampling rates are between 6 and 11 s). The number of samples of IWV or ILW is about 48 million. The number of samples with cloud occurrence (group CF4) is about 24 million. The number of samples of the group CF1 is about 7 million. The number of samples of CF2 is 6 million, and the number of samples of CF3 is about 10 million. The number of samples of rain (ILW $\geq$ 400 g/m^2) is about 4 million. The number of samples of IWV, rain, CF1, CF2, CF3 and CF4 are shown in Figure 1 as a function of month. Most critical might be the statistics of thick supercooled liquid water clouds (CF2) in summer.

The climatologies of IWV, ILW, CF1, CF2, CF3 and CF4 above Bern from 2004 to 2016 are shown in Figure 2. Generally, there are strong seasonal variations in Figure 2. The error bars indicate the standard deviation of the monthly mean from year to year. Next, we use the climatological curves for computation of the diurnal cycles with respect to the monthly means. IWV varies from 8 kg/m^2 in winter to 24 kg/m^2 in summer. This annual oscillation of IWV is due to the seasonal change in air temperature and the Clausius–Clapeyron equation. The seasonal cycle of ILW contains an annual and a semi-annual component where the latter is a bit stronger. Cloud fraction of thin liquid water clouds (CF1) is fairly constant during the year except during winter months when CF1 presents higher values. Thick supercooled liquid water clouds (CF2) almost disappear during the summer months when the air temperature is high. Thick warm liquid water clouds (CF3) are strongly reduced during the winter. Cloud fraction of all liquid water clouds (CF4) increases slightly during winter, and a semi-annual oscillation is present.

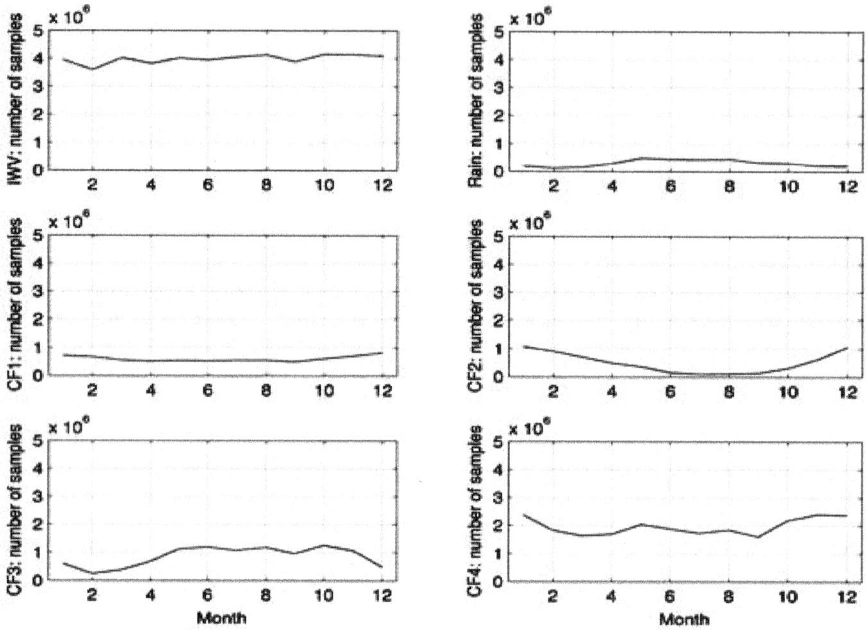

Figure 1. Number of samples of integrated water vapour (IWV), rain (ILW $\geq$ 400 g/m^2) and cloud fraction (CF) of four cloud categories (defined in the text) as a function of month observed by the TROWARA radiometer in Bern from 2004 to 2016.

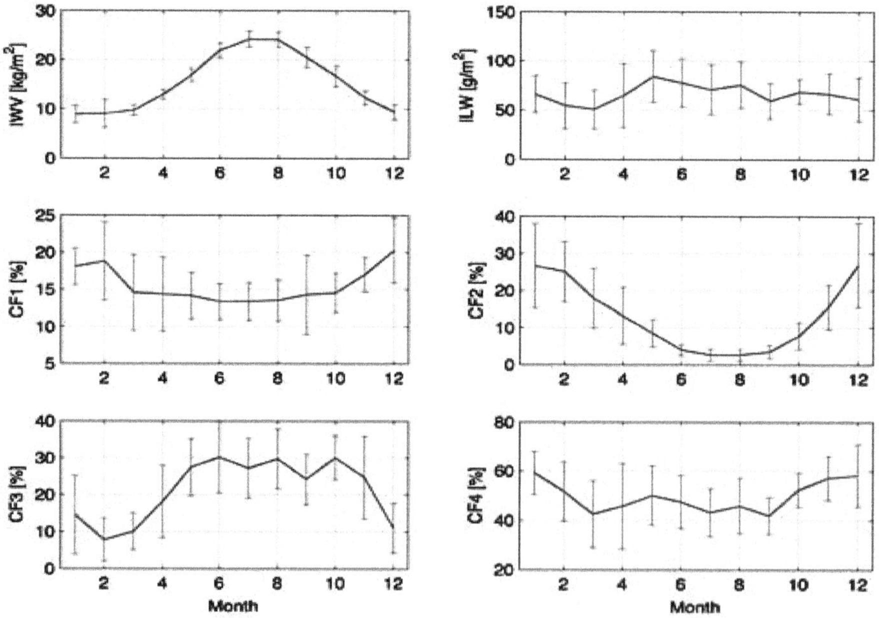

Figure 2. Climatologies of integrated water vapour (IWV), integrated liquid water (ILW) and cloud fraction (CF) of four cloud categories (defined in the text) observed by the TROWARA radiometer in Bern from 2004 to 2016. The error bars indicate the standard deviation of the monthly mean from year to year.

3.2. Spectral Behaviour of the Time Series

The normalized FFT power spectra of the complete time series of fluctuations provide information on the occurrence and the strength of the diurnal and semi-diurnal cycles with respect to the annual oscillation. Only in the case of ILW, the semi-annual oscillation is stronger than the annual oscillation, and so we normalize ILW by the power of the semi-annual oscillation. The FFT power spectra mainly provide information on the phase-locked fluctuations, e.g., phase-locked to the daily or annual cycle of solar radiation. Thus, many intermittent short-term fluctuations may average out by taking the spectrum over the time interval 2004–2016. This is the reason why we show next the amplitude spectrum obtained by a wavelet-method.

Figure 3 shows the normalized FFT power spectra in a blue colour for the fluctuations of IWV, ILW, CF1, CF2, CF3 and CF4 above Bern from 2004 to 2016. The annual oscillation is the blue spike at 1/(365 day), which is near 0 cycles/day. Compared to the power of the annual oscillation in IWV, the diurnal cycle in IWV power is about 400 times smaller. This means that the amplitude of the diurnal cycle is about 0.4 kg/m^2, which is 20 times smaller than the annual cycle of IWV, which is about 8 kg/m^2 in Figure 2. Furthermore, the amplitude of the diurnal cycle is about five times larger than the amplitude of the semi-diurnal cycle in IWV. The magnitudes of the annual and diurnal cycle in IWV are consistent with those of [25].

The amplitudes of the diurnal and semidiurnal oscillation in IWV are known to 5 σ-level confidence. This is indicated by the red line, which lies at a 5 σ distance above the yellow line, which is the mean power of the blue spectra sampled over 1000 frequency grid points. Here, we assumed that the mean power is equal to the noise σ.

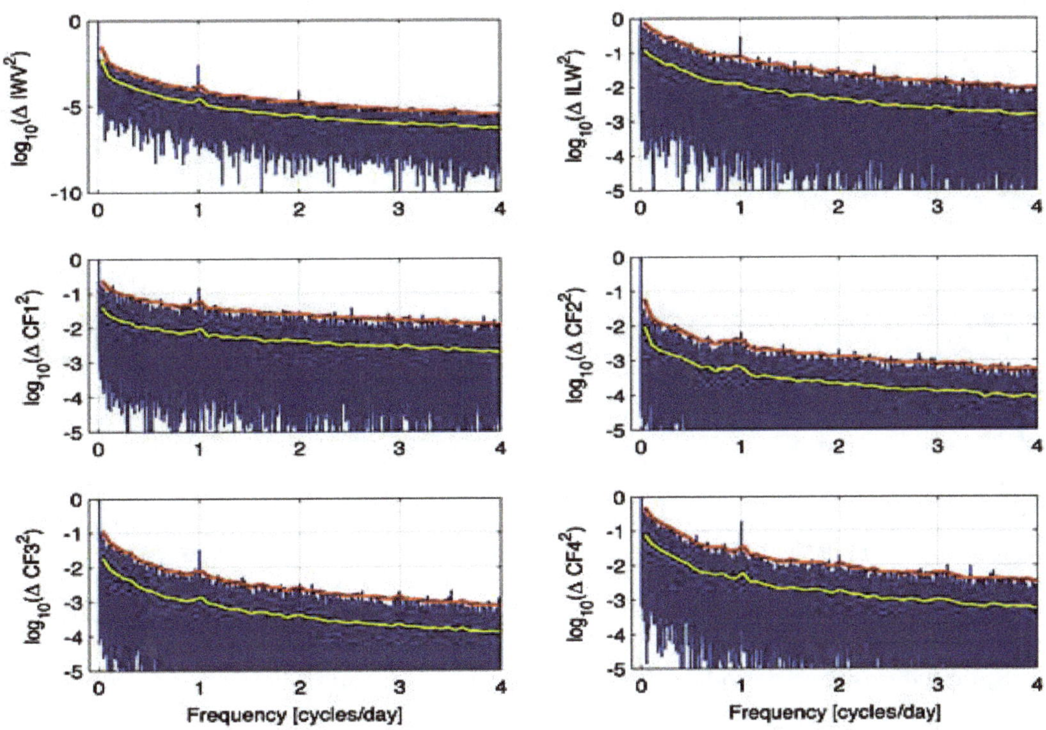

Figure 3. Fast Fourier Transform (FFT) power spectra (blue line) of the temporal fluctuations of IWV, ILW and CF in Bern for the time interval from 2004 to 2016. The spectra are normalized by the power of the maximum (power of the annual or semi-annual oscillation). The yellow line is the mean of the spectrum averaged over 1000 frequency grid points. The red line is the five sigma level of confidence. A significant diurnal cycle shows up as a blue spike at 1 cycle/day in each parameter. A significant semidiurnal cycle (at 2 cycles/day) is present for IWV and CF4.

In the case of ILW, we see a peak of the diurnal cycles that is rather close to those of the semi-annual oscillation, which is the dominant oscillation. There are several other significant oscillations with peaks closely above the red line, e.g., the semi-diurnal oscillation. In the cases of CF1, CF2, CF3 and CF4, we find significant diurnal cycles. The semi-diurnal cycle is well present for CF2 and CF4.

As a supplement to the FFT power spectra, we derive amplitude spectra by means of the fast response bandpass filter that takes care of phase-unlocked oscillations, which may persist only about time intervals of a few wave periods. Figure 4 shows the short-term variability of IWV, ILW, CF1, CF2, CF3 and CF4. An unresolved small peak is seen at the position of the diurnal cycle, which is marked by the red vertical line. The amplitude of the diurnal cycle in IWV is about 1 kg/m^2, which is larger than those obtained by the FFT power spectrum. This is reasonable since the phase of the diurnal cycle may change in time, and phase-unlocked intermittent oscillations with a one day period may occur as well. In case of the thin liquid water clouds (CF1), we can see that the maximal amplitudes are reached for periods smaller than one day. This is likely since small and thin clouds have a short life time and short horizontal scales. In the cases of IWV, ILW, CF3 and CF4, there are relatively strong oscillations with periods from two to ten days, which might be related to changes in synoptic weather patterns. The peak amplitude of IWV is around the seven-day period, whereas, for clouds, the peak is for shorter periods.

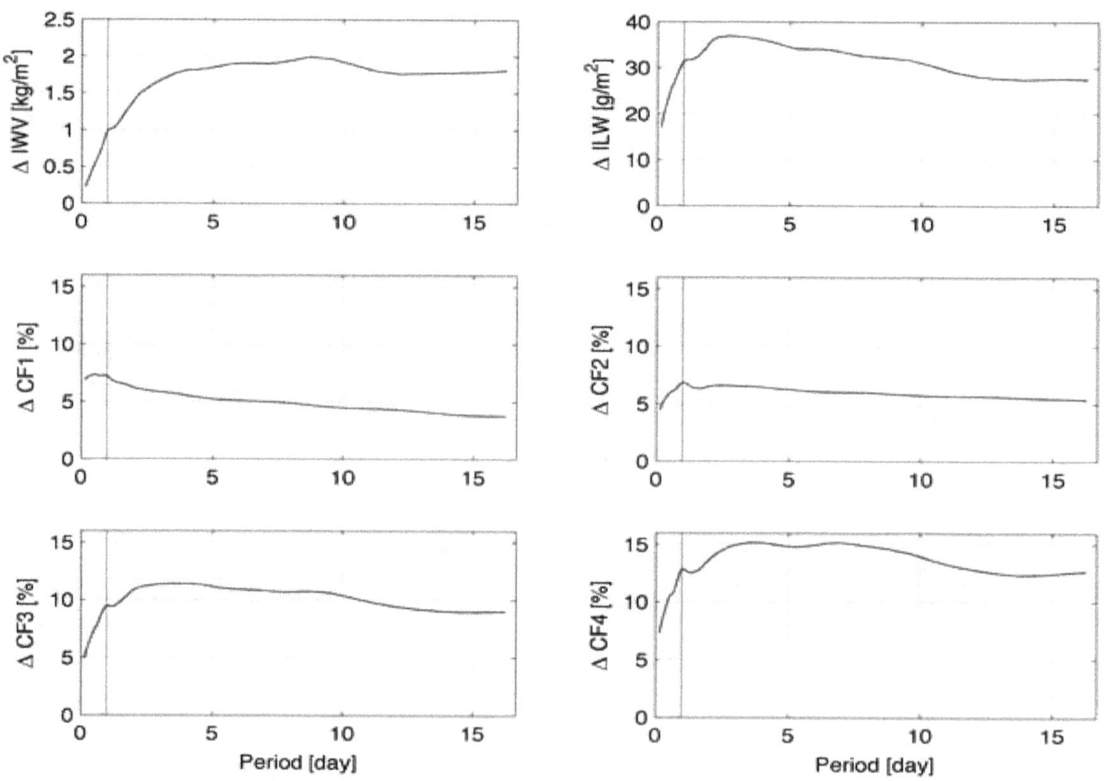

Figure 4. Amplitude spectra of IWV, ILW and CF over Bern for the time interval 2004–2016 obtained by a fast response bandpass filter. The red line marks the position of the diurnal cycle. The short-term variability of fluctuations with periods <10 days is high for ILW and CF.

3.3. Seasonal Dependence and Annual Mean of the Diurnal Cycle in IWV

Figure 5 shows the absolute and the relative diurnal cycles in IWV with respect to the monthly mean $\langle \mathrm{IWV} \rangle$. The subtraction of the monthly mean is necessary for the intercomparison of the seasonal curves since the monthly mean strongly varies from 8 kg/m^2 in winter to 24 kg/m^2 in summer. The absolute diurnal cycle is shown in the upper panel and is defined as $\Delta\mathrm{IWV} = \mathrm{IWV} - \langle \mathrm{IWV} \rangle$. The relative diurnal cycle is shown in the lower panel and is defined as $\Delta\mathrm{IWV} = (\mathrm{IWV} - \langle \mathrm{IWV} \rangle)/\langle \mathrm{IWV} \rangle$.

The seasonal curves of the diurnal cycles are given in color while the black dots denote the annual mean of the diurnal cycle. The amplitude of the mean diurnal cycle is 0.41 kg/m^2. Morland, J. et al. [25] obtained a quite similar curve for the mean diurnal cycle in IWV using measurements of TROWARA from 2003 to 2007. They found a mean amplitude of 0.32 kg/m^2. The phase of the diurnal cycle with a maximum around 19:00 LT in Figure 5 was also found by [25]. In addition, they compared the TROWARA results with the mean diurnal cycle of a GPS station in Bern, and they found an excellent agreement for the amplitude and the phase of the diurnal cycle.

Morland, J. et al. [25] suggested that evaporation of soil moisture into the atmosphere may explain the shape of the diurnal cycle in IWV. Accumulation of the evaporated water in the atmosphere during the daytime leads to the maximum of IWV in the evening while accumulated condensation of water vapour during the nighttime induces the IWV minimum in the morning hours. The hydrological atlas of Switzerland shows that the daily evaporation rate ranges from about 0.1 kg/m^2 per day in winter to about 3.8 kg/m^2 per day in summer in the Swiss plateau, which is dominated by agriculture [34]. Since the surroundings of Bern also have forests and a river, we expect higher evaporation rates of about 0.5 kg/m^2 per day in winter [34]. The annual mean of the rate of change of IWV in Figure 5 is about 0.82 kg/m^2 from the morning to the evening, which lies within the range of values of the evaporation rate in Bern. Thus, we regard the diurnal cycle of insolation, evaporation and condensation as the main reason for the observed diurnal cycle in IWV. Other factors such as diurnal variations of the surface wind vector and the vertical mixing rate of moisture lead to additional modifications of the diurnal cycle in IWV.

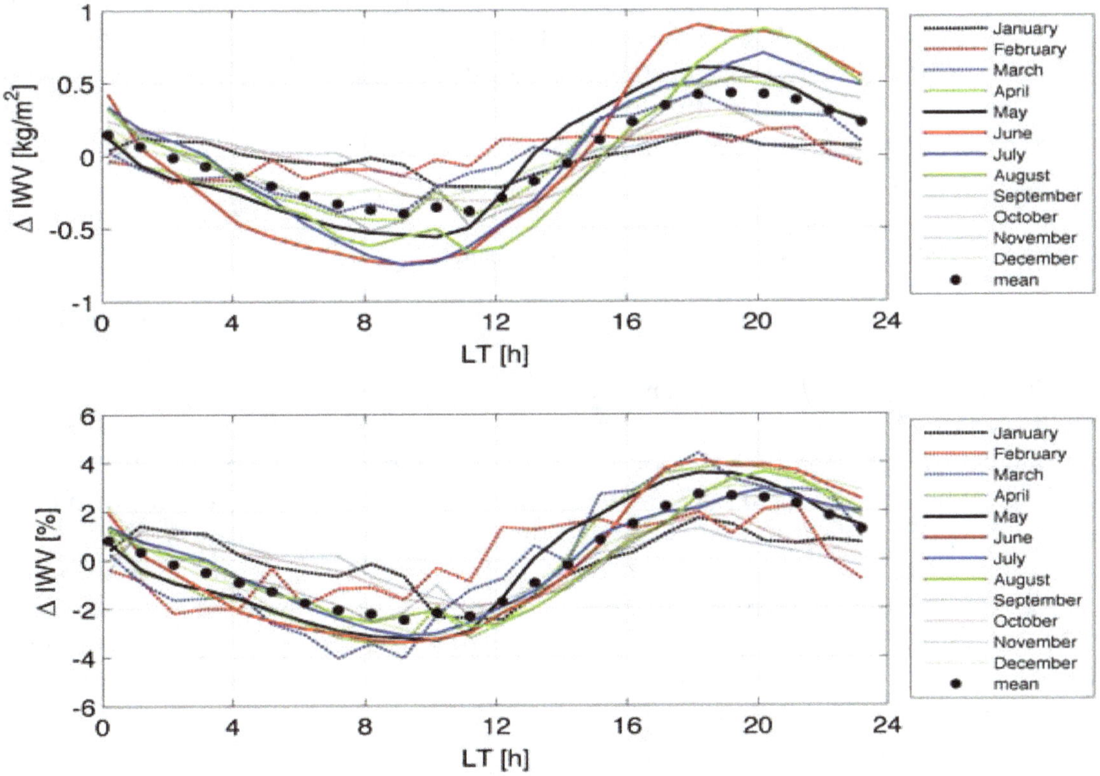

Figure 5. Seasonal dependence of the diurnal cycle in ΔIWV as a function of local time over Bern for the time interval 2004–2016. The upper panel shows the absolute diurnal cycle (ΔIWV = IWV − monthly mean of IWV) while the lower panel shows the relative diurnal cycle with respect to the monthly mean and in percent. The black dots indicate the annual mean of the seasonal curves.

Analyzing a GPS ground station network in Spain, Ortiz de Galisteo, J.P. et al. [35] found an amplitude of about 0.35 kg/m^2 for the mean diurnal cycle in IWV of all stations. The time of the maximum is at 18:30 UTC (which is close to local time in Spain). These values agree well with TROWARA in Bern. However, we find a significant difference if we look at the time of the minimum of IWV. It is about 10:00 LT for TROWARA in Bern while the IWV minimum is about 05:00 LT for the GPS network in Spain.

Ortiz de Galisteo, J.P. et al. [35] found a semi-diurnal oscillation in IWV with an amplitude of 0.13 kg/m^2, which is consistent with our FFT power spectra in Figure 3. Although the mean diurnal cycle of the Spanish stations agrees quite well with TROWARA in Bern, the Spanish stations showed a remarkable variation amongst themselves. Ortiz de Galisteo, J.P. et al. [35] explained that the shape of the diurnal cycle depends on factors that cause condensation, evaporation and moisture transport. The diurnal cycle in solar radiation is possibly most important for the observed diurnal cycle in IWV since evaporation of water from the surface, increase of air temperature and the growth of the planetary boundary layer during the daytime are closely related to absorption of solar radiation. On the other hand, the decrease of air temperature during the night leads to enhanced condensation of water vapour and to the observed minimum of IWV in the early morning hours. The time lag of the IWV minimum and maximum with respect to those of the air temperature might be related to the slow accumulation processes of liquid water on the ground and water vapour in the air.

Dai, A. et al. [36] retrieved diurnal cycles in IWV from a GPS station network in North America and found amplitudes of 1.0–1.8 kg/m^2 in the summer season and weaker in other seasons, which is greater than the 0.8 kg/m^2 June and August amplitudes of TROWARA in Figure 5. There is a large variability in the diurnal IWV cycles of the North American stations, but there are several stations that show an increase in IWV during the afternoon hours and a maximum around 16:00–19:00 LT, which is similar to that of TROWARA in Bern. The time of the IWV minimum is often late in the morning hours from 07:00–10:00 LT. Generally, Dai, A. et al. [36] reported relative diurnal IWV cycles with amplitudes less than 5%, while Figure 5 shows maximal relative amplitudes of 4%. In the percentage scale, the seasonal curves come closer together and the strong absolute diurnal cycles of the summer months do not differ much from those of March, April and May. The smallest relative amplitudes (<2%) occur in winter from November to January in Bern.

3.4. Seasonal Dependence and Annual Mean of the Diurnal Cycle in ILW

Figure 6 shows the absolute and the relative diurnal cycles in ILW with respect to the monthly mean. We define ΔILW in the same manner as ΔIWV as described in the subsection above. Compared to IWV, the diurnal cycle in ILW is more variable from month to month and relative amplitudes of 15 to 25% are reached especially during October, November and January. These strong diurnal cycles are connected with a break up of a cloud layer (stratus) at about 10:00 LT before noon and a relatively clear sky in the afternoon, which is consistent with our daily weather experience in Bern in late autumn and winter. The annual mean of the diurnal cycle (black dots) shows a minimum of about -10%, which occurs around noon. We suggest that this minimum is connected with the maximum evaporation or loss of cloud droplets at noon. During nighttime, a relative maximum of 9% is reached at around 03:00 LT.

Wood, R. et al. [37] reported diurnal ILW amplitudes of about 15–35% over the subtropical and tropical oceans observed by the Tropical Rainfall Measuring Mission Microwave Imager (TMI). The time of the ILW maximum is in the early morning (at 03:00 LT) and the time of the ILW minimum is at 15:00 LT, which is a bit later than for our measurements in Bern. They explain that the diurnal cycle in ILW is mainly driven by cloud solar absorption. Wood, R. [37] emphasize that the ILW measurements provide important constraints for models simulating the diurnal cycle of clouds. Snider, J.B. et al. [17] performed surface-based radiometric observations of water vapour and cloud liquid in the temperate zone and in the tropics. They found a sub-daily maximum of ILW at 05:00 LT and a minimum at 14:00 LT. The relative sub-daily variation of ILW was about 30%. In contrast with our study, Roebeling, R.A. et al. [16] reported diurnal variations in ILW values of the SEVIRI satellite experiment and ground-based microwave radiometers in France and England, which show increasing ILW values toward local solar noon.

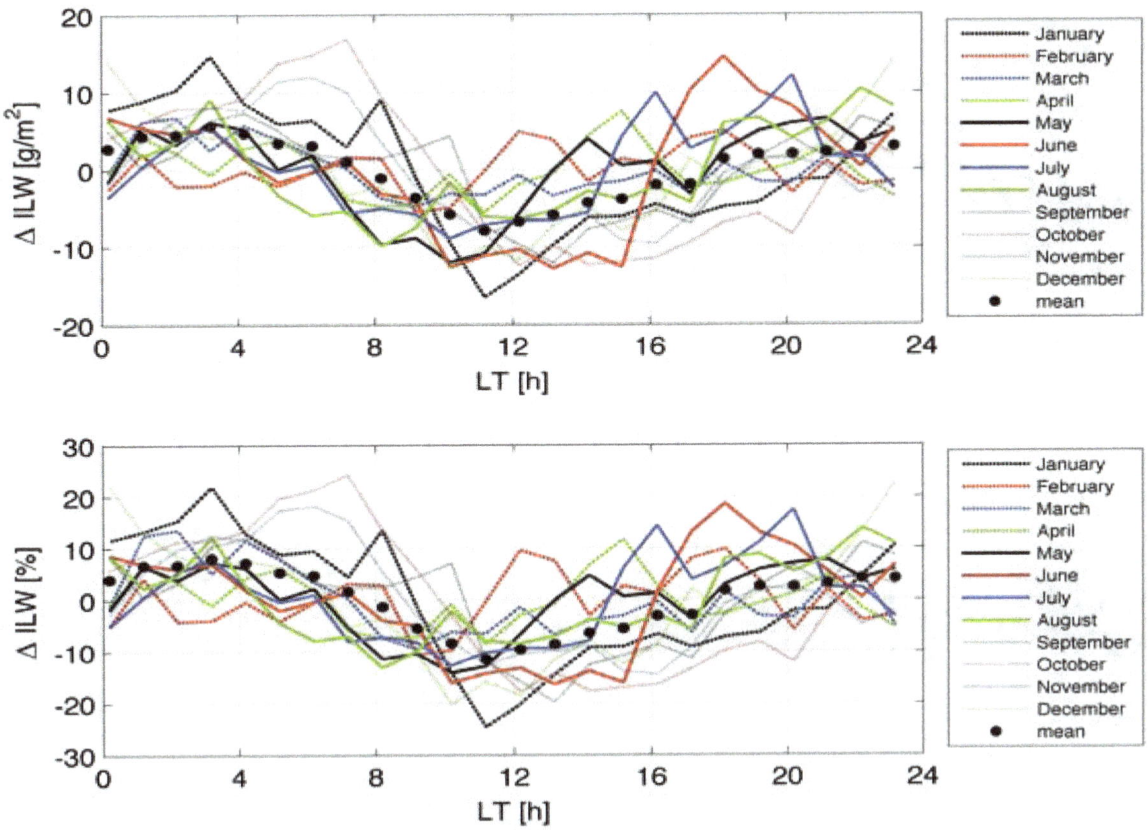

Figure 6. Seasonal dependence of the diurnal cycle in ΔILW as a function of local time over Bern for the time interval 2004–2016. The upper panel shows the absolute diurnal cycle (Δ ILW = ILW − monthly mean of ILW) while the lower panel shows the relative diurnal cycle with respect to the monthly mean and in percent. The black dots indicate the annual mean of the seasonal curves.

The seasonal curves for June to August in Figure 6 show local maxima from 16:00–20:00 LT. We suggest that this effect could be due to diurnal water vapour convection during summer. This phenomenon was indicated by the increase of IWV during afternoon in Figure 5. Schlemmer, L. et al. [38] performed idealized cloud-resolving simulations for the study of mid-latitude diurnal convection over land. They found an increase of specific cloud water content from 15:00 to 21:00 LT at 2–3.5 km altitude. They explained that convection and evaporation determine the moisture content of the lower troposphere. Then, the moisture content regulates the timing and intensity of the diurnal convection.

3.5. Seasonal Dependence and Annual Mean of the Diurnal Cycle in CF

In the case of cloud fraction, we present and discuss only the diurnal variations $\Delta CF = CF - \langle CF \rangle$. Figure 7 shows the diurnal cycle in CF1 (thin liquid water clouds) in the upper panel and those of CF2 (thick supercooled liquid water clouds) in the lower panel. The diurnal cycle in ΔCF1 has a maximum of 2% at 10:00 LT. Generally, the sub-daily variation is less than 5%. The climatology of CF2 showed that the thick supercooled liquid water clouds mainly occur during winter. The lower panel of Figure 7 shows that the curves of December and January have a strong diurnal variation with a fast decrease of ΔCF2 at 10:00 LT. This means that the occurrence of the thick supercooled liquid water clouds is favored during the nighttime and in the morning hours until 10:00 LT. Possibly, a part of the clouds are depleted by insolation during the daytime. The amplitude of the diurnal cycle in CF2 is about 4% in December and January.

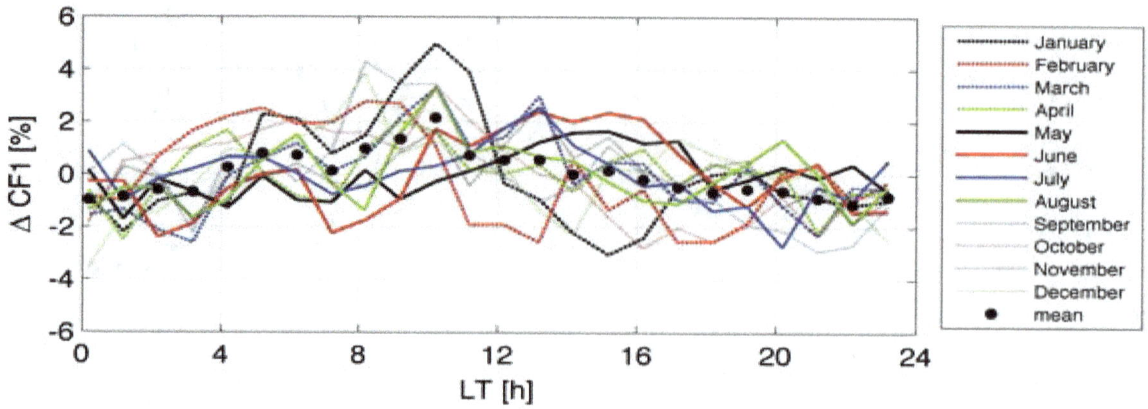

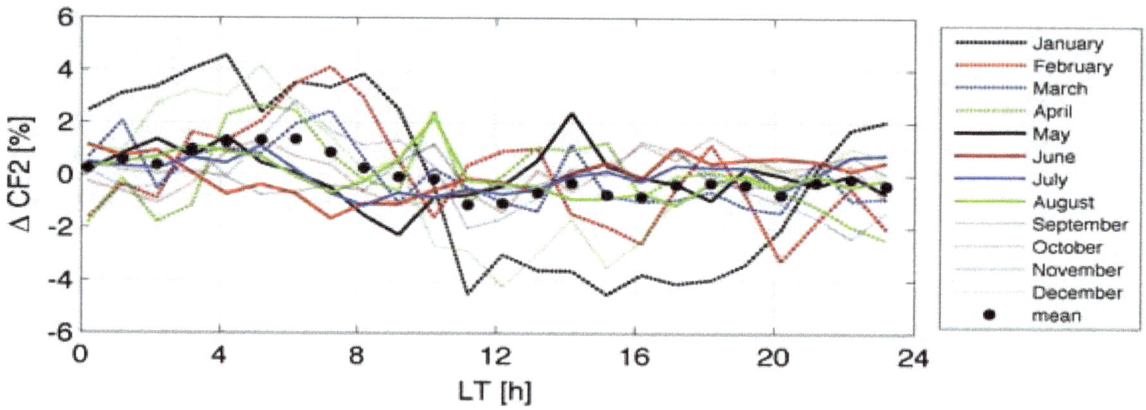

Figure 7. Seasonal dependence of the diurnal cycle in cloud fraction ΔCF1 (thin liquid water clouds) and ΔCF2 (supercooled thick liquid water clouds) as a function of local time over Bern for the time interval 2004–2016. The panels show ΔCF = (CF − monthly mean of CF). The black dots indicate the annual mean of the seasonal curves.

Figure 8 shows the diurnal cycle in CF3 (thick warm liquid water clouds) in the upper panel and those of CF4 (all liquid water clouds) in the lower panel. The strongest diurnal variation is found for CF3 and CF4 in October and November with deviations of about ±10% for CF4. Maximum values are reached at 07:00 LT and then a decrease starts towards the minimum, which is reached at 16:00 LT. Our daily experience with the cloud cover above Bern during autumn supports this objective measurement. Often, the clouds disappear around noon in autumn. A similar diurnal variation of cloud cover was derived by [13] from the ISCCP-C2 cloud climatology for the cloud category of maritime non-convective low-level clouds. "Maritime climate" may fit to Switzerland since it has a west coast climate and many lakes. The study of [13] showed that each cloud category has a different diurnal cycle. Min, M. and Zhang, Z. [39] presented a sinusoidal-like diurnal cycle in cloud fraction (five-year mean of SEVIRI observations over the southeast Atlantic). They found a slow 20% decrease of CF starting after sunrise and lasting until the evening.

In the case of CF3 in June, there seems to be an increase in the late afternoon, which might be connected to diurnal convection and cloud formation as described by [38].

The behavior of the annual mean of the diurnal cycle in Figure 8 (black dots) is a bit similar to those of ILW in Figure 6. Cloud solar absorption may explain the decrease of CF and the slightly negative values during the daytime.

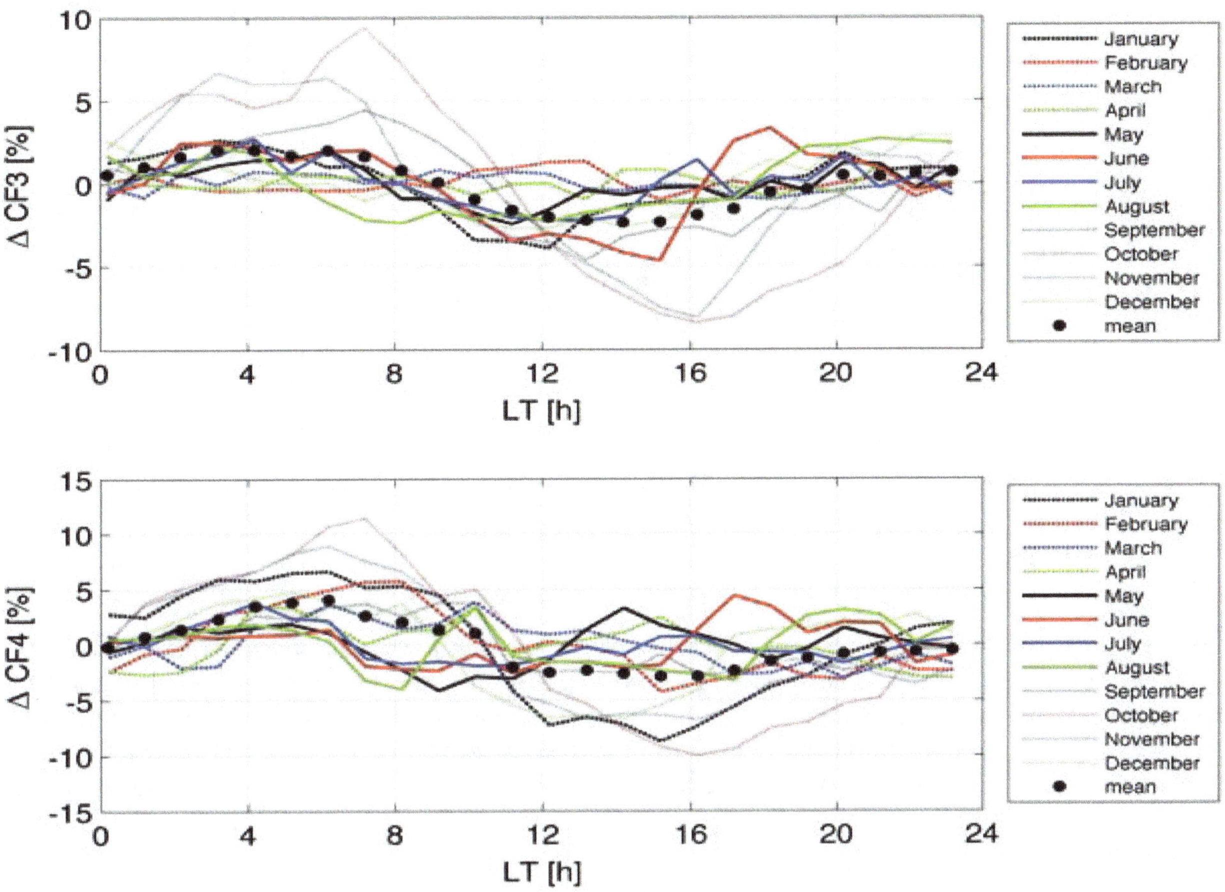

Figure 8. Seasonal dependence of the diurnal cycle in cloud fraction ΔCF3 (warm thick liquid water clouds) and ΔCF4 (all liquid water clouds) as a function of local time over Bern for the time interval 2004–2016. The panels show ΔCF = (CF − monthly mean of CF). The black dots indicate the annual mean of the seasonal curves.

3.6. Seasonal Variation and Diurnal Cycle in Rain Fraction

Analogously to cloud fraction, one can define rain fraction (RF), which is a measure of the occurrence of rain droplets in the measurements of TROWARA. Here, rain or rain droplets occur if the ILW measurements of TROWARA are greater than or equal to 400 g/m^2. At the edges of a time interval of rain, ILW increases or decreases within a short time from a small value to a high value or vice versa so that the choice of the threshold (e.g., 300, 400, or 500 g/m^2) plays a marginal role for the calculation of rain fraction [32].

Figure 9a shows the seasonal variation in rain fraction which varies from about 4% in winter to about 11% in summer. Figure 9b depicts the diurnal cycle in $\Delta RF = RF - \langle RF \rangle$. Rain fraction is enhanced by a few percent in the late afternoon during the summer months of June and July. This diurnal cycle in rain fraction might be connected to diurnal convection and precipitation over land during summer as described by [38].

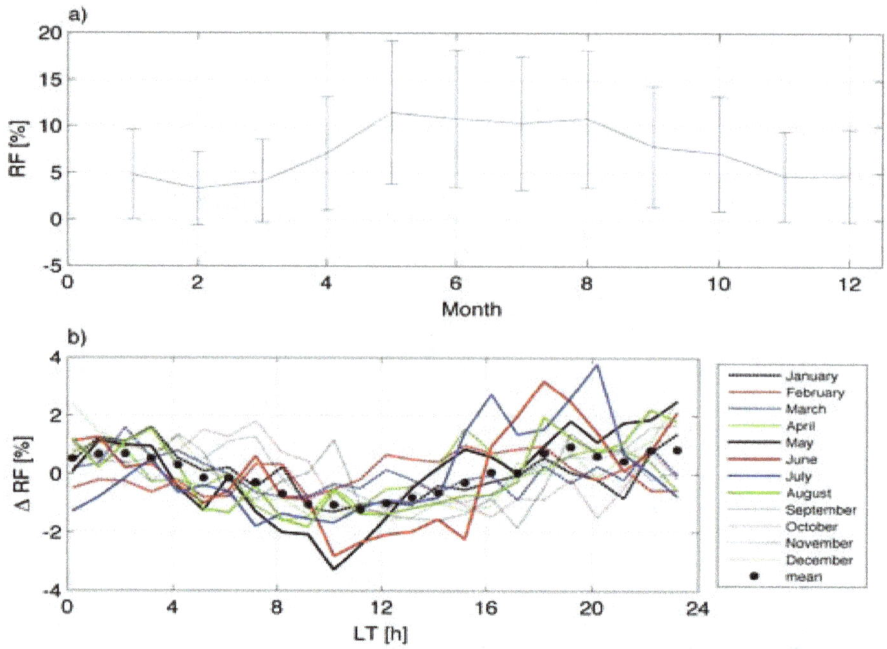

Figure 9. (**a**) seasonal variation in rain fraction (defined in the text) observed by the TROWARA radiometer in Bern from 2004 to 2016. The error bars indicate the standard deviation of the monthly mean from year to year. (**b**) seasonal dependence of the diurnal cycle in rain fraction ΔRF (for ILW as a function of local time over Bern for the time interval 2004–2016. The panel shows ΔRF = (RF − monthly mean of RF). The black dots indicate the annual mean of the seasonal curves.

4. Conclusions

The TROpospheric WAter RAdiometer (TROWARA) continuously measured cloud fraction (CF), integrated liquid water (ILW) and integrated water vapour (IWV) in Bern in Switzerland from 2004 to 2016. For our study, we derived hourly means from the TROWARA data sampled every 10 s. We presented and discussed the diurnal cycles in cloud fraction (CF), integrated liquid water (ILW) and integrated water vapour (IWV) for different seasons and the annual mean. Furthermore, we divided CF into four categories: thin liquid water clouds (CF1), thick supercooled liquid water clouds (CF2), thick warm liquid water clouds (CF3) and all liquid water clouds (CF4).

The amplitude of the mean diurnal cycle in IWV is 0.41 kg/m^2. The sub-daily minimum of IWV is at 10:00 LT, while the maximum of IWV occurs at 19:00 LT. The relative amplitudes of the diurnal cycle in ILW are up to 25% in October, November and January, which is possibly related to a breakup of the cloud layer at 10:00 LT. The minimum of ILW occurs at 12:00 LT, possibly explained by the maximum loss of cloud droplets due to maximum insolation at noon. In the case of cloud fraction of liquid water clouds (CF4), maximum values of +10% are reached at 07:00 LT and then a decrease starts towards the minimum of −10%, which is reached at 16:00 LT in autumn. This breakup of cloud layers in the late morning and early afternoon hours during autumn seems to be typical for the weather in Bern. A similar behavior is observed for CF3 so that we conclude that mainly thick warm liquid water clouds are responsible for the described diurnal variation in autumn. Finally, the TROWARA observations show that rain fraction is enhanced in the late afternoon hours in June and July.

The study showed that long-term measurements of a microwave radiometer equipped with an additional infrared channel objectively provide information on the diurnal cycle in six atmospheric water parameters. This information is of great interest for cross-validations with satellite data, high-resolution reanalyses and model simulations.

Acknowledgments: The study was supported by Swiss National Science Foundation under Grant No. 200021-165516. We are grateful to all technicians and scientists of the Institute of Applied Physics for designing, building and operating the TROWARA instrument over the last two decades. We thank the reviewers for valuable comments and corrections.

Author Contributions: Klemens Hocke carried out the spectral analysis. Francisco Navas-Guzmán and Christian Mätzler took care of the radiometer. All authors contributed to the interpretation of the data set.

References

1. Boucher, O.; Randall, D.; Artaxo, P.; Bretherton, C.; Feingold, G.; Forster, P.; Kerminen, V.M.; Kondo, Y.; Liao, H.; Lohmann, U.; et al. *Clouds and Aerosols, Climate Change 2013: The Physical Science Basis. Contribution of Working Group I to the Fifth Assessment Report of the Intergovernmental Panel on Climate Change*; Stocker, T.F., Qin, D., Plattner, G.-K., Tignor, M., Allen, S.K., Boschung, J., Nauels, A., Xia, Y., Bex, V., Midgley, P.M., Eds.; Cambridge University Press: Cambridge, UK; New York, NY, USA, 2013; Section 7, pp. 571–658.
2. Held, I.; Soden, B.J. Water vapor feedback and global warming. *Annu. Rev. Energy Environ.* **2000**, *25*, 441–475.
3. Jiang, J.H.; Su, H.; Zhai, C.; Perun, V.S.; Del Genio, A.; Nazarenko, L.S.; Donner, L.J.; Horowitz, L.; Seman, C.; Cole, J.; et al. Evaluation of cloud and water vapor simulations in CMIP5 climate models using NASA A-Train satellite observations. *J. Geophys. Res. Atmos.* **2012**, *117*, doi:10.1029/2011JD017237.
4. Gettelman, A.; Kay, J.E.; Shell, K.M. The Evolution of Climate Sensitivity and Climate Feedbacks in the Community Atmosphere Model. *J. Clim.* **2012**, *25*, 1453–1469.
5. Biasutti, M.; Yuter, S.E.; Burleyson, C.D.; Sobel, A.H. Very high resolution rainfall patterns measured by TRMM precipitation radar: Seasonal and diurnal cycles. *Clim. Dyn.* **2012**, *39*, 239–258.
6. Stubenrauch, C.J.; Chédin, A.; Rädel, G.; Scott, N.A.; Serrar, S. Cloud Properties and Their Seasonal and Diurnal Variability from TOVS Path-B. *J. Clim.* **2006**, *19*, 5531–5553.
7. Wylie, D. Diurnal Cycles of Clouds and How They Affect Polar-Orbiting Satellite Data. *J. Clim.* **2008**, *21*, 3989–3996.
8. Hong, G.; Heygster, G.; Notholt, J.; Buehler, S.A. Interannual to Diurnal Variations in Tropical and Subtropical Deep Convective Clouds and Convective Overshooting from Seven Years of AMSU-B Measurements. *J. Clim.* **2008**, *21*, 4168–4189.
9. Labow, G.J.; Herman, J.R.; Huang, L.K.; Lloyd, S.A.; DeLand, M.T.; Qin, W.; Mao, J.; Larko, D.E. Diurnal variation of 340 nm Lambertian equivalent reflectivity due to clouds and aerosols over land and oceans. *J. Geophys. Res. Atmos.* **2011**, *116*, doi:10.1029/2010JD014980
10. Dai, A.; Trenberth, K.E. The Diurnal Cycle and Its Depiction in the Community Climate System Model. *J. Clim.* **2004**, *17*, 930–951.
11. Dirmeyer, P.A.; Cash, B.A.; Kinter, J.L.; Jung, T.; Marx, L.; Satoh, M.; Stan, C.; Tomita, H.; Towers, P.; Wedi, N.; et al. Simulating the diurnal cycle of rainfall in global climate models: Resolution versus parameterization. *Clim. Dyn.* **2012**, *39*, 399–418.
12. Bergman, J.W.; Salby, M.L. The Role of Cloud Diurnal Variations in the Time-Mean Energy Budget. *J. Clim.* **1997**, *10*, 1114–1124.
13. Bergman, J.W.; Salby, M.L. Diurnal Variations of Cloud Cover and Their Relationship to Climatological Conditions. *J. Clim.* **1996**, *9*, 2802–2820.
14. Zhang, Y.; Long, C.N.; Rossow, W.B.; Dutton, E.G. Exploiting diurnal variations to evaluate the ISCCP-FD flux calculations and radiative-flux-analysis-processed surface observations from BSRN, ARM, and SURFRAD. *J. Geophys. Res. Atmos.* **2010**, *115*, doi:10.1029/2009JD012743.
15. Westwater, E.R.; Crewell, S.; Matzler, C. Frontiers in surface-based microwave and millimeter wavelength radiometry. In Proceedings of the IEEE International Geoscience and Remote Sensing Symposium (IGARSS), Anchorage, AK, USA, 20–24 September 2004; Volume 2, pp. 1268–1272.
16. Roebeling, R.A.; Deneke, H.M.; Feijt, A.J. Validation of Cloud Liquid Water Path Retrievals from SEVIRI Using One Year of CloudNET Observations. *J. Appl. Meteorol. Climatol.* **2008**, *47*, 206–222.
17. Snider, J.B.; Hazen, D.A. Surface-based radiometric observations of water vapor and cloud liquid in the temperate zone and in the tropics. *Radio Sci.* **1998**, *33*, 421–432.
18. Considine, G.; Curry, J.A.; Wielicki, B. Modeling cloud fraction and horizontal variability in marine boundary layer clouds. *J. Geophys. Res. Atmos.* **1997**, *102*, 13517–13525.

19. Peter, R.; Kämpfer, N. Radiometric Determination of Water-Vapor and Liquid Water and Its Validation with Other Techniques. *J. Geophys. Res. Atmos.* **1992**, *97*, 18173–18183.
20. Morland, J. *TROWARA—Tropospheric Water Vapour Radiometer. Radiometer Review and New Calibration Model*; IAP Research Report 2002-15; Institut für Angewandte Physik, Universität Bern: Bern, Switzerland, 2002.
21. Ingold, T.; Peter, R.; Kämpfer, N. Weighted mean tropospheric temperature and transmittance determination at millimeter-wave frequencies for ground-based applications. *Radio Sci.* **1998**, *33*, 905–918.
22. Mätzler, C.; Morland, J. Refined Physical Retrieval of Integrated Water Vapor and Cloud Liquid for Microwave Radiometer Data. *IEEE Trans. Geosci. Remote Sens.* **2009**, *47*, 1585–1594.
23. Brocard, E. Ground-Based Remote Sensing of the Troposphere in the Thermal Infrared. Ph.D. Thesis, Philosophisch-Naturwissenschaftliche Fakultät, Universität Bern, Bern, Switzerland, 2008.
24. Cossu, F.; Hocke, K.; Mätzler, C. A 10-Year Cloud Fraction Climatology of Liquid Water Clouds over Bern Observed by a Ground-Based Microwave Radiometer. *Remote Sens.* **2015**, *7*, 7768–7784.
25. Morland, J.; Collaud Coen, M.; Hocke, K.; Jeannet, P.; Mätzler, C. Tropospheric water vapour above Switzerland over the last 12 years. *Atmos. Chem. Phys.* **2009**, *9*, 5975–5988.
26. Hocke, K.; Kämpfer, N.; Gerber, C.; Mätzler, C. A complete long-term series of integrated water vapour from ground-based microwave radiometers. *Int. J. Remote Sens.* **2011**, *32*, 751–765.
27. Hocke, K.; Navas Guzmán, F.; Cossu, F.; Mätzler, C. Cloud Fraction of Liquid Water Clouds above Switzerland over the Last 12 Years. *Climate* **2016**, *4*, doi:10.3390/cli4040048.
28. Brocard, E.; Schneebeli, M.; Morland, J.; Mätzler, C. *Detection of Supercooled Cloud Layers with Microwave and Infrared Radiometry and Additional Tools*; IAP Research Report 2006-02-MW; Institut für Angewandte Physik, Universität Bern: Bern, Switzerland, 2006.
29. Mätzler, C.; Rosenkranz, P.W.; Cermak, J. Microwave absorption of supercooled clouds and implications for the dielectric properties of water. *J. Geophys. Res. Atmos.* **2010**, *115*, doi:10.1029/2010JD014283.
30. Hirsch, E.; Agassi, E.; Koren, I. Determination of optical and microphysical properties of thin warm clouds using ground based hyper-spectral analysis. *Atmos. Meas. Tech.* **2012**, *5*, 851–871.
31. Bernet, L.; Navas-Guzmàn, F.; Kämpfer, N. The effect of cloud liquid water on tropospheric temperature retrievals from microwave measurements. *Atmos. Meas. Tech. Discuss.* **2017**, *2017*, 1–35.
32. Cossu, F. Study of Atmospheric Water over Bern by Means of Ground-Based Observations and Numerical Simulations. Ph.D. Thesis, Philosophisch-Naturwissenschaftliche Fakultät, Universität Bern, Bern, Switzerland, 2015.
33. Studer, S.; Hocke, K.; Kämpfer, N. Intraseasonal oscillations of stratospheric ozone above Switzerland. *J. Atmos. Sol. Terr. Phys.* **2012**, *74*, 189–198.
34. Menzel, L.; Lang, H.; Rohmann, M. Mean Annual Actual Evaporation 1973–1992. Plate 4.1 in Hydrological Atlas of Switzerland 1999. Available online: http://www.hydrologischeratlas.ch/en/products/printed-issue/evaporation/plate-4-1 (accessed on 31 August 2017).
35. Ortiz de Galisteo, J.P.; Cachorro, V.; Toledano, C.; Torres, B.; Laulainen, N.; Bennouna, Y.; de Frutos, A. Diurnal cycle of precipitable water vapor over Spain. *Q. J. R. Meteorol. Soc.* **2011**, *137*, 948–958.
36. Dai, A.; Wang, J.; Ware, R.H.; Van Hove, T. Diurnal variation in water vapor over North America and its implications for sampling errors in radiosonde humidity. *J. Geophys. Res. Atmos.* **2002**, *107*, doi:10.1029/2001JD000642.
37. Wood, R.; Bretherton, C.S.; Hartmann, D.L. Diurnal cycle of liquid water path over the subtropical and tropical oceans. *Geophys. Res. Lett.* **2002**, *29*, doi:10.1029/2002GL015371.
38. Schlemmer, L.; Hohenegger, C.; Schmidli, J.; Bretherton, C.S.; Schär, C. An Idealized Cloud-Resolving Framework for the Study of Midlatitude Diurnal Convection over Land. *J. Atmos. Sci.* **2011**, *68*, 1041–1057.
39. Min, M.; Zhang, Z. On the influence of cloud fraction diurnal cycle and sub-grid cloud optical thickness variability on all-sky direct aerosol radiative forcing. *J. Quant. Spectrosc. Radiat. Transf.* **2014**, *142*, 25–36.

5

Gaps Analysis and Requirements Specification for the Evolution of Copernicus System for Polar Regions Monitoring: Addressing the Challenges in the Horizon 2020–2030

Estefany Lancheros [1,*], Adriano Camps [1,*], Hyuk Park [1], Pierre Sicard [2,†], Antoine Mangin [2], Hripsime Matevosyan [3] and Ignasi Lluch [3]

1 Universitat Politècnica Catalunya—BarcelonaTech & IEEC, Campus Nord, 08034 Barcelona, Spain; park.hyuk@tsc.upc.edu
2 ACRI Bâtiment Le Grand Large, Quai de la Douane—2 ème éperon, 29200 Brest, France; psicard@argans.eu (P.S.); antoine.mangin@acri.fr (A.M.)
3 Skolkovo Institute of Science and Technology, Moscow 143026, Russia; hripsime.matevosyan@skolkovotech.ru (H.M.); ignasi.lluchicruz@skolkovotech.ru (I.L.)
* Correspondence: estefany@tsc.upc.edu (E.L.); camps@tsc.upc.edu (A.C.)

† Current address: ARGANS, 260 routes du Pin Montard, BP234, 06904 Sophia-Antipolis CEDEX, France.

Abstract: This work was developed as part of the European H2020 ONION (Operational Network of Individual Observation Nodes) project, aiming at identifying the technological opportunity areas to complement the Copernicus space infrastructure in the horizon 2020–2030 for polar region monitoring. The European Earth Observation (EO) infrastructure is assessed through of comprehensive end-user need and data gap analysis. This review was based on the top 10 use cases, identifying 20 measurements with gaps and 13 potential EO technologies to cover the identified gaps. It was found that the top priority is the observation of polar regions to support sustainable and safe commercial activities and the preservation of the environment. Additionally, an analysis of the technological limitations based on measurement requirements was performed. Finally, this analysis was used for the basis of the architecture design of a potential polar mission.

Keywords: Earth Observation (EO); satellite; sensors; platform; microwave radiometer; SAR; GNSS-R; optical sensors; polar; weather; ice; marine

1. Introduction

Copernicus is a program that powers the European Earth Observation (EO) capacity to meet the user needs and be highly competitive globally. Copernicus addresses six thematic services: land, marine, atmosphere, climate change, emergency management and security. Each service relies on a product portfolio that is derived from space and in situ infrastructure. The European Space Agency (ESA) has developed the space segment, a series of missions called the Sentinels, specifically tailored to the operational needs of the Copernicus program. Additionally, the Sentinels' missions are supported by contributing missions, such as the Earth Explorer missions by the ESA and the Meteorological Satellites (EUMETSAT) and include missions from European Union (EU) and non-EU member states.

Sentinel-1 is equipped with a C-band Synthetic Aperture Radar (SAR) for land, ocean and emergency services. This is based on a constellation of two polar orbiting satellites, in the same orbital plane with a 180° orbital phase difference. Currently, Sentinel 1-A and Sentinel 1-B are operational satellites, and Sentinel 1-C and Sentinel-1D are future missions to ensure data continuity. The first Sentinel-1 satellite was Sentinel-1A, and it was launched on 3 April 2014. Sentinel-1B was launched on 25 April 2016. Sentinel-1C will be launched in 2021 and Sentinel-1D in 2023. Each Sentinel-1 is expected to have at least seven years of lifetime.

Sentinel-2 is equipped with a Multi-Spectral Imaging (MSI) sensor, to cover the land and emergency services of Copernicus. Constituted by the A/B/C/D series, at present, there are two satellites in polar orbit (Sentinel-2A and Sentinel-2B). Sentinel 2-A was launched on 23 June 2015, and Sentinel 2-B was launched on 7 March 2017. Planned missions to provide data continuity are Sentinel-2C and Sentinel-2D, which will be launched in 2021 and 2022.

Sentinel-3 is equipped with seven instruments for land- and ocean-monitoring services. The two multispectral sensors are named the Ocean and Land Color Imager (OLCI) and Sea and Land Surface Temperature Radiometer (SLSTR). It also has a Synthetic aperture Radar Altimeter (SRAL) that requires a micro-wave radiometer for water vapor correction, a Doppler Orbitography and Radio-positioning Integrated by Satellite (DORIS), a Laser Retro-Reflector (LRR) and a GPS receiver for orbitography correction. It is composed by four satellites series (A/B/C/D), with two operational polar orbit satellites at the present time. Sentinel-3A and Sentinel-3B were launched on 16 February 2016 and 25 April 2018, respectively. The future Sentinel-3C and Sentinel-3D will be launched in 2023 and 2024.

Sentinel-5p, also known as Sentinel-5 Precursor, is equipped with a Tropospheric Monitoring Instrument (TROPOMI). This mission brings support for atmospheric services. It is based on a single satellite in polar orbit, launched on 13 October 2017.

Sentinel-4 and Sentinel-5 are planned as hosted payloads for a mission operated by EUMETSAT, to ensure the atmospheric and climate change services of Copernicus. Sentinel-4 is a spectrometer called the Ultra-violet, Visible and Near-infrared sounder (UVN), which will be onboard the Meteosat Third Generation-Sounder (MTG-S). MTG-S is composed of two series (MTG-S1 and MTG-S2), in geostationary orbit. MTG-S1 and MTG-S2 are scheduled for launch in 2023 and 2031, respectively. Sentinel-5 is a sounder called the Ultra-violet, Visible and Near-infrared Sounder (UVNS) onboard the MetOp-Second Generation (MetOp-SG, with the series A1, A2 and A3), in polar orbit. MetOp-SG A1/A2/A3 will be launched in 2021, 2028 and 2035, respectively.

Sentinel-6, also called the Joint Altimetry Satellite Oceanography Network-Continuity of Service (JASON-CS), will be developed and implemented through a partnership between EUMETSAT, ESA, National Aeronautics and Space Administration (NASA), and National Oceanic and Atmospheric Administration (NOAA). A radar altimeter package like the one in Sentinel-3 will be equipped in two Sun-synchronous series satellites (JASON-CS-A and JASON-CS-B) with a seven-year lifetime each. Currently, the launches of JASON-CS A and B are planned for 2020 and 2025, respectively.

In recent years, European Commission (EC) has led the Horizon 2020 program-supporting mission aligned with major EU policy priorities. In the context of Copernicus, the priorities are to contribute to the evolution of its services and to satisfy the end-user needs. The H2020 ONION (Operational Network of Individual Observation Nodes) project played an important role in defining the technological EO requirements based on the user needs and future measurement gaps of the Copernicus system in the horizon 2020–2030. Each use case is linked to a Copernicus service, and they are integrated by a set of measurements required to fulfil the users' needs. The measurements are the geophysical product estimated from satellite acquisitions. The measurements with gaps are the measurements detected with an observation gap (in terms of spatial resolution, and/or revisit time, and/or accuracy, and/or temporal continuity, and/or data latency) in the Copernicus space infrastructure in the time period from 2020–2030. The main objective of the ONION project was to place the user requirements at the center of the design process, as well as to identify solutions to meet these needs. This project has helped to understand the challenges for the evolution of the new Copernicus missions.

From the knowledge of the end-user needs, this project has provided an important scientific basis to address the measurement requirements, the instrumentation and remote sensing technologies that have to be explored to cover the next decade of the measurement gaps of the Copernicus system, where monitoring of the polar regions is an emerging need, with improved revisit time and latency time for marine weather forecast and sea ice monitoring use cases.

The methodology used in this work is described in Figure 1. First, the top 10 use cases were ranked according to the end-user needs [1], and end-user requirements were defined [2]. Second, a database of the future Copernicus instruments and contributing missions was generated to analyze the measurement gaps in the horizon 2020–2030. The gaps were detected based on the ability of these sensors to monitor each measurement defined in the use cases. Measurement gaps were analyzed in terms of the spatial resolution, revisit time, precision and temporal continuity, as well as the data latency for products requiring near real-time data (Section 2). Based on the results of the gap analysis in the time frame 2020–2030, monitoring of the polar regions arose as the top emerging need. Accordingly, Section 3 describes the importance of observing the polar regions. Section 4 presents the potential instrumentation required to cover the emerging needs, based on the measurement characteristics. Section 5 presents a discussion based on the limitation of current technologies and the challenges addressed to next generation of the sensors, to ensure all the measurements with gaps in the polar region are covered by the Copernicus space segment. Finally, the conclusions are presented.

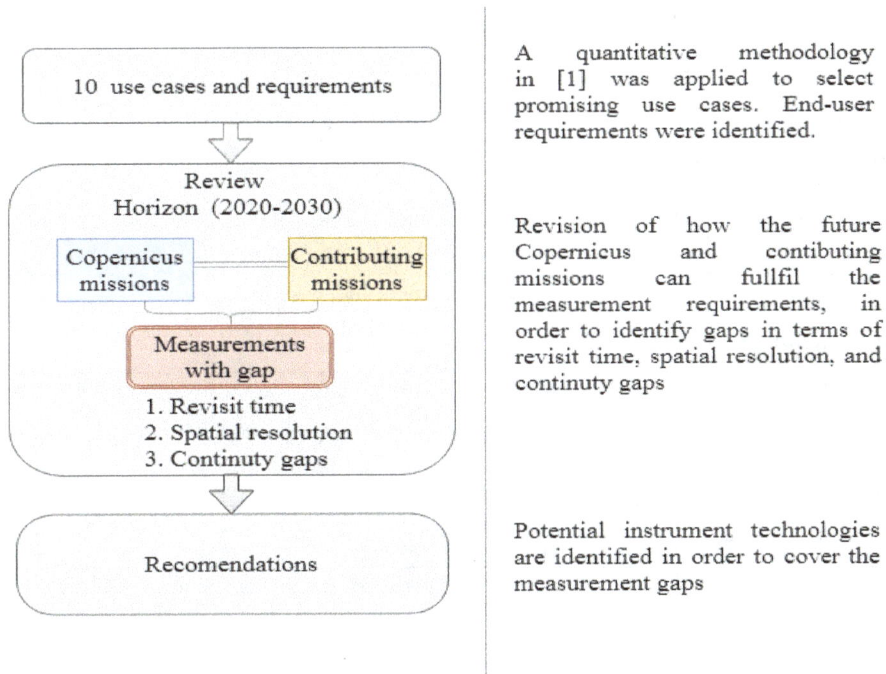

Figure 1. Methodology applied to define the end-user requirements and measurement gaps.

2. Requirements Specifications and Measurements Gaps

This study focuses on the identification of the EO measurements gaps in the time frame from 2020–2030, to complement the Copernicus space infrastructure, based on the top 10 use cases [1]. These use cases are not satisfied by the existing Copernicus infrastructure, and they were generated through a quantitative methodology that involved the prioritization of 38 EO data needs (the complete list of the identified needs with their description is presented in Table 1), 96 products across the six Copernicus services, 63 stakeholders, 131 measurements and 48 uses cases, which were scored. The top 10 use cases were defined as (1) marine weather forecast, (2) sea ice monitoring (extent and thickness), (3) fishing pressure and fish stock assessment, (4) land for infrastructure status assessment, (5) agriculture and forestry (hydric stress), (6) land for mapping (risk assessment), (7) sea ice melting emissions, (8) atmosphere for weather forecast, (9) climate for ozone layer and UV and (10) natural habitat and protected species monitoring.

Table 1. Description of the identified user needs [1].

Need Name	Need Description
Agriculture, Rural Development and Food Security	Estimates of crop production, water satisfaction index, early warning of harvest shortfalls.
Air Quality and Atmospheric Composition	The quality of air that one directly breathes at the surface.
Alerting Service	Alert of an ongoing crisis.
Animal Migration Maps	Track for animal migration.
Assessment of Renewable Energies Potential	Provide meteorological (cloud, water vapor) and atmospheric (aerosol, ozone) data; and solar irradiance maps.
Basic Maps	Base layer information with key geographical features.
Biodiversity Assessment	Vegetation indices, information on habitat deterioration, evolution of vegetation parameters.
Climate Evolution	Assess long-term climate evolution.
Climate Forcing	Monitoring human-forced climate change.
Climate Policy Development	Informing policy development to protect citizens from climate-related hazards such as high-impact weather events.
Communication/Reporting Resources	Context/supporting and justifying operations.
Crisis and Damage Mapping	Updated (24 h) geographical information.
Emission and Surface Flux Assessment	Anthropogenic emissions, greenhouse gases.
Fish Stock Management	Analysis and forecasting of fish stocks.
Forest Resources Assessment	Deforestation rates, forest intactness.
In-Field Data Collection	Locally-sampled information.
Infrastructure Status Assessment	Roads, railroads, buildings, power lines, pipelines and others.
Inland Water Management Maps	Measure quantity, quality (acidity) and track for algae.
Land Degradation and Desertification Assessment	Degradation risk index, degradation hot spots, etc.
Maintenance information	Estimation of the required ship maintenance date
Marine Operations Safety	Oil spill combat, ship routing, weather forecasting, defense, search and rescue
Mining	Focused on information for the mining industry
Mitigation and Adaptation	Improving planning of mitigation and adaptation practices for key human and societal activities.
Ocean Color Maps	Track for algae, bloom, toxicity, "red tide" and acidity.
Oil and Gas Assessment	Focused on information retrieval for the oil and gas industry.
On time operation	Optimized routing and ship speed.
Ozone Layer and UV	Archive and forecast information on ozone layer and UV.
Ports Monitoring	Monitoring of ports and facilitate traffic management.
Refugee Support Mapping	Snapshot of temporary settlements and internally-displaced people.
Ship Positioning Mapping	Monitoring ship positions and information.
Situation Mapping	After crisis mapping.
Solar Radiation	The amount of solar radiation coming to Earth.
Thematic Mapping	Focused on the spatial variation of a theme.
Urban and Regional Development	Monitoring of settlements, land losses or gain.
Water quality	Water quality and pollution both in high seas and coast.
Water Resources	Erosion risk maps, average water available for watershed.
Weather Forecast	Climate monitoring, ice seasonal forecast.

These 10 use cases require 75 measurements. The results of the observation gap in the time frame from 2020–2030 presented 20 measurements with gaps that correspond only to eight use cases (see Table 2). The marine weather forecast use case has six measurements with gaps; sea ice monitoring (extent and thickness) involves 13 measurements with gaps; fishing pressure and fish stock assessment has five measurements with gaps; land for infrastructure status assessment use case will be met by future Copernicus infrastructure and contributing missions (the measurements require high spatial resolution with a long revisit time); the agriculture and forestry (hydric stress) use case implies four measurements with gaps; the land for mapping (risk assessment) use case presents only one measurement with observation gaps (this measurement is associated with the agriculture and forestry (hydric stress) use case); the sea ice melting emission use case involves six measurements with gaps (these measurements are associated with sea ice monitoring (extent and thickness) use case); the atmosphere for weather forecast use case has only one measurement with gaps, and this measurement is associated with the marine weather forecast use case; the climate for ozone layer and UV use case will be satisfied by future Copernicus infrastructure and contributing missions; and natural habitat and protected species monitoring has only one variable with an observation gap (this measurement is associated with the agriculture and forestry (hydric stress) use case).

In most cases, each measurement is associated with different use cases. There are four use cases that involve all the measurements with gaps (marine weather forecast, sea ice monitoring (extent and thickness), fishing pressure and fish stock assessment and agriculture and forestry (hydric stress)). In this way, if only four use cases are addressed, they cover all the 20 measurements with gaps. These four use cases that involve all the measurements are congruent with respect to a recent survey into the state and the health of the European EO services industry [3], where the results indicated that the Copernicus data and services do not fully respond to the customer needs. Principally, it reported that the agriculture, maritime and fishery domains are of major importance for the European EO market. For these reasons, the next sections focus on the analysis of these four use cases.

Marine weather forecast covers measurements such as wave and wind parameters. This information is of predominant importance to a wide variety of activities, from tourism to fishing, oil and gas exploration and exploitation. Results of the measurement gap analysis focused on "marine weather forecast" showed that revisit time gap to be reduced to 3 h and data latency <1 h. Table 2 shows the Copernicus and contributing missions in the time frame from 2020–2030, which are capable of measuring the variables defined for the marine weather forecast use case. These future missions will provide high horizontal spatial resolution for ocean measurements, but the provision of appropriate sea-state products and the adequacy of EO observations in near- or real-time (<1 h) are not satisfied. This translates into a system mission that shall support existing and planned EU infrastructure to reduce the revisit time for "marine weather forecast" use case to 3 h and data latency to less than 1 h.

The use case "sea ice monitoring" covers a wide range of measurements that are of high relevance to marine operations and to understand global climate change. This use case requires providing real-time sea ice data and improving the precision of ice thickness measurements, as well as increasing the operational monitoring capability of polar regions. On the one hand, the Arctic and Antarctic are the parts of the globe with better revisit time statistics as most EO missions are in polar Low Earth Orbits (LEO) and fly over the poles 14–15 times per day. On the other hand, polar regions represent a blind zone for instruments flying in a geostationary orbit and are poorly covered by some narrow-swath nadir-pointing instruments such as radar altimeters (e.g., SRAL/Sentinel-3 and Poseidon-4/Sentinel-6). This means that for instruments in polar LEO orbits (typically Sun-Synchronous Orbits, SSO) with off-side acquisition capacity, the coverage of the polar regions is limited by the swath of the instrument and by the number of satellites considered. The latency of the data is also an issue for near real-time products. Moreover, a very small subset of EU or cooperating missions can provide sea ice thickness, sea ice type, sea ice concentration, sea ice cover, sea ice drift and extent at the resolutions required by end-users. Therefore, sea ice monitoring with a short revisit time, short latency time and high spatial resolution are the main requirements, the mission shall support sea ice products (e.g., sea ice type, cover, extent, drift, thickness and iceberg tracking) with a revisit time <3 h and latency time < 1 h [4].

The use case "fishing pressure and fish stock assessment" promises also new prospects for the system in the maritime domain and would benefit from a reduction in access and revisit times for the provision of appropriate oceanic conditions (e.g., sea surface temperature, ocean chlorophyll concentration) and fishing pressure (e.g., vessel tracking). Ocean chlorophyll concentrations are related to the presence of planktonic life, which is the base of the marine food chain. In this regard, this parameter brings information on the health and productivity of marine ecosystems. Therefore, these data are valuable and help to develop strategies for sustainable and productive commercial fishery. Another important measurement to cover is Sea Surface Temperature (SST) in this time frame.

The SST over polar regions in a global context is essential for climate modeling, weather forecast, as well as for the fishing and maritime industry. Missions with optical instruments (in the visible and infrared domains) provide information on the sea surface composition (e.g., ocean chlorophyll concentration, color dissolved organic matter) and sea surface temperature. One of the main difficulties of these types of techniques is that they are directly impacted by cloud coverage and depend on solar illumination conditions. The current Copernicus infrastructure provides about 24–48 h of latency for those measurements. From the side of contributing missions in geostationary orbits, these provide data every 30 min (shorter revisit time), but with a coarse resolution of 5 km. Geostationary satellites are essential in equatorial and mid-latitude areas, but for high latitudes, image distortion and atmospheric effects are too large for effective use.

In the context of the fishing pressure and fish stock assessment, data on fish farming cages (density) and vessel's identity and location (position, speed and direction) are potentially valuable for emergency and management services. For instance, the provision of these observations will help to improve ship routing services, offshore operations and search and rescue operations, thus contributing to marine safety. This type of data can be provided by using an Automatic Identification System (AIS). The ESA has promoted the use of AIS systems on Satellites (SAT-AIS) [5] through of Advanced Research in Telecommunications Systems (ARTES) program. In the horizon 2020–2030, only the NORSAT-2 and Triton-2 missions carrying AIS have been planned [6].

The "agriculture and forestry (hydric stress)" use case is based on methods enabling precision agriculture, efficient irrigation, fire prevention, forest protection and impacts on hydrological basins, supporting agronomic research and production, assessment of population food security and sovereignty and environmental impact evaluation. Soil moisture is a key parameter for the hydrological cycle, meteorology, climatology and agriculture production. The role of soil moisture for meteorology lies in the global transfer of water and energy between the Earth's surface and the atmosphere. In the agricultural context, the amount of soil moisture is an important element affecting production and plant growth. Surface soil moisture can be estimated with a high spatial resolution by the Advanced Scatterometter (ASCAT) on the meteorological Operational (MetOp) mission and SAR-C on the Sentinel-1. However, the accuracy, revisit time and temporal resolution are insufficient to meet the user requirements. New developments in the miniaturization of cameras in the visible and infrared bands with high-resolution data make new techniques available for remote observation of crops [7,8]. However, for precision agriculture applications, the use of remote sensing can be limited because of inadequate spatial, temporal and thematic products tailored to the needs of farmers. Future satellites carrying sensors in the thermal infrared band present coarse spatial and temporal resolutions and are also limited to clear sky conditions. Due to these limitations, thermal imagery is not useful at the plot scale for precise irrigation monitoring. There is also an emerging need to consider L-band microwave radiometers (with high spatial resolution) to support crop condition monitoring.

Table 2. Gap analysis results over the Copernicus space segment in the horizon 2020–2030.

Measurements	Use Cases	Requirements [2]	(2020–2030) Copernicus Instruments/Mission [9]	(2020–2030) Contributing Instruments/Mission [10]	Gap
Ocean surface currents	1. Marine weather forecast 2. Sea ice monitoring 7. Sea ice melting emissions	Spatial resolution: 1–25 km Revisit time <3 h 0.5 m/s and 10° accuracy	SAR-C/Sentinel-1 SRAL/Sentinel-3 Poseidon-4/Sentinel-6	Karin/SWOT SWIM/CFOSAT SAR-2000 S.G/CSG SAR / HRWS SAR-X / TSX-NG SAR-X/PAZ	Revisit time <3 h Latency time <1 h
Dominant wave direction	1. Marine weather forecast 2. Sea ice monitoring	Spatial resolution: 1–15 km Revisit time <3 h 10° accuracy			Revisit time <3 h Latency time <1 h
Significant wave height	Marine weather forecast Sea ice monitoring	Spatial resolution: 1–25 km Revisit time <3 h 0.1 m accuracy			Revisit time <3 h Latency time <1 h
Wind speed over sea surface (horizontal)	1. Marine weather forecast 2. Sea ice monitoring 8. Atmospheric for weather forecast	Spatial resolution: 1–10 km Revisit time <3 h 0.5 m/s accuracy Latency time <1 h		ASCAT/MetOp SCA/MetOp-SG Karin/SWOT SAR-2000 S.G/CSG SAR/HRWS SAR-X/TSX-NG SAR-X/PAZ	Revisit time <3 h Latency time <1 h
Sea ice type	2. Sea ice monitoring 7. Sea ice melting emissions	Spatial resolution: 10 m Revisit time <3 h 0.25/classes accuracy Latency time <1 h			Revisit time <3 h Latency time <1 h
Iceberg tracking	2. Sea ice monitoring	Spatial resolution: 10 m Revisit time <3 h 5% accuracy Latency time <1 h			Revisit time <3 h Latency time <1 h
Sea ice cover	2. Sea ice monitoring 7. Sea ice melting emissions	Spatial resolution: 12 km–10 m Revisit time <3 h 5% accuracy Latency time <1 h		SAR-2000 S.G/CSG SAR/HRWS SAR-X/TSX-NG SAR-X/PAZ MSI/Earth-CARE FLORIS/FLEX	Revisit time <3 h
Sea ice extent	2. Sea ice monitoring	Spatial resolution: 12 km–10 m Revisit time <3 h 5% accuracy Latency time <1 h			Revisit time <3 h Latency time <1 h
Sea ice drift	2. Sea ice monitoring	Spatial resolution: 10 m Revisit time <3 h 0.5 m/s and 10° accuracy			Revisit time <3 h Latency time <1 h
Sea ice thickness	2. Sea ice monitoring 7. Sea ice melting emissions	Spatial resolution: 1 cm (vertical) Revisit time <3 h 1 cm accuracy Latency time <1 h		KARIN/SWOT SAR-2000 S.G/CSG SAR/HRWS SAR-X/TSX-NG SAR-X/PAZ	Revisit time <3 h Latency time <1 h

Table 2. *Cont.*

Measurements	Use Cases	Requirements [2]	[2020–2030] Copernicus Instruments/Mission [9]	Contributing Instruments/Mission [10]	Gap
Atmospheric pressure over sea surface	1. Marine weather forecast 8. Atmospheric for weather forecast	Spatial resolution: 1–25 km Revisit time <3 h 5% accuracy Latency time <1 h	OLCI/Sentinel-3	CPR/Earth-CARE	Revisit time <3 h Latency time <1 h
Sea surface temperature	1. Marine weather forecast 2. Sea ice monitoring 3. Fishing pressure and and fish stock assessment 7. Sea ice melting emissions	Spatial resolution: 1–10 km Revisit time <3 h 0.3 k accuracy Latency time <1 h	SLSTR/Sentinel 3	SEVERI/MSG MSI/Earth-CARE IASI and AVHRR/MetOp METimage, IASI-NG/MetOp-SG FCI/MTG-I IRS/MTG-S	Revisit time <3 h Latency time <1 h
Ocean chlorophyll concentration	3. Fishing pressure and fish stock assessment	Spatial resolution: 1 km Revisit time <72 h 0.05 mg/m^3 accuracy	OLCI/Sentinel-3	3MI/MetOp-SG METimage/MetOp-SG	Latency time <1 h
Ocean imagery and water leaving radiance	1. Sea ice monitoring 3. Fishing pressure and fish stock assessment	Spatial resolution: 1 km Revisit time <72 h 5% accuracy Latency time <1 h	OLCI/Sentinel-3 SAR-C/Sentinel-1	AVHRR/3/MetOp-A/B/C SAR/RADARSAT-2	Latency time <1 h
Color dissolved organic matter	3. Fishing pressure and and fish stock assessment	Spatial resolution: 1 km Revisit time <72 h 5% accuracy	OLCI/Sentinel-3	3MI /MetOp-SG METimage/MetOp-SG FLORIS/FLEX	Latency time <1 h
Detection of water stress in crops	5. Agriculture and forestry (hydric stress)	Spatial resolution: 2–7 m Revisit time <24 h 5% accuracy Latency time <1 h	SLSTR / Sentinel-3	SEVERI/MSG MSI/Earth-CARE IASI and AVHRR/MetOp METimage, IASI-NG /MetOp-SG FCI/MTG-I IRS/MTG-S	Spatial resolution <7 m Latency time <1 h
Estimation of crop evapotranspiration	5. Agriculture and forestry (hydric stress)	Spatial resolution: 1–10 m Revisit time <24 h			
Soil moisture at the surface	5. Agriculture and forestry (hydric stress) 6. Land for mapping: risk assessment 10. Natural habitat and protected species 2. Sea ice monitoring	Spatial resolution: 10 km Revisit time <24 h 0.01 m^3/m^3 accuracy Latency time <1 h	Sentinel-1	ASCAT/MetOp SAR-2000 SG/CSG SEVERI/MSG SAR-P/BIOMASS FCI/MTG-I MSI/Earth-CARE	Accuracy <0.01 m^3/m^3 Latency time <1 h
Crop growth & condition	5. Agriculture and forestry (hydric stress)	Spatial resolution: 2 km Revisit time <24 h Latency time <1 h	N/A	N/A	Spatial resolution: 2 km Revisit time <24 h Latency time <1 h
Monitoring system—vessels and fish farming cages position tracking	3. Fishing pressure and fish stock assessment	Spatial resolution: 1 km Revisit time <72 h (cloud free) 5% accuracy Latency time <1 h	Sentinel-1	NAIS/NORSAT-2 E-SAIL/Triton-2	from 2025

3. Sea Ice Monitoring and Marine Weather Forecast over Polar Regions as an Emerging Need for Future Copernicus Missions

Over the past few decades, the polar regions have been subjected to significant changes. Total sea ice extent has decreased, and it has thinned [11]. Arctic sea ice melts, and it is increasingly influencing human activities, as some Arctic marine routes have gone from being covered by sea ice to being navigable during part of the year. In this direction, the Arctic and Northern Ocean have been considered as interesting areas to extend the commercial operations related to fishing, oil and gas.

As shown in Table 2, 65% of the measurements with gaps correspond to sea ice monitoring, and marine weather forecasts represent around 30%. Here, there is potential to cover these use cases. The sea ice monitoring (extent and thickness) and marine weather forecast use cases would benefit enormously from the improvement of the latency time, revisit time and accuracy. In response to the end-user needs, instrumentation and remote sensing technologies have to be explored to cover the future measurement gaps of the Copernicus system. In this way, we focus on EO over polar regions, where there is a high priority to monitor the previously-mentioned domains.

New remote sensing opportunities will be explored to provide real time data to ensure navigation safety, to increase the operational monitoring capability on sea ice to understand climate change and for marine weather forecast information to a wide variety of activities such as fishing, oil and gas operation. In this direction, the measurements and requirements to cover are detailed in the next section.

4. Instrumentation and Remote Sensing Technologies Required to Cover the Future Measurement Gaps over Polar Regions

The measurements with gaps that have been detected are of dominant importance to a wide variety of activities, from marine traffic, fishery and the environment. Therefore, in this section, these measurements are analyzed in terms of the available instrument technologies to cover the detected gaps. Additionally, the operational limitations of the potential instruments are identified from the viewpoint of measurement requirements. Table 3 summarizes the technologies identified to ensure that the measurements with gaps are covered. The state of the art of the instrumentation is presented and the technology limitations from the perspective of the measurement requirements.

4.1. Ocean Surface Currents

This variable presents a gap in the required revisit time. The required revisit time is 3 h, and the required spatial resolution is 1–25 km, with a speed accuracy of 5 m/s and 10° in direction. A federation of three missions, Sentinel-3 (radar altimeter SRAL) and a specific combination with a constellation of Global Navigation Satellite Systems-Reflectometry (GNSS-R) instruments [12,13], microwave radiometer, altimeter radar [14] and SAR on small platforms, could improve the accuracy and the revisit time efficiently.

4.2. Dominant Wave Direction and Significant Wave Height

These variables can be measured with radar altimeters and SAR. In the time frame 2020–2030, the instruments capable of measuring these variables are Poseidon-4 (Sentinel-6/JASON), SRAL (Sentinel-3) and SAR-C (Sentinel-1), but the 3-h revisit time is not satisfied. The current Copernicus infrastructure delivers about 24–48 h of latency for the corresponding measurements.

4.3. Wind Speed

This variable presents gaps in the revisit time (<3 h) and latency time (<1 h). The required revisit time is 3 h at 10-km spatial resolution, with an accuracy of 0.5 m/s. Typical resolutions for microwave radiometer winds are about 25 km [15] or 12.5 km [16] over oceans; the measurement range is 0–50 m/s [17], and the accuracy is from about 2–10 m/s (depending on the rain flag) [18].

Microwave radiometers infer wind speed from frequencies near 6.8, 10.7, 19 and 37 GHz. Another technology that can be used to measure wind speed is SAR. The current Copernicus infrastructure has a constellation of two SARs, but the revisit time is between one and two days for high latitudes and the high resolution dataset (1 km) [19]. Another solution is to infer the wind vector using radar scatterometers, with a 10-km spatial resolution over the oceans.

Table 3. Mapping of the potential technologies to cover measurements with gaps. The technology limitations from the perspective of the measurement requirements are presented.

Technology Type	Measurements	Instrument Limitations
GNSS-R	Sea ice thickness [a]	Accuracy (~20 cm) [12,20]
	Dominant wave direction [b]	Coarse spatial resolution (~25 km) [21]
	Wind speed over the sea surface (horizontal) [a]	Coarse spatial resolution (~25 km) Accuracy (2 m /s) [12]
	Significant wave height [b] Sea ice cover [b] ocean surface currents [b]	No specific limitation [22]
Microwave sounder (50–60 GHz)	Atmospheric pressure (over sea surface) [c]	Coarse spatial resolution (20 km, at 400 km altitude) [23]
Microwave radiometer (X-, K-, Ka-, W-bands)	Wind speed over sea surface (horizontal) [b] Sea ice cover [b] Sea ice type [a] Sea ice drift [a] Sea surface temperature [a]	Coarse spatial resolution (~25 km) [24] Accuracy: (1.5-m/s ocean wind speed) [25] (0.5 K for SST) [25] (from 10%–20% for sea ice data) [26]
L-band microwave radiometer	Sea ice cover [b]	Coarse spatial resolution
	Soil moisture at the surface [c]	Coarse spatial resolution
	Sea ice thickness [a]	Accuracy [27,28]
AIS decoder	Monitoring system: vessels [c]	No specific limitation [5]
Cloud radar (oxygen band)	Atmospheric pressure (over sea surface) [c]	Narrow swath (2 km) [29,30]
Radar scatterometer	Wind speed over sea surface (horizontal) [c] Sea ice type [a] Sea-ice cover [a]	Accuracy: Wind speed (<2 m/s) [31]
Radar altimeter (SAR)	Ocean surface currents [3] Significant wave height [b] Dominant wave direction [b] Sea ice type [b] Sea ice cover [b] Wind speed over sea surface (horizontal) [a]	Narrow coverage (nadir-pointing) Long-term analysis and narrow coverage
SAR	Ocean surface currents [c] Iceberg tracking [c] Sea ice drift [c] Sea ice extent [c] Sea ice type [c] Sea ice cover [c] Dominant wave direction [b] Dominant wave period [b] Significant wave height [b] Sea ice thickness [a] Wind speed over sea surface [a]	Narrow coverage (<600 km) Long-term analysis and narrow coverage [32]
LiDAR	Sea ice thickness [b]	Long-term analyses, narrow coverage, Cloud sensitive
Multispectral radiometer VIS/NIR/TIR	Ocean chlorophyll concentration [c] (λ center: 442.5, 490, 510, 560 nm) Ocean imagery and water leaving radiance [c] (λ Centre: 485, 560, 660, 2100 nm) Color Dissolved Organic Matter (CDOM) [c] (λ center: 442.5, 490, 510, 560, 665 nm) Sea surface temperature [c] (λ center: 3.7, 4.05, 8.55, 11, 12 μm) Sea ice cover [a] (λ center: 640, 1610 nm)	Cloud sensitive, Daylight only
Hyperspectral radiometer (VIS/NIR)	CDOM [c] Sea ice cover [b]	Cloud sensitive, Daylight only
Spectrometer/sounder IR	Sea ice cover [c] Sea surface temperature [c]	Cloud sensitive, Daylight only

[a] Marginal relevance; [b] medium relevance; [c] high relevance .

The use of novel techniques using Signals of Opportunity (SoOp), such as those from Direct Broadcast Satellite (DBS) television signals at the Ku- or X-band [33], can be potentially exploited in the future to measure precipitation and winds over the sea surface at higher revisit times. These signals are potentially sensitive to detecting fluctuations of the sea surface roughness and light precipitation. As compared to GNSS-R systems, the spatial resolution will be better (higher frequency) and the Signal-to-Noise Ratio (SNR) higher as transmitters transmit more power. In this regard, a receiver of signals of opportunity at the Ku- or X-band can provide cost benefits and high quality data, but these techniques have yet to be developed.

4.4. Sea Ice Type

Sea ice type is a critical parameter of marine weather forecast to improve and to understand climate change. The requirement for this measurement is a 3-h revisit time with 10 km as the spatial resolution. Previous experiments demonstrated the sensitivity of microwave scatterometers to discriminate sea ice type [34]. According to the gap detected in the revisit time, this instrument will be evaluated in conjunction with other solutions such as microwave imagers, SAR and SAR altimeters [35].

4.5. Sea Ice Cover

This variable requires a revisit time of 3 h, with a minimum requirement of 12 km of spatial resolution and an accuracy of 5%. In the time frame 2020–2030, Sentinel-1 with a C-band SAR and Sentinel-3 with an SRAL radar altimeter are the Copernicus missions capable of measuring this variable. At high latitudes, Sentinel-1 presents a revisit time <1 day, and its utilization period is from 2014–2030. MetOp is a contributing Copernicus mission that can provide sea ice cover data thanks to the Micro-Wave Imager (MWI, instrument operated by ESA and EUMETSAT), with global coverage once per day and a spatial resolution of 25 km. The utilization period for MWI is from 2022–2043. The Sentinel-1 and MetOp missions do not meet the 3-h revisit time at a 12-km horizontal spatial resolution. In this regard, the formulation of potential technologies to measure sea ice cover is necessary. In this way, microwave radiometers, SAR, SAR altimeters and GNSS-R are technologies capable of measuring this variable.

Microwave radiometers can measure this variable using the 19.35-, 37- and 90-GHz channels (e.g., the Special Sensor Microwave, SSM/I ; the Special Sensor Microwave-imager/sounder, SSMIS, with a spatial resolution of 25 km) [36]; or the 18.7- and 89-GHz ones (the Advanced Microwave Scanning Radiometer for Earth Observation System, AMSR-E, with a spatial resolution of 12.5 km) [37]. However, in order to achieve the high spatial resolution required, the antennae of these instruments should be enlarged. Currently, dual-polarization GNSS-R has emerged as a promising technique to measure sea ice cover and thickness. Experiments with the mission TechDemoSat-1 have demonstrated the capability and high accuracy of this instrument to conduct these types of measurements in [22,38,39].

4.6. Sea Ice Extent

Sea ice extent is an important parameter to understand the global climate. Previous works have shown that Ku-band scatterometer data are able to measure sea ice extent [40,41]. The required revisit time is 3 h with 10 km as minimal spatial resolution and 5% accuracy. Scatterometers have a wide swath and a reduced revisit time, which makes them a valuable technology for this application. However, the main limitation is the coarse resolution of the data generated by this sensor. In this sense, many studies have been conducted to develop algorithms to improve the spatial resolution of the data, combining data from multiple passes of the satellite [42,43] or from different sensors [44] (e.g., microwave imager, SAR and scatterometer).

4.7. Iceberg Tracking

Iceberg monitoring is important for climate studies and for navigation safety. In the time frame 2020–2030, the Copernicus EO infrastructure to support iceberg tracking will be SAR-C/Sentinel-1,

SRAL/Sentinel-3 and SRAL/Sentinel-4 (JASON). Contributing Copernicus missions that are capable of this are ASCAT/MetOp and Scaterometer (SCA) on MetOp-SG. However, these missions are not enough to meet the user requirements, as iceberg tracking presents a gap in the revisit time (Table 2). Nowadays, the technologies measuring this parameter are the scatterometer, the SAR and the radar altimeter (in SAR processing). The main limitation of SAR and altimeters is their narrow swath, long revisit time and the long time to analyze the data (long latency). Scatterometer data are also valuable due to their coverage, but they present a coarse spatial resolution. The instruments that will be studied to cover the gaps in this measurement are SAR, SAR altimeters and microwave scatterometers.

4.8. Sea Ice Drift

Nowadays, the Copernicus Marine Environment Monitoring Service offers daily sea ice drift data over the Arctic and Antarctic from active (Sentinel-1, ASCAT) and passive microwave sensors (SSM/I, AMSR-E) and optical passive (the Advanced Very High Resolution Radiometer /3, AVHRR/3). With the data combined from different sensors, a global dataset of the sea ice drift is obtained with a spatial resolution of 10 km. The data latency is 5 h. In this regard, the data latency has to be improved down to <1 h and the revisit time down to <3 h. Potential sensing technologies are: microwave radiometers, SAR, multispectral optical and microwave scatterometers.

4.9. Sea Ice Thickness

In the time frame from 2020–2030, the sea ice thickness variable presents a gap in the revisit time (24 h) and in the vertical spatial resolution of 1 cm with 0.1-cm accuracy [2]. Ice thickness maps with a resolution of 50 km and up to 50–60 cm in thickness are produced with the Microwave Imaging Radiometer using Aperture Synthesis (MIRAS), L-band microwave radiometer onboard the Soil Moisture and Ocean Salinity (SMOS) mission of the ESA, with an accuracy mission of 0.5–6 cm [45]. Another technology is the LiDAR altimeter, but its narrow swath and high cost do not compensate its benefits in terms of the variables with a gap. GNSS-R [39] and SAR [46] technologies also are capable of measuring this variable.

4.10. Atmospheric Pressure over the Sea Surface

Atmospheric pressure oscillations produce SST variations, and its monitoring provides a better understanding of the factors disturbing climate variations, contributing to accurate marine weather forecasts. Microwave radiometers can infer this measurement using several channels in the 50–60-GHz frequency bands. Among these, future microwave sounders in the time frame from 2020–2030 to measure the atmospheric pressure over the sea surface are: the Advanced Microwave Sounding Unit-A (AMSU-A), the Advanced Technology Microwave Sounder (ATMS) and nanosatellites, such as the Microwave Radiometer Technology Acceleration (MiRaTA) [23], and the Earth Observation Nanosatellite–MicroWave (EON-MW) missions [23]. AMSU-A is a whisk broom line scanner instrument in a SSO, with a global coverage of twice per day. It has 14 channels in the oxygen band (50–60 GHz). It is operated by NOAA, NASA and EUMETSAT, and the utilization period is from 2006–2024 on MetOp. ATMS on the Join Polar Satellite System (JPSS) is a cross-track scanning microwave sounder in a Sun-synchronous orbit, with a global coverage of twice per day. It is able to measure atmospheric pressure (over the sea surface). Its utilization period is from 2018–2021. It is also operated by NASA, NOAA and EUMETSAT. ATMS is the functional equivalent to AMSU-A with improved coverage thanks to a swath of 2600 km. The EON-MW mission has been proposed by MIT Lincoln Laboratory to extend the JPSS. It aims at demonstrating ATMS quality on a low-cost CubeSat platform in order to mitigate the gaps in weather observations. This mission is scheduled for launch in 2018–2019. The Infrared (IR) sounder/spectrometer also can measure this variable. The IR sounder on a CubeSat is feasible (6U). A good example of the compact from of this type of sensor is the Earth Observation Nanosatellite Infrared (EON-IR) on the CIRAS (CubeSat Infrared Atmospheric Sounder) mission [47]. In this regard, a constellation of microwave sounders or IR sounders as the payload could be considered, in order to cover the low revisit time (<3 h). These data will be complementary to those obtained by sensors onboard large platforms.

4.11. Sea Surface Temperature

The SST can be measured by microwave radiometer imagers [48] and infrared sounders [47]. Microwave radiometer imagers at frequencies of 6–7 and/or 11 GHz with coarse spatial resolution can also provide global SST data. Currently, a spatial resolution of 25 km can be achieved using the microwave imagers' (Tropical Rainfall Measuring Mission Microwave Imager, or TMI; Wind Microwave Radiometer, WindSAT; Global precipitation measurement Microwave Imager, GMI; AMSR-E; and AMSR-2) data fusion technique [24]. Infrared radiometers are capable of measuring this variable over cloud-free areas with high spatial resolution (1–4 km) using wavelength in the 10–12 μm range. Microwave radiometers can improve the coverage in polar regions because microwave signals penetrate the clouds. The challenge is an achievement of a relatively small footprint (<10 km) at low frequency bands (6.8 GHz).

4.12. Surface Soil Moisture

The soil moisture is a key parameter to understand the water cycle, and in cryosphere regions, it provides information about the freeze-thaw cycles. Several remote sensing technologies have been proposed to estimate soil moisture. Two criteria were selected to perform the technology categorization. The first one is related to the region of the spectrum (optical or microwave). Optical instruments acquire soil moisture measurements using the Thermal Infrared (TIR). Microwave instruments use signals in the L-, S- and C-bands. L-band is the main frequency band to acquire soil moisture due to its large sensitivity and its direct relationship with the soil water content [49]. The second criterion is related to the way of measuring: passive vs. active microwave instruments.

Microwave sensors do not rely on Sun illumination and are able to work in all weather and illumination conditions. This particular characteristic is especially important in polar regions that have long dark periods in winter and where it is cloudy most of the time. This feature also makes microwave sensors more suitable than optical sensors in this region. Several missions have been launched with active microwave instruments, which can be grouped into two main families: SAR and radar scatterometers. Current SAR instruments for the C-band are Sentinel-1A and Sentinel-1B, RADARSAT-2, the Radar Imaging Satellite-1 (RISAT-1)and Gao Fen-3 (GF-3). They provide dual polarization and multi-polarization data. Soil moisture estimation by Sentinel-1 is derived from the Advanced Synthetic Aperture Radar (ASAR) algorithm [50]. Operational SAR sensors in the S-band are HJ-1A/B/C and in the L-band the one on the Advanced Land Observing Satellite-2 (ALOS-2). The main limitations of these instruments are: the narrow swath, the dependence on the vegetation cover and the surface roughness and the speckle noise that makes SAR images appear very noisy. The main limitation of the SAR system for soil moisture is also the lower accuracy, as compared to passive microwave data.

The SMOS mission of the ESA was the first satellite dedicated to providing global soil moisture data [51,52]. SMOS-derived soil moisture products have an accuracy of 0.04 m^3/m^3 at a spatial resolution ranging from 35–50 km, as well as a revisit time of 1–3 days. Before SMOS, soil moisture measurements were performed using passive microwave radiometers [53] at 7 and 10 GHz (AMSR; AMSR-2; AMSR-E; the Multi-frequency Scanning Microwave Radiometer, MSMR). However, at these frequencies, soil moisture measurements are more affected by the vegetation cover. On the other hand, the L-band offers additional advantages such as less atmospheric attenuation than at higher frequencies and additional smaller water content effects (up to at least 5 kg/m). GNSS-reflectometry is another potential technology, so far having modest accuracy, that can be implemented even in CubeSats for soil moisture measurements. Data fusion between microwave radiometry, optical and SAR can improve the spatial resolution and accuracy of soil moisture measurements, as has been demonstrated in the SMOS and the Soil Moisture Active-Passive (SMAP) missions [54].

4.13. Monitoring System: Vessel and Fish Farming Cage Position Tracking

Norsat-2 and Triton-2 are contributing missions under the Advanced Research in Telecommunications Systems (ARTES) program by ESA. Spaceborne SAR and AIS can also be considered complementary systems to improve the security and surveillance services for maritime navigation.

5. Discussion

The aim of this section is to establish the promising technologies and to address the technological challenges, to ensure they satisfy the measurement requirements for the observation gaps detected in the Copernicus space segment. As shown in Table 3, the measurements with gaps detected in this study can be monitored by different types of sensors. According to the state of the art of the potential payloads to cover the gaps of the Copernicus space infrastructure in 2020–2030, the next generation of instruments require overcoming challenges and new technological developments in order to meet the end-user requirements.

GNSS-R is a promising technology to detect surface currents, significant wave height, sea ice cover, horizontal wind speed, dominant wave direction and sea ice thickness. The advantage of this technology is that it can process data in real time and onboard, through the use of Delay-Doppler Mapping (DDM); in this way, the latency time can be improved. However, the next generation of GNSS-R sensors should improve the spatial resolution to <10 km and the accuracy to <1 cm and <0.5 m/s, in order to meet the end-user requirements. In this regard, a precise clock module is required to reduce the errors in the retrieval computation.

For the passive microwave, a spatial resolution of <25 km is feasible by increasing the size of the antenna and would be suitable on CubeSats, as well as meet the spatial resolution requirement for atmospheric pressure over the sea surface. Microwave radiometers on small platforms demand the use of inflatable antennas in order to improve the spatial resolution and meet the end-user requirements, covering a variety of measurements between sea ice parameters, ocean conditions and sea surface temperature.

Future cloud radar, radar altimeters, LiDAR and SAR are required to improve the coverage and adopt the use of a new concept to allow a wide-swath. For the the next generation of scatterometer, it is imperative to improve the accuracy for horizontal wind speed over the sea surface <0.5 m/s with a spatial resolution <10 km, in order to meet the user requirements.

Optical sensors (such multispectral, hyperspectral radiometers and spectrometers) can monitor many variables with gaps, but the data can only be acquired in daylight and clear sky, a limiting factor for observing polar regions, where it is very often cloudy and the dark period is long. However, they are a good complement to microwave sensors, where the fusion of data could result in better products in terms of spatial resolution and accuracy.

6. Conclusions

This work is the first study reviewing the instrument requirements based on the end-users needs. The gaps of the future European EO spaceborne infrastructure were identified to select the potential technological areas for complementing and refining the Copernicus infrastructure. The exploration of the requirements and future European EO technology led to detecting and identifying the measurements with gaps and the emerging need to monitor the polar regions. Based on these results, future instruments, services and technology areas to upgrade were identified. Some of the most important characteristics, from which all the Copernicus services can benefit, are latency time and lower revisit time. Specifically, reductions of both revisit time from 24 h to 3 h and product delivery in the marine services from 24–48-h to 1 h would support, e.g., marine services, enabling polar navigation, enhanced marine real-time weather forecast, oil and gas exploration and oil spill remediation.

Author Contributions: E.L. and A.C. identified the measurement with gaps. P.S., A.M., H.M. and I.L. defined the user requirements. E.L. and A.C. identified the potential technologies to cover the measurement gaps. E.L. wrote the paper. A.C., H.P. and P.S. revised it.

References

1. Matevosyan, H.; Lluch, I.; Poghosyan, A.; Golkar, A. A Value-Chain Analysis for the Copernicus Earth Observation Infrastructure Evolution: A Knowledgebase of Users, Needs, Services, and Products. *IEEE Geosci. Remote Sens. Mag.* **2017**, *5*, 19–35.10.1109/MGRS.2017.2720263. [CrossRef]
2. World Meteorological Organization Observational Requirements and Capabilities. Available online: https://www.wmo-sat.info/oscar/requirements (accessed on 21 June 2018).
3. A Survey into the State & Health of the European EO Services Industry. Available online: http://earsc.org/news/results-eo-industry-survey-september-2017 (accessed on 29 May 2018).
4. Lancheros, E.; Park, H.; Camps, A.; Di Simone, A.; Matevosyan, H.; Lluch, I. Analysis of the potential of small satellites to cover the sea ice data products gap. In Proceedings of the 2017 IEEE International Geoscience and Remote Sensing Symposium (IGARSS), Fort Worth, TX, USA, 23–28 July 2017; pp. 1173–1176. [CrossRef]
5. Tobehn, C.; Schönenberg, A.; Bolea, A.; Coppola, D.; Ginesi, A.; Ginati, A.; Bal, L.; Sciberras, L. Joint EMSA/ESA initiative for innovative AIS satellites, technologies, applications and services. In Proceedings of the 64th International Astronautical Congress 2013, Beijing, China, 23–27 September 2013; Volume 5, pp. 4077–4086.
6. Sat-AIS: Satellite-Based Automatic Identification System. Available online: https://esamultimedia.esa.int/docs/telecom/SAT-AIS-factsheet{_}WEB.pdf (accessed on 28 May 2018).
7. Selva, D.; Krejci, D. A survey and assessment of the capabilities of Cubesats for Earth observation. *Acta Astronaut.* **2012**, *74*, 50–68. [CrossRef]
8. Cosine: HyperScout. Available online: http://hyperscout.nl/ (accessed on 28 May 2018).
9. European Space Agency: Copernicus Observing the Earth. Available online: http://www.esa.int/Our_Activities/Observing_the_Earth/Copernicus/Overview4 (accessed on 9 July 2018).
10. Copernicus Space Component Data Access: Contributing Missions. Available online: https://spacedata.copernicus.eu/web/cscda/missions (accessed on 9 July 2018).
11. Lang, A.; Yang, S.; Kaas, E. Sea ice thickness and recent Arctic warming. *Geophys. Res. Lett.* **2017**, *44*, 409–418. [CrossRef]
12. Wickert, J.; Cardellach, E.; Martin-Neira, M.; Bandeiras, J.; Bertino, L.; Andersen, O.B.; Camps, A.; Catarino, N.; Chapron, B.; Fabra, F.; et al. GEROS-ISS: GNSS REflectometry, Radio Occultation, and Scatterometry Onboard the International Space Station. *IEEE J. Sel. Top. Appl. Earth Obs. Remote Sens.* **2016**, *9*, 1552–4581. [CrossRef]
13. Park, H.; Camps, A.; Pascual, D.; Kang, Y.; Onrubia, R.; Querol, J.; Alonso-Arroyo, A. A Generic Level 1 Simulator for Spaceborne GNSS-R Missions and Application to GEROS-ISS Ocean Reflectometry. *IEEE J. Sel. Top. Appl. Earth Obs. Remote Sens.* **2017**, *10*, 4645–4659. [CrossRef]
14. Klemas, V. Remote Sensing Techniques for Studying Coastal Ecosystems: An Overview. *J. Coast. Res.* **2011**, 10.2112/JCOASTRES-D-10-00103.1. [CrossRef]
15. Halpern, D.; Hollingsworth, A.; Wentz, F. ECMWF and SSM/I Global Surface Wind Speeds. *J. Atmos. Oceanic Technol.* **1994**, *11*, 779–788.<0779:EASGSW>2.0.CO;2. [CrossRef]
16. Wentz, F.J.; Ricciardulli, L.; Gentemann, C.; Meissner, T.; Hilburn, K.; Scott, J. The accuracy of Preliminary WindSat Vector Wind Measurements: Comparisons With NDBC Buoys and QuikSCAT. Available online: http:\www.remss.com/missions/windsat (accessed on 29 May 2018).
17. Wentz, F.J.; Ricciardulli, L.; Gentemann, C.; Meissner, T.; Hilburn, K.A.; Scott, J. Remote Sensing Systems Coriolis WindSat. Available online: http:\www.remss.com/missions/windsat (accessed on 29 May 2018).
18. Freilich, M.H.; Vanhoff, B.A. The accuracy of preliminary WindSat vector wind measurements: Comparisons with NDBC buoys and QuikSCAT. *IEEE Trans. Geosci. Remote Sens.* **2006**, *44*, 622–637. [CrossRef]
19. La, T.V.; Khenchaf, A.; Comblet, F.; Nahum, C. Comparison Of Inversion Models of Wind Speed Retrieval From C-Band Sentinel-1 and X-Band TerraSAR-X Data; In Proceedings of the International Geoscience and Remote Sensing Symposium (IGARSS), Beijing, China, 10–15 July 2016; pp. 2261–2264. [CrossRef]

20. Li, W.; Cardellach, E.; Fabra, F.; Rius, A.; Ribó, S.; Martín-Neira, M. First spaceborne phase altimetry over sea ice using TechDemoSat-1 GNSS-R signals. *Geophys. Res. Lett.* **2017**, *44*, 8369–8376. [CrossRef]
21. Schiavulli, D.; Nunziata, F.; Migliaccio, M.; Frappart, F.; Ramilien, G.; Darrozes, J. Reconstruction of the Radar Image from Actual DDMs Collected by TechDemoSat-1 GNSS-R Mission. *IEEE J. Sel. Top. Appl. Earth Obs. Remote Sens.* **2016**, *9*, 4700–4708. [CrossRef]
22. Yan, Q.; Huang, W. Sea ice detection from GNSS-R Delay-Doppler Map. In Proceedings of the 2016 17th International Symposium on Antenna Technology and Applied Electromagnetics, ANTEM, Montreal, QC, Canada, 10–13 July 2016; doi:10.1109/ANTEM.2016.7550123. [CrossRef]
23. Blackwell, W.J. New Small Satellite Capabilities for Microwave Atmospheric Remote Sensing (Conference Presentation). In Proceedings of the SPIE, CubeSats and NanoSats for Remote Sensing , San Diego, CA, USA, 7 December 2016; Volume 997807, doi:10.1117/12.2238675.
24. Remote Sensing Systems: Microwave OI SST Product Description. Available online: http://www.remss.com/measurements/sea-surface-temperature/oisst-description/ (accessed on 29 May 2018).
25. Robinson, I.S. *Measuring the Oceans from Space: The Principles and Methods Of Satellite Oceanography*; Springer Praxis Books in Geophysical Sciences: Berlin, Germany, 2004; p. 669.
26. Ocean and Sea Ice: Global Sea Ice Concentration Data Record Relase 1.2 (SSMI/SSMIS). Available online: http://www.osi-saf.org/?q=content/global-sea-ice-concentration-data-record-ssmissmis (accessed on 29 May 2018).
27. Pațilea, C.; Heygster, G.; Huntemann, M.; Spreen, G. Combined SMAP/SMOS Thin Sea Ice Thickness Retrieval. *Cryosphere Discuss.* **2017**, 1–16. [CrossRef]
28. Universität Bremen: Sea Ice Remote Sensing: SMOS Thin Ice Thickness—Data Browser. Available online: https://seaice.uni-bremen.de/databrowser/#p=smos (accessed on 21 June 2018).
29. Oceanic Surface Air: Via Oxygen Band Radar. Available online: http://technology.nasa.gov//t2media/tops/pdf/LAR-TOPS-102.pdf (accessed on 30 May 2018).
30. Peral, E.; Tanelli, S.; Haddad, Z.; Sy, O.; Stephens, G.; Im, E. Raincube: A proposed constellation of precipitation profiling radars in CubeSat. In Proceedings of the International Geoscience and Remote Sensing Symposium (IGARSS), Milan, Italy, 26–31 July 2015; Volume 2015, pp. 1261–1264. [CrossRef]
31. OSISAF: MetOp-A ASCAT Coastal Winds. Available online: http://www.osi-saf.org/?q=content/MetOp-ascat-coastal-winds (accessed on 29 May 2018).
32. Balz, T.; Hammer, H.; Auer, S. Potentials and limitations of SAR image simulators—A comparative study of three simulation approaches. *ISPRS J. Photogramm. Remote Sens.* **2015**, *101*, 102–109. [CrossRef]
33. Ribo, S.; Arco, J.C.; Oliveras, S.; Cardellach, E.; Rius, A.; Buck, C. Experimental results of an x-band PARIS receiver using digital satellite TV opportunity signals scattered on the sea surface. *IEEE Trans. Geosci. Remote Sens.* **2014**, *52*, 5704–5711. [CrossRef]
34. Otosaka, I.; Rivas, M.B.; Stoffelen, A. Bayesian sea ice detection with the ERS scatterometer and sea ice backscatter model at C-band. *IEEE Trans. Geosci. Remote Sens.* **2018**, *56*, 2248–2254. [CrossRef]
35. Shen, X.Y.; Zhang, J.; Zhang, X.; Meng, J.M.; Ke, C.Q. Sea Ice Classification Using Cryosat-2 Altimeter Data by Optimal Classifier–Feature Assembly. *IEEE Geosci. Remote Sens. Lett.* **2017**, *14*, 1948–1952. [CrossRef]
36. Comiso, J.C.; Nishio, F. Trends in the sea ice cover using enhanced and compatible AMSR-E, SSM/I, and SMMR data. *J. Geophys. Res. Oceans* **2008**.10.1029/2007JC004257. [CrossRef]
37. Beitsch, A.; Kern, S.; Kaleschke, L. Comparison of SSM/I and AMSR-E sea ice concentrations with ASPeCt ship observations around antarctica. *IEEE Trans. Geosci. Remote Sens.* **2015**, *53*, 1985–1996. [CrossRef]
38. Yan, Q.; Huang, W.; Moloney, C. Neural Networks Based Sea Ice Detection and Concentration Retrieval from GNSS-R Delay-Doppler Maps. *IEEE J. Sel. Top. Appl. Earth Obs. Remote Sens.* **2017**, *10*, 3789–3798. [CrossRef]
39. Fabra, F.; Cardellach, E.; Rius, A.; Ribó, S.; Oliveras, S.; Nogués-Correig, O.; Belmonte Rivas, M.; Semmling, M.; D'Addio, S. Phase altimetry with dual polarization GNSS-R over sea ice. *IEEE Trans. Geosci. Remote Sens.* **2012**, *50*, 2112–2121. [CrossRef]
40. Belmonte Rivas, M.; Verspeek, J.; Verhoef, A.; Stoffelen, A. Bayesian sea ice detection with the advanced scatterometer ASCAT. *IEEE Trans. Geosci. Remote Sens.* **2012**, *50*, 2649–2657. [CrossRef]
41. Remund, Q.P.; Long, D.G. A decade of QuikSCAT scatterometer sea ice extent data. *IEEE Trans. Geosci. Remote Sens.* **2014**, *52*, 4281–4290. [CrossRef]

42. Long, D.G.; Hardin, P.J.; Whiting, P.T. Resolution Enhancement of Spaceborne Scatterometer Data. *IEEE Trans. Geosci. Remote Sens.* **1993**, *31*, 700–715. [CrossRef]
43. Remund, Q.P.; Long, D.G. Sea ice extent mapping using Ku band scatterometer data. *J. Geophys. Res. Oceans* **1999**, *104*, 11515–11527. [CrossRef]
44. Multisensor Analyzed Sea Ice Extent. 2010. Available online http://dx.doi.org/10.7265/N5GT5K3K (accessed on 7 June 2018).
45. Maab, N. Remote Sensing of Sea Ice Thickness Using SMOS Data. Ph.D. Thesis, University of Hamburg, Hamburg, Germany, 2013; doi10.17617/2.1737721.
46. Laxon, S.W.; Giles, K.A.; Ridout, A.L.; Wingham, D.J.; Willatt, R.; Cullen, R.; Kwok, R.; Schweiger, A.; Zhang, J.; Haas, C.; et al. CryoSat-2 estimates of Arctic sea ice thickness and volume. *Geophys. Res. Lett.* **2013**. [CrossRef]
47. Pagano, T.S.; Rider, D.; Teixeira, J.; Aumann, H.H.; Rud, M.; Pereira, J.; Furlong, D.; Mamula, D. The CubeSat Infrared Atmospheric Sounder (CIRAS), Pathfinder for the Earth Observing Nanosatellite-Infrared (EON-IR). In Proceedings of the AIAA/USU Conference on Small Satellites, Pasadena, CA, USA, 22–24 March 2016.
48. Coppo, P.; Ricciarelli, B.; Brandani, F.; Delderfield, J.; Ferlet, M.; Mutlow, C.; Munro, G.; Nightingale, T.; Smith, D.; Bianchi, S.; et al. SLSTR: A high accuracy dual scan temperature radiometer for sea and land surface monitoring from space. *J. Mod. Opt.* **2010**, *57*, 1815–1830. [CrossRef]
49. Njoku, E.G.; Entekhabi, D. Passive microwave remote sensing of soil moisture. *J. Hydrol.* **1996**, *184*, 101–129. [CrossRef]
50. Dostálová, A.; Doubková, M.; Sabel, D.; Bauer-Marschallinger, B.; Wagner, W. Seven Years of Advanced Synthetic Aperture Radar (ASAR) Global Monitoring (GM) of Surface Soil Moisture over Africa. *Remote Sens.* **2014**, *6*, 7683–7707. [CrossRef]
51. Kerr, Y.H.; Waldteufel, P.; Wigneron, J.P.; Delwart, S.; Cabot, F.; Boutin, J.; Escorihuela, M.J.; Font, J.; Reul, N.; Gruhier, C.; et al. The SMOS Mission: New Tool for Monitoring Key Elements ofthe Global Water Cycle. *Proc. IEEE* **2010**, *98*, 666–687. [CrossRef]
52. Font, J.; Camps, A.; Borges, A.; Martin-Neira, M.; Boutin, J.; Reul, N.; Kerr, Y.H.; Hahne, A.; Mecklenburg, S. SMOS: The challenging sea surface salinity measurement from space. *Proc. IEEE* **2010**, *98*, 649–665. [CrossRef]
53. Owe, M.; de Jeu, R.; Holmes, T. Multisensor historical climatology of satellite-derived global land surface moisture. *J. Geophys. Res.* **2008**, *113*, F01002. [CrossRef]
54. Piles, M.; Entekhabi, D.; Camps, A. A change detection algorithm for retrieving high-resolution soil moisture from SMAP radar and radiometer observations. *IEEE Trans. Geosci. Remote Sens.* **2009**. [CrossRef]

6

Large Anomalies in the Tropical Upper Troposphere Lower Stratosphere (UTLS) Trace Gases Observed during the Extreme 2015–16 El Niño Event by using Satellite Measurements

S. Ravindrababu [1], M. Venkat Ratnam [2], Ghouse Basha [2], Yuei-An Liou [1,*] and N. Narendra Reddy [2]

1 Center for Space and Remote Sensing Research, National Central University, No. 300, Jhongda Rd., Jhongli Dist., Taoyuan City 32001, Taiwan; baburavindra595@gmail.com

2 National Atmospheric Research Laboratory, Gadanki 517112, India; vratnam@narl.gov.in (M.V.R.); mdbasha@narl.gov.in (G.B.); narendra.paperwork@gmail.com (N.N.R.)

* Correspondence: yueian@csrsr.ncu.edu.tw

Abstract: It is well reported that the 2015–16 El Niño event is one of the most intense and long lasting events in the 21st century. The quantified changes in the trace gases (Ozone (O_3), Carbon Monoxide (CO) and Water Vapour (WV)) in the tropical upper troposphere and lower stratosphere (UTLS) region are delineated using Aura Microwave Limb Sounder (MLS) and Atmosphere Infrared Radio Sounder (AIRS) satellite observations from June to December 2015. Prior to reaching its peak intensity of El Niño 2015–16, large anomalies in the trace gases (O_3 and CO) were detected in the tropical UTLS region, which is a record high in the 21st century. A strong decrease in the UTLS (at 100 and 82 hPa) ozone (~200 ppbv) in July-August 2015 was noticed over the entire equatorial region followed by large enhancement in the CO (150 ppbv) from September to November 2015. The enhancement in the CO is more prevalent over the South East Asia (SEA) and Western Pacific (WP) regions where large anomalies of WV in the lower stratosphere are observed in December 2015. Dominant positive cold point tropopause temperature (CPT-T) anomalies (~5 K) are also noticed over the SEA and WP regions from the high-resolution Constellation Observing System for Meteorology, Ionosphere and Climate (COSMIC) Global Position System (GPS) Radio Occultation (RO) temperature profiles. These observed anomalies are explained in the light of dynamics and circulation changes during El Niño.

Keywords: El Niño 2015–16; trace gases; Upper Troposphere Lower Stratosphere

---

1. Introduction

The Upper Troposphere and Lower Stratosphere (UTLS) is one of the important regions of the Earth's atmosphere and crucial for the Earth's energy balance [1]. The redistribution of the water vapour (WV), ozone (O_3) and other chemical species in the UTLS region has a direct impact on the Earth's radiation budget. WV acts as a major source of cooling in the upper troposphere whereas O_3 is the main source for the warming in the lower stratosphere (LS). These WV and O_3 changes in the UTLS region can strongly influence the temperature structure of the atmosphere and hence atmospheric transport [2–6]. The variability in the UTLS trace gases is strongly influenced by several dominant atmospheric oscillations such as quasi-biennial oscillation (QBO) and El Niño Southern Oscillation (ENSO). ENSO is a dominant mode of the inter-annual variability of the tropical troposphere climate. El Niño is a warm phase of the ENSO, which is having a strong impact on the global atmosphere. In general, during El Niño, enhancement of atmospheric convection and increase of sea surface temperatures (SSTs) in the tropical Eastern Pacific (EP) or Central Pacific (CP) generally observed [7].

The opposite signatures are noticed during the La Niña periods. Due to increase of the convection, lots of large scale atmospheric waves released into the atmosphere vertically. These waves indirectly strengthen the Brewer Dobson Circulation (BDC), increasing the upward motion in the tropics and downward motion in poles in the stratosphere [8]. Due to these circulation changes in the El Niño period, tropical troposphere generally warms while the tropical lower stratosphere (LS) cools [8,9]. The warm troposphere allows more WV into the LS in El Niño events. The negative zonal mean O_3 and temperature anomalies in the LS are also evident during the El Niño period. These anomalies are attributed to the strengthening tropical upwelling of the BDC in El Niño period [10,11]. Due to the warm, high troposphere temperatures in El Niño period, drought conditions generally observed over South East Asia (SEA) region with lot of biomass burning and forest fires [12–14]. In addition, more carbon emissions observed in El Nino period over SEA region, which causes high concentrations of CO in the troposphere [12–14]. Several previous observational and modelling studies are reported on the atmospheric composition (mainly on O_3, WV and CO) changes during the El Niño periods [12–16].

The recent El Niño event in 2015–16 was one of the strongest El Niño events in the 21st century satellite era [7]. It is also only one strongest boreal summer El Nino event in the Microwave Limb Sounder (MLS) record [17]. During this event, several unusual atmospheric changes happened and were reported well. For example, strong drought conditions were observed over SEA especially over the Indonesia region and caused a record amount of forest fires and biomass burning in September and October [18,19]. The carbon emissions that occurred over SEA in 2015 are the largest one since 1997 [18]. Due to these forest fires, a huge amount of carbon was released into the atmosphere in the form of CO_2 and CO [18–20]. These trace gases have strong influence on global atmospheric chemical budget. By using Greenhouse gases Observing SATellite (GOSAT) data, Parker et al. [21] reported the strong enhancement of CO_2 and CH_4 over the Indonesian region. Whitburn et al. [22] observed 3 times higher amounts of ammonia during this period than in the previous seven years based on satellite measurements acquired in 2008–2015. It is also well reported that the second half of 2015 witnessed massive propagation of Rossby waves into the tropics [23,24].

It was also reported that during the 2015–16 El Niño event, the large anomalies of WV and ice in the LS were observed in December 2015 over SEA and WP regions [7,25]. This enhancement in WV was either due to the warm tropopause temperatures [7] or due to the 2015–16 QBO disruption [25]. However, very recently, Diallo et al. [9] argued that the interplay between the El Nino event and the QBO disruption made an important contribution to the change in the LS WV anomalies during this event. Another study by Garfinkel et al. [26] discussed the impact of ENSO on the LS temperature and WV and suggested that the impact is nonlinear in boreal spring whereas linear in boreal winter. It is well known that most of the WV enters the LS through the tropical tropopause layer (TTL) and the temperature in the TTL controls the WV entering the LS [27–30]. However, the trace gases such as CO and O_3 variability in the UTLS region during the 2015–16 El Niño event have not been well investigated. O_3 is a strong radiative gas and plays a significant role on the radiative forcing of the atmosphere along with CO_2 [31]. The inter-annual variability of O_3 at tropopause level is strongly dominated by ENSO [32]. The strength of the BDC and SSTs in the equatorial pacific has strong impact on the O_3 anomalies in ENSO events. CO is important for global warming and it acts as a precursor for CO_2 and tropospheric O_3 and a major sink for OH radicals [33]. It is well established that the strong enhancement of the CO concentrations during the El Niño events [7–9]. In the present study, we present the observed large anomalies of these trace gases in the tropical UTLS region during the recent strong El Nino event of 2015–16.

2. Data base and Methodology

In the present study, we used version 4.2 level 2 profiles of O_3, CO and WV data from Aura MLS provided by the Jet Propulsion Laboratory along with CO data from Atmosphere Infrared Radio Souder (AIRS) in order to see the vertical changes in CO during the year 2015. The high resolution temperature profiles were obtained from the Constellation Observing System for Meteorology, Ionosphere and

Climate (COSMIC) Global Position System (GPS) Radio Occultation (RO) are used for tropopause changes. Multivariate ENSO Index (MEI) data obtained from http://www.esrl.noaa.gov/psd/enso/mei were used as an ENSO index [34,35].

2.1. MLS Data

Earth Observing System (EOS) MLS is one of the four instruments aboard NASA's Aura satellite. The Aura MLS gives around 3500 vertical profiles per day and it crosses the Equator at ~01:40 and ~13:40 local time [36]. In the present work, version 4.2 Aura MLS data of O_3, CO and WV were used. The vertical resolution of the WV is in the range 2.0 to 3.7 km from 316 to 0.22 hPa and the along-track horizontal resolution varies from 210 to 360 km for pressure greater than 4.6 hPa. For ozone, vertical resolution is ~2.5 km and the along-track horizontal resolution varies between 300 and 450 km. CO is retrieved from radiance measurements of two bands in the MLS 240 GHz radiometer. Vertical resolution for CO is in the range 3.5–5 km from the upper troposphere to the lower mesosphere and the useful range is 215–0.0046 hPa. The horizontal resolution for CO is about 460 km at 100 hPa and 690 km at 215 hPa. The precision and systematic uncertainty for WV and O_3 are ±10–40%, ±10–25% and ±0.02–0.04 ppmv, ±0.02–0.05 ppmv ±5–10%, respectively. The accuracy of CO at 100 hPa is ±19 ppbv and ±30%. The data was collected for the period from January 2006 to December 2017. Profiles of O_3, CO and WV were used to construct a monthly mean 2.5° × 2.5° degree gridded data set for the study period. We used data from 2006 to 2014 for background climatology for each individual month. To get zonal mean of trace gases in each month during the reporting period, all the available MLS profiles within the latitude bands that is, for 10°N–10°S. More details about MLS version 4 level 2 data can be found in Livesey et al. [34].

2.2. Atmosphere Infrared Radio Souder Data

AIRS is one of six instruments onboard Aqua, which is part of NASA's EOS of satellites launched into Earth orbit on 4 May 2002. Version 6 Level 2 data of CO is utilized in the present study. AIRS CO is retrieved with horizontal resolution of 45 km at nadir, in a swath of width about 1600 km. This orbit gives global coverage in the tropics every 2 days. The retrieval uses a cloud-clearing methodology providing the CO with sensitivity that peaks around 500 hPa, with ~0.8–1.2 degrees of freedom of signal for 50–70% of scenes.

2.3. COSMIC GPS RO Data

The temperature profiles obtained from the COSMIC GPS RO were utilized for the present study. The GPS RO data were downloaded from the COSMIC Data Analysis and Archive Centre (CDAAC) website (http://cosmic-io.cosmic.ucar.edu/cdaac/index.html). COSMIC GPS RO is a joint Taiwan–U.S. mission, which is a constellation of six microsatellites equipped with GPS receivers [37]. These satellites were launched in early 2006 and started providing data from April 2006. It provides 2000–2500 occultations for a day over the entire globe. There are different vertical resolutions available for COSMIC GPS RO data but for the present study we used 200 m resolution temperature (atmPrf) profiles available at CDAAC website only which is freely available for the public use. Note that these data are validated with a variety of techniques, including GPS radiosonde data and matched very well, particularly in the UTLS region [38–40]. A comprehensive introduction to the RO method for remote sensing of the atmosphere and ionosphere was presented by Liou et al. [41]. Deployment of COSMIC was presented by Fong et al. [42] and its constellation spacecraft system performance after one year in orbit was presented by Fong et al. [43]. Later, follow-on mission of COSMIC, that is COSMIC-2 and its deployment were introduced by Fong et al. [44] and [45], respectively. The suitability and importance of the GPS RO data for different topics of the atmospheric research (e.g., gravity wave studies [46–50], tropopause structure [51], tropical cyclones [52,53], etc.) were well reported in the literature.

3. Results and Discussion

3.1. Unusual Behaviour of Trace Gases in the Tropical UTLS Region during the 2015–16 El Niño Event

The monthly mean time series of observed O_3 at 100 and 82 hPa and CO at 146 and 100 hPa over the equatorial region (averaged over 10°N and 10°S) along with MEI index from January 2006 to December 2017 are shown in Figure 1. O_3 shows significant seasonal variations with high values during Northern Hemisphere (NH) summer months and low values during NH winter months (Figure 1a). However, during the year 2015 a drastic change in O_3 mixing ratio at both pressure levels is noticed particularly during the summer months of July and August in 2015 as compared to the other years. The MEI index also shows the gradual increase in its strength and reaches record high values (>2 MEI index value) in September 2015 (Figure 1c). After that, the MEI index maintains its value of ~2 till May 2016. It is well known that the ENSO dominates the inter-annual variability in the O_3 at tropopause level [10,32]. The CO (Figure 1b) also shows a strong enhancement in September, October and November 2015. The values are very high during 2015 as compared to the other years. The O_3 and CO show significant decrease and increase. The relative percentage changes in the O_3 and CO with respect to the averaged period from 2006 to 2014 are presented in the Section 4, respectively.

The monthly climatological (2006–2014) means of O_3 (CO) mixing ratio at 146, 100 and 82 hPa (146, 100 and 68 hPa) in the tropics (20°N–20°S) along with the seasonal change in the year 2015 are illustrated in Figure 2. Three latitude bands within the tropical latitudes, namely equator (averaged over 10°N–10°S), Northern Hemisphere (NH) (averaged over 11°N–20°N) and Southern Hemisphere (SH) (averaged over 11°S–20°S) are selected to obtain the variability in the O_3 mixing ratio. We can see the clear seasonal variability in the climatological O_3 mixing ratio at both the pressure levels over the tropics.

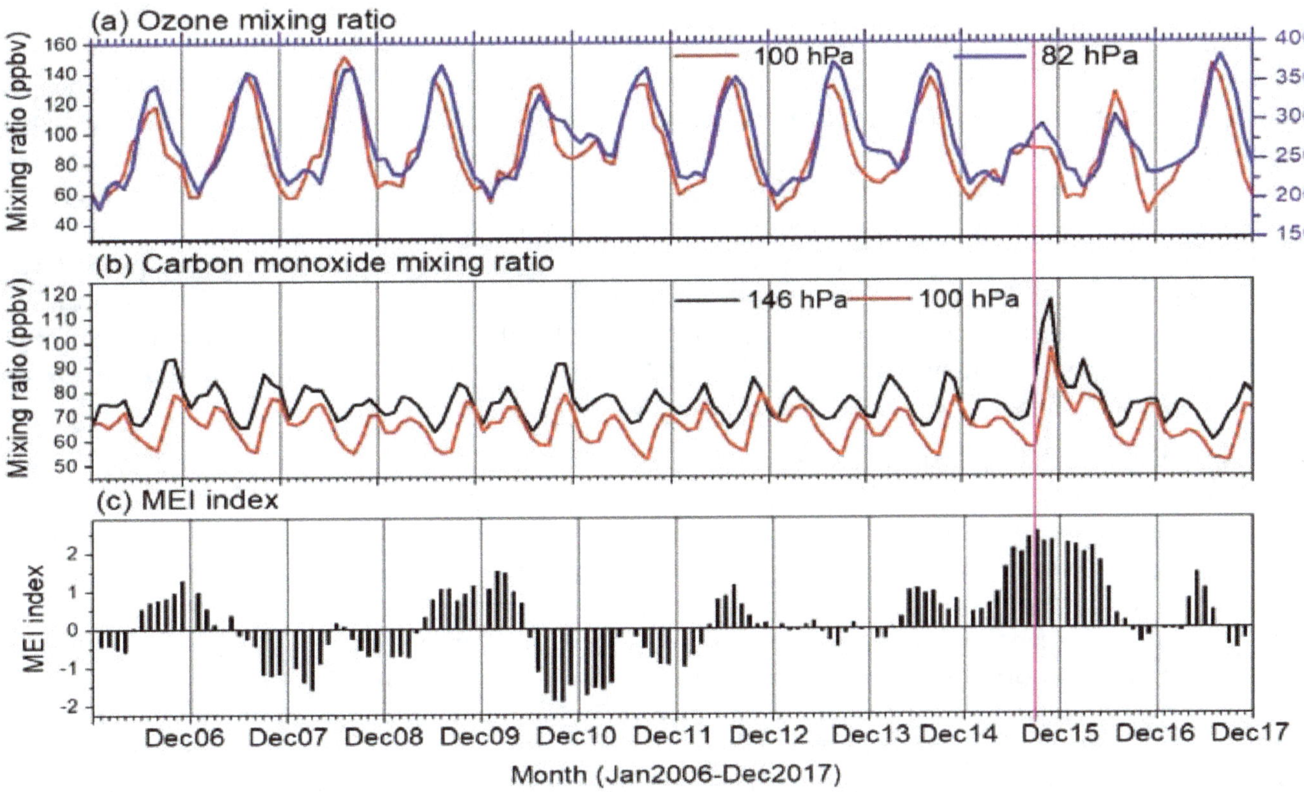

Figure 1. Time series of observed monthly mean (**a**) ozone at 100 and 82 hPa, (**b**) carbon monoxide at 100 and 146 hPa and (**c**) Multivariate ENSO Index (MEI) from January 2006 to December 2017.

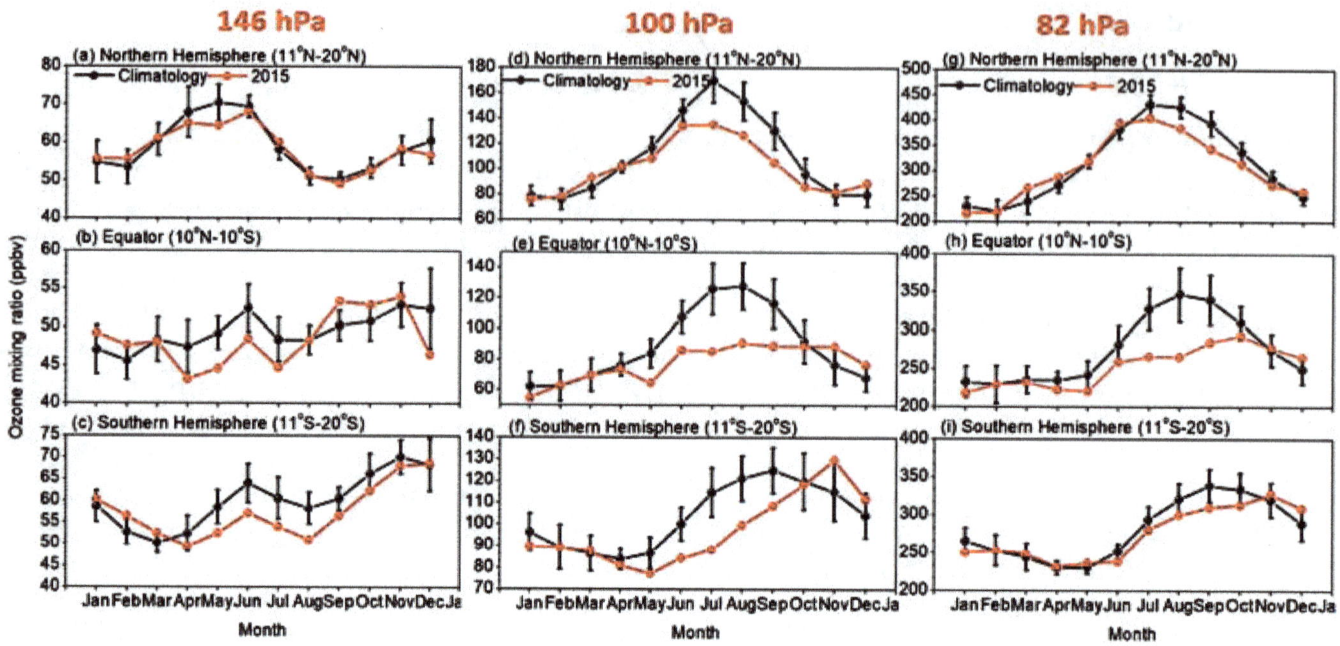

Figure 2. Annual cycle of ozone mixing ratio observed at 146 hPa (**a–c**), 100 hPa (**d–f**) and at 82 hPa (**g–i**) averaged over different latitude bands. Black colour line shows monthly climatology of ozone mixing ratio calculated by using MLS data from January 2006 to December 2017 and red colour line shows the monthly mean of ozone mixing ratio during 2015. Vertical bars indicate standard deviations of the measurements.

Note that over the equator (Figure 2e,h) and SH (Figure 2f), the O_3 shows a quite different seasonal change in the year 2015 compared to the climatology. A clear drop in the O_3 mixing ratio in the equator from June to September is observed (Figure 2e,f). In 2015, from June to December, the seasonal change is completely disappeared in the tropical regions at 100 hPa and 82 hPa as seen in Figure 2e,h, respectively. In NH, the climatological and the year 2015 O_3 values show a clear distinct picture. At 100 (82) hPa, the O_3 value from January to May follows the climatological pattern and then in June it starts to deviate from it. The difference is greater at 100 hPa as compared to 82 hPa. From these results, it is clear that O_3 shows a significant drop in the tropics from June to September 2015. Similar unusual changes are noticed in the CO mixing ratio over the tropics during this period but a little bit later (Figure 3). Figure 3 shows that, in the entire tropics, CO shows a strong enhancement in the months from September to December at both pressure levels (146 and 100 hPa) as compared to the climatology. The monthly mean values of O_3 and CO over the tropics clearly indicate that strong and unusual changes occurred in the UTLS region during the 2015–16 El Niño event.

To quantify the changes in the trace gases during the year 2015, we have estimated the monthly anomalies by subtracting the climatology from the individual monthly mean trace gas concentrations. The time series of O_3 and CO anomalies observed at different levels in the UTLS region over the 10°N–10°S region is shown in Figure 4. Strong negative anomalies in O_3 are observed at 82 and 100 hPa but not at 146 hPa (Figure 4a). The O_3 decrease is high at 82 hPa (~80 ppbv) as compared to 100 hPa (40 ppbv). This decrease is very high from September to December 2015 (peaking in November) with large negative anomalies and continued in 2016 with less magnitude. This might be due to the strengthening of the Brewer Dobson circulation (BDC), that is, strong enhancement of tropical upwelling during El Niño period [9–11]. Previous studies clearly demonstrated the impact of the ENSO on the inter-annual variability of O_3 at the tropopause level [32]. These inter-annual variations in the O_3 anomalies are linked with the SSTs in the equatorial Pacific Ocean and are explained by the strength of the BDC. In general, during the El Niño period, the tropical upwelling increases whereas in La Niña events, the tropical upwelling decreases. Note that previous studies focused on the O_3 and other trace gases changes mainly in the boreal winter (mainly in December when the intensity of the El Niño becomes peak). However, the observed high decrease of O_3 in the present study is in July and August that belong to boreal summer. These results matched well with those reported by Tweedy et al. [17] and Diallo et al. [11]. They clearly demonstrated that the changes in the O_3 anomalies in NH are due to the meridional advection in northern subtropics altered by boreal summer ENSO events and in SH due to the tropical upwelling [17].

The structural changes in the BDC due to 2015–16 El Niño event is also one of the reasons for the O_3 anomalies observed in the tropical LS during this event [11]. Overall, instead of chemical reactions, the transport processes (due to BDC) are the major possible reason for the presently observed O_3 anomalies. The CO anomalies (20–35 ppbv) show a substantial increase in the UTLS (mainly 146 and 100 hPa) from September to December 2015 (peaking in November) as compared to the whole MLS data period (Figure 4b). Note that the observed CO anomalies in 2015 are a record high in the 21st century. From Figure 4, it is evident that the O_3 drop was high from June to September 2015 (peaking in August) while strong enhancement in CO was observed from September to December 2015 (peaking in November).

More significantly, the maximum CO anomalies are noticed in the troposphere in October whereas in the LS (~100 and 68 hPa), it was in November (Figures 7 and 11). In the following section, we investigate the spatial distributions of these anomalies in 2015.

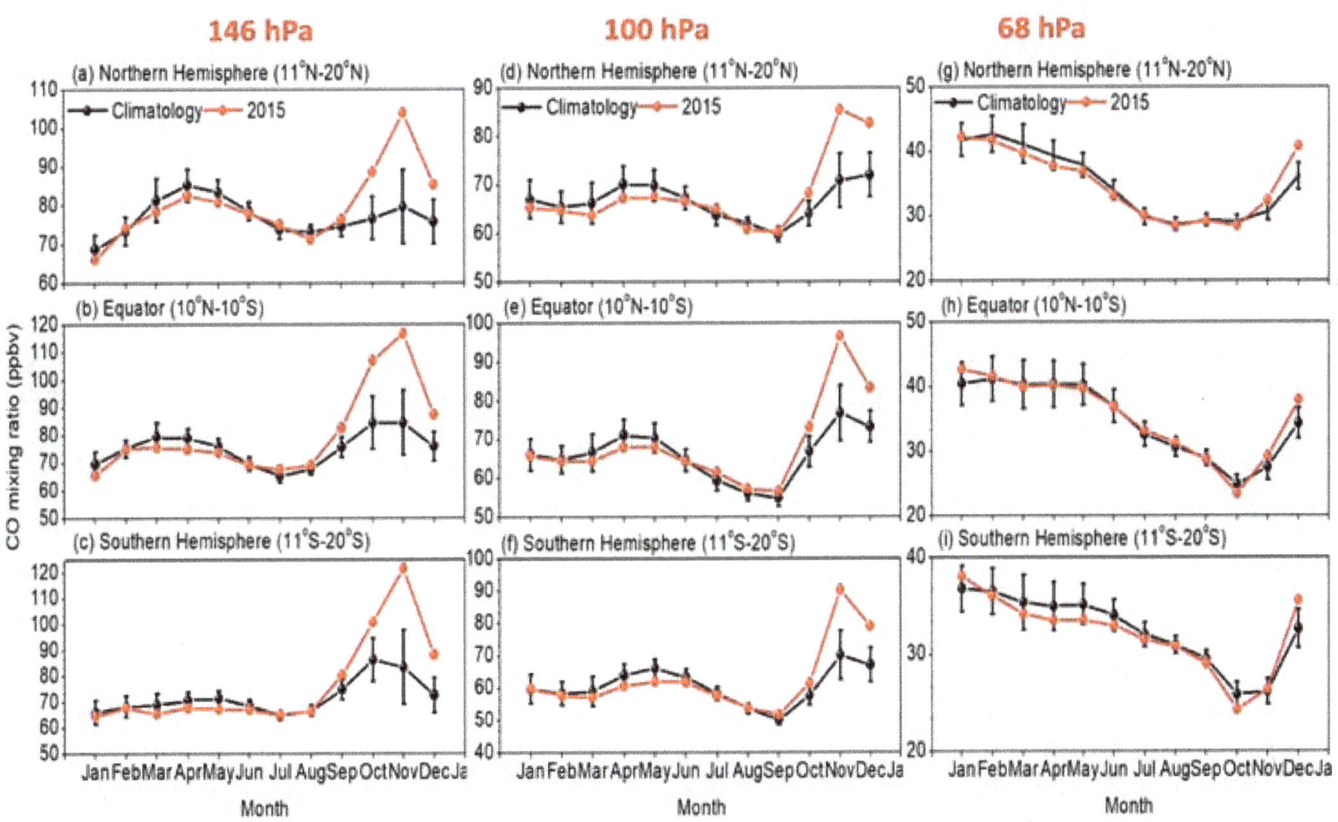

Figure 3. Same as Figure 2 but for the monthly mean of carbon monoxide at 146 hPa 100 hPa and 68 hPa.

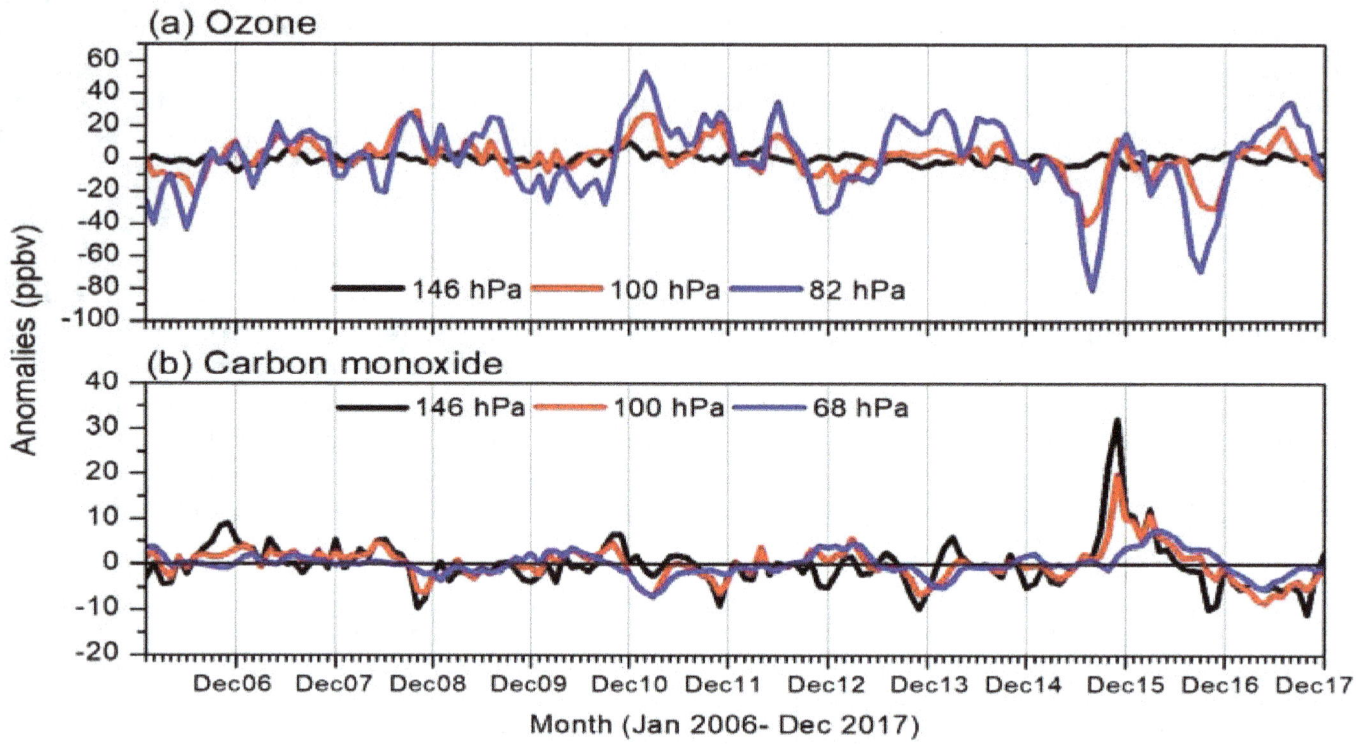

Figure 4. Time series of de-seasonalized anomalies of (**a**) ozone and (**b**) carbon monoxide observed over Equatorial region (10°N and 10°S) from January 2006 to December 2017. Different colours indicate different pressure levels.

3.2. Spatial Distributions of Trace Gas Anomalies during the 2015–16 El Niño Event

Figure 5 shows the spatial distribution of O_3 at 82 hPa in July 2015 and CO at 100 hPa in October 2015 as well as respective climatologies. Significant differences are noticed in the trace gas distributions in 2015 over the tropical region. Compared to the climatology, minimum O_3 (100 ppbv) is noticed over the SEA and Atlantic regions but with a relatively zonally uniform feature in 2015. High values of CO (200 ppbv) are observed, particularly over SEA in 2015. These observations clearly indicate the huge difference between climatology and 2015. Figure 6 illustrates the spatial distribution of anomalies of O_3 at 82 hPa and CO at 100 hPa with respect to the climatology in July and October 2015. Distinct characteristics in the trace gas anomalies are perceived between these two months. In the O_3, strong negative anomalies (~200 ppbv) are observed over the entire tropical region except some parts of the WP region in July 2015. No such changes are observed for O_3 in October 2015 over the tropical region. However, the observed O_3 decrease is much higher and recorded strong decrease in its concentrations (~200 ppbv) in the recent decade. The O_3 and temperatures in the Tropical Tropopause Layer (TTL) are linked both dynamically and radiatively. It is reported that the O_3 perturbations have a positive radiative feedback, with negative O_3 anomalies locally cooling the TTL and positive O_3 anomalies locally warming the TTL (Gilford et al., 2016). We also observed the negative cold point tropopause temperature (CPT) anomalies from COSMIC GPS RO data in July and August 2015 over most of the tropical equatorial region except over SEA and WP regions (Figures are not shown). The observed CPT anomalies are well correlated with the O_3 anomalies in July and August. Large anomalies of CO (~150 ppbv) are observed in October 2015 and the prominent enhancement is observed over SEA region. The observed increase of CO mixing ratio is much higher compared to the previous El Niño events in the 21st century. This clearly indicates the transport of lower troposphere air into the tropopause level in October 2015.

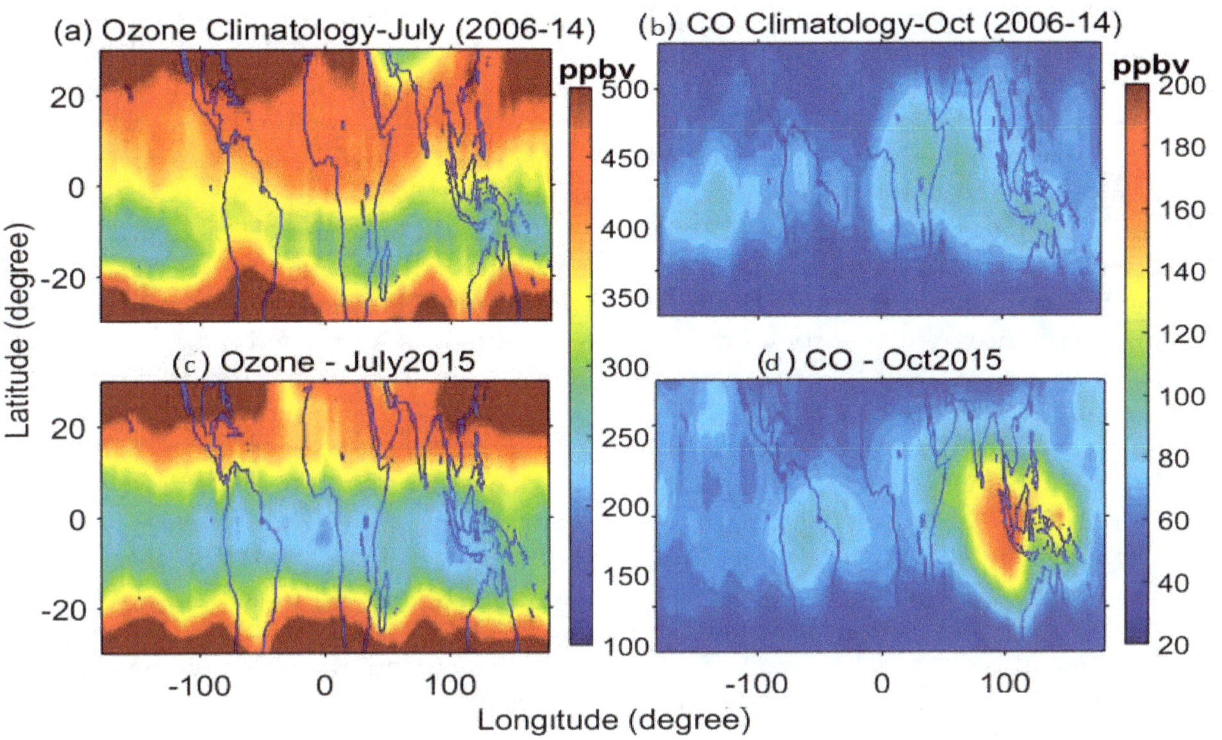

Figure 5. Background climatology (averaged 2006–2014) of (**a**) ozone mixing ratio at 82 hPa in July and (**b**) carbon monoxide at 100 hPa in October. (**c**,**d**) same as above but for the year 2015.

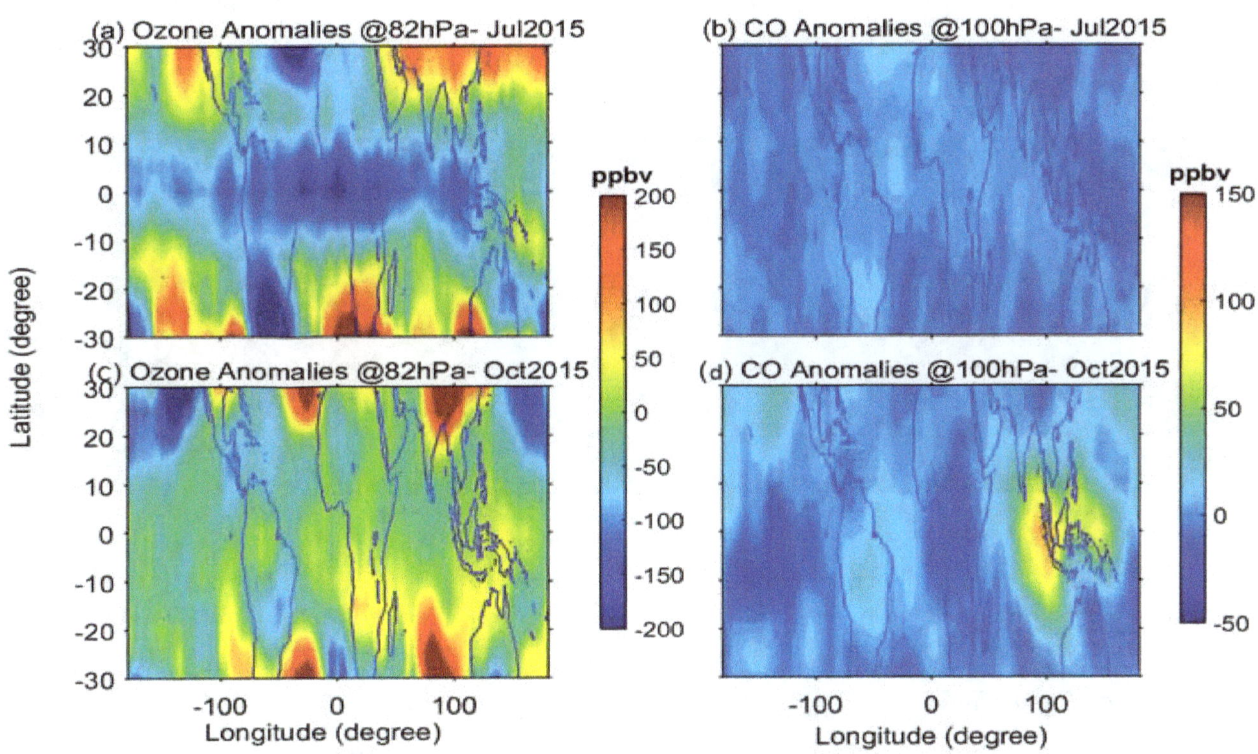

Figure 6. Spatial distribution of ozone anomalies observed at 82 hPa in (**a**) July 2015 and (**c**) October 2015. (**b**,**d**) same as (**a**,**c**), respectively but for carbon monoxide anomalies observed at 100 hPa. Anomalies are calculated by removing the background climatology (January 2006–December 2014) of individual months.

The 2015 El Niño induced drought conditions, which further allowed active biomass burning and forest fires to spread rapidly in the SEA region in September and October [18,19]. These carbon emissions, which occurred in 2015 are the largest emissions since 1997 [18]. The effects of these fires are clearly seen in the enhancement of CO concentrations in the UTLS region. Due to these emissions, CO at the tropopause level is increased in 2015. The observed values of decreasing O_3 (200 ppbv) and increasing CO (~150 ppbv) are high values recorded in 2015 and the results clearly indicate the unusual strong enhancement/decrease of trace gases (CO and O_3) in the UTLS region. The carbon emissions in September and October over maritime SEA play an important role in the enhancement of CO in the UTLS region. Based on ground based and satellite measurements, it is well reported that the strong enhancement of carbon emissions over Indonesia happened in September–October 2015 [18–20]. It is also evident that the overall emissions from the tropical Asian biomass burning in 2015 were almost three times the 2001–2014 average [54]. In the present study, we also observed the strong increase in the CO concentrations in most of the troposphere even up to the 100 hPa over SEA and WP regions in October 2015 (Figure 7). MLS data is available from 215 hPa and has only 4 pressure levels (215, 146,100 and 68 hPa) in the troposphere to the UTLS region. Based on MLS data, it is difficult to see the vertical change in the CO. To avoid this we have utilized AIRS observed monthly mean CO data over SEA region from January to December 2015. Height-time cross section of CO over the SEA and WP regions observed from AIRS measured CO data from January to December 2015 is shown in Figure 7. It is clear that the vertical transport of the CO into the UTLS region is clearly seen in the Height-time cross section of the CO. The anomalies in these trace gases clearly demonstrated the unusual changes that are occurring in the UTLS region before the El Niño becomes strong and strengthened during the 2015–16 winter. It was also clear from the study that localized strong carbon emissions over SEA play a crucial role on the large enhancement of zonal mean CO. In the 'State of the Climate 2015' [54], it is clearly shown that the biomass burning in Indonesia region led to increasing of the CO, aerosols and tropospheric O_3 in 2015. Our satellite measurement results are also well matched with their results. In the next section, we explore the changes in tropopause temperature and WV in the LS during this event.

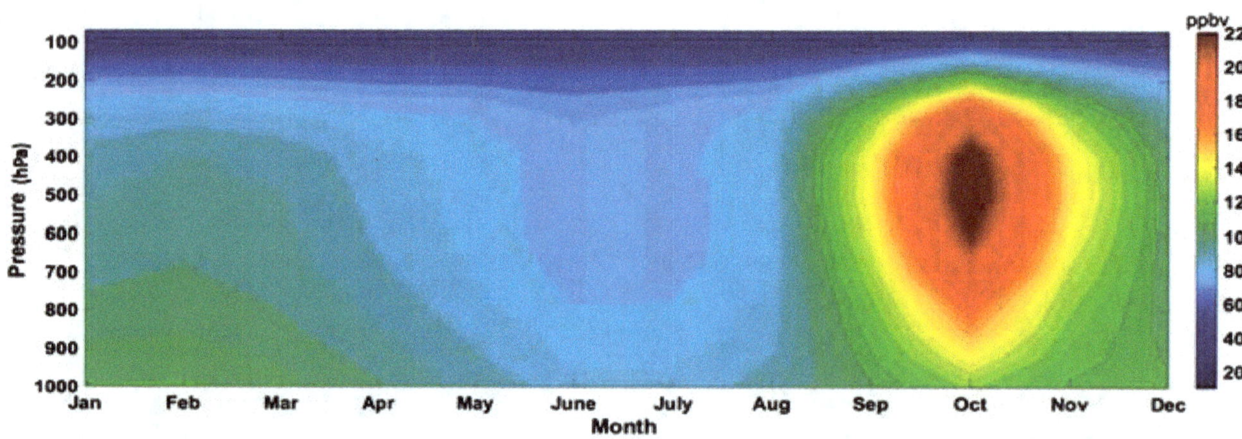

Figure 7. Pressure time cross section of monthly mean carbon monoxide observed over Southeast Asia and Western Pacific region (averaged over 10°N–10°S/85°E–140°E) from January to December 2015. To obtain this, AIRS satellite measured carbon monoxide data is utilized.

3.3. Spatial Variability of Cold Point Tropopause Temperature during December 2015

Changes in the CPT-T are crucial for the understanding the lower stratosphere WV changes [27,52]. It has been reported warm CPT-T (~3.5 K) over the SEA and WP regions in December 2015 from the reanalysis data sets [7]. In this study, we have utilized COSMIC GPS RO data for estimating the CPT temperatures. Figure 8 shows the spatial variability in the CPT-T in December 2015 along with the background climatology of CPT-T. The coldest CPT-T (~185 K) is situated over the central pacific (CP) (Figure 8b) in 2015 while background climatological minimum CPT-T is located over the WP region (Figure 8a). It clearly indicates the shifting of the minimum CPT-T region towards CP with extension towards EP in December 2015. The spatial distribution of CPT-T anomalies shows two strong warm anomaly areas over SEA/WP and Atlantic Ocean regions in December (Figure 8c). However, the strongest warm anomalies of ~5 K are observed over the SEA and WP regions. The height-longitudinal cross section of COSMIC temperature also shows the minimum temperature over eastern pacific region (Figure 8d). The thickness of the minimum temperature layer was much lower over the Indian Ocean and WP regions than the EP region.

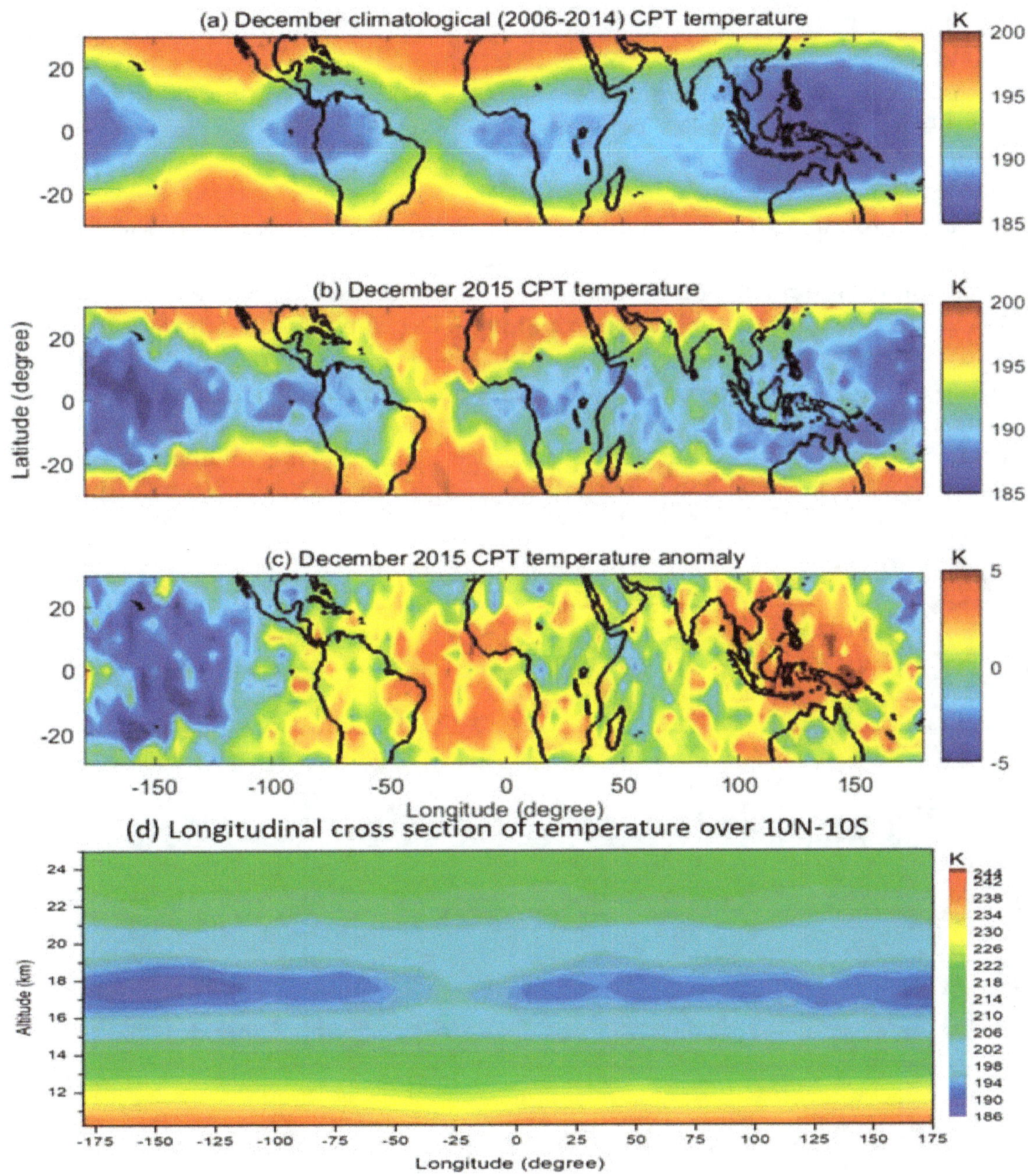

Figure 8. Spatial distribution of COSMIC GPS RO observed cold point tropopause temperature (**a**) December climatological (2006–2014) CPT, (**b**) December 2015 CPT and (**c**) December 2015 CPT anomaly. (**d**) Height-longitudinal cross section of COSMIC temperature over 10°N–10°S in December 2015.

The observed results matched well with the previous report by using reanalysis data sets [7] but with higher CPT-T anomalies during the same period. However, this warm CPT-T is formed well before December (figure not shown). The formation of warm CPT-T anomalies initiated during October gradually increased from November and reached the maximum warm CPT-T anomaly (~5 K) in December 2015 over the SEA and WP regions.

3.4. Water Vapour Changes during the 2015–16 El Nino

WV has major consequences for the radiative and heat transport in the atmosphere. Even very small changes in the LS WV could affect the surface climate [6]. It also plays an important role in the distribution of O_3 in the LS as an important contributor for long term change in the LS temperatures [4,5,55]. Debate is still going on the impact of El Niño 2015–16 and QBO disruption in 2016 on the WV concentrations in the LS [7,9,25]. However, in the present study we do not focus on which one is having more impact on changes in the WV concentrations in the LS. Instead, we focus only on the quantification of WV concentrations with respect to the background climatology within the tropical UTLS region. Recently, Avery et al. [7] clearly demonstrated the increasing of WV in the LS over the WP region along with warm CPT-T in December 2015. In the present study we tried to examine the zonally averaged changes in the WV at different pressure levels in the tropical UTLS region. Figure 9 shows zonally averaged WV mixing ratios at 146, 100 and 82 hPa over different latitudes. Black colour line represents climatology and red colour line represents year 2015. Strong increase of WV at 146 hPa was clearly noticed from July to December over all the latitudes (Figure 9a–c). But at 100 and 82 hPa, strong enhancement was observed from October to December 2015. There was a little drop in the WV compared to the background climatology during the summer months (June–August) of 2015 over the equator (Figure 9e,f). From the Figure 9, it is clear that the enhancement of the WV in 2015 started from October and it continues after that. The zonal mean anomalies reveal that the enhancement is high over equatorial latitudes as compared to the other latitudes.

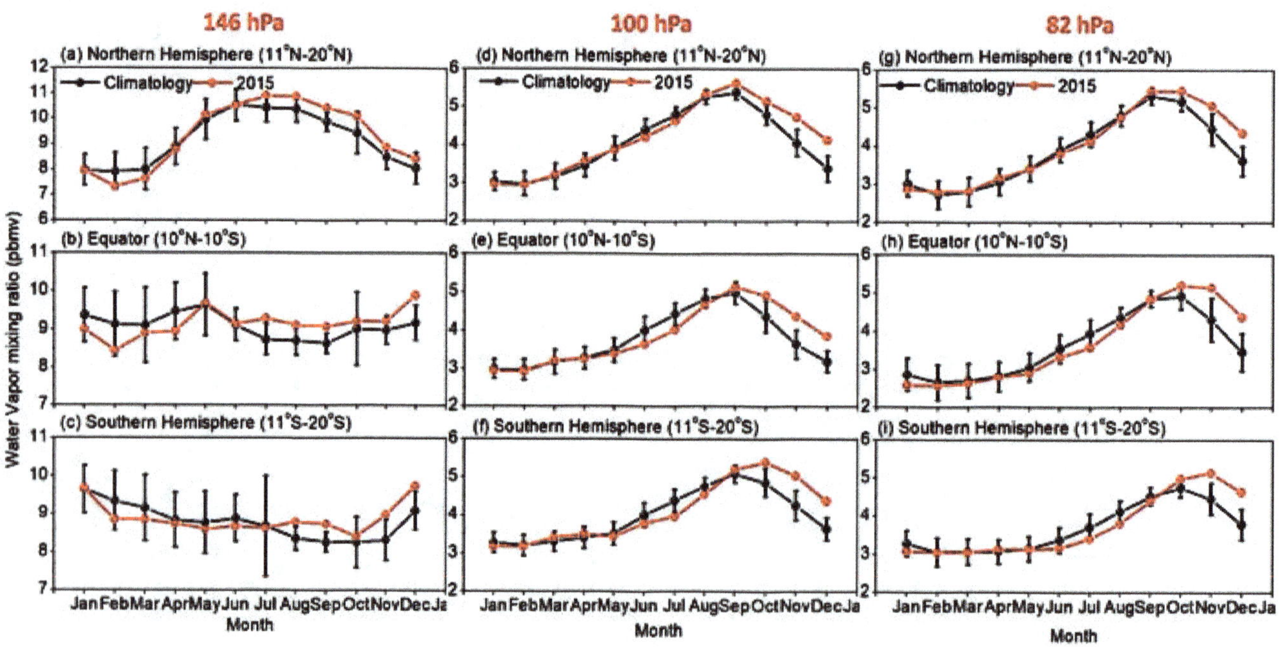

Figure 9. Same as Figure 2 but for the monthly mean of water vapor.

The WV enhancement in December 2015 at 82 hPa was little high over the equator (averaged over 10°S–10°N) compared to the NH and SH. (Figure 10). Strong El Niño events like 2015–16 El Niño cause the warming even up to the cold point tropopause allowing more WV to enter the stratosphere [26]. In a recent paper, Garfinkel et al. [26] suggested that the impact of ENSO events on the LS temperature and WV is nonlinear in boreal spring whereas linear in boreal winter. They also clearly mentioned that the strong El Niño events led to warming over Indo-WP region that subsequently warms the CPT and moistens the tropical LS [26]. The observed zonal mean anomalies clearly indicate the strong decrease of WV in the LS in 2016. At 82 hPa, the decrease was quite higher over equator compared to NH and SH. This record loss of WV was reported well in an earlier study reported in 'State of the Climate 2016' [56]. However, in their study, it is reported that the WV anomalies were found at 82 hPa only. In the present study, we tried to see the changes in the tropical UTLS region by estimating the WV anomalies at different pressure levels based on different latitude bands. The loss of WV in the LS in 2016 is strongly correlated with the large negative anomalies of CPT-T [25]. From the study by Tweedy et al. [25], it is evident that the decrease in global WV in the LS in December 2016 is the lowest in the record (1992–2016) [Figure 7 of [25]]. Our results also matched well with their results, except that the higher drop in the WV is observed over equatorial latitudes (>1 ppmv).

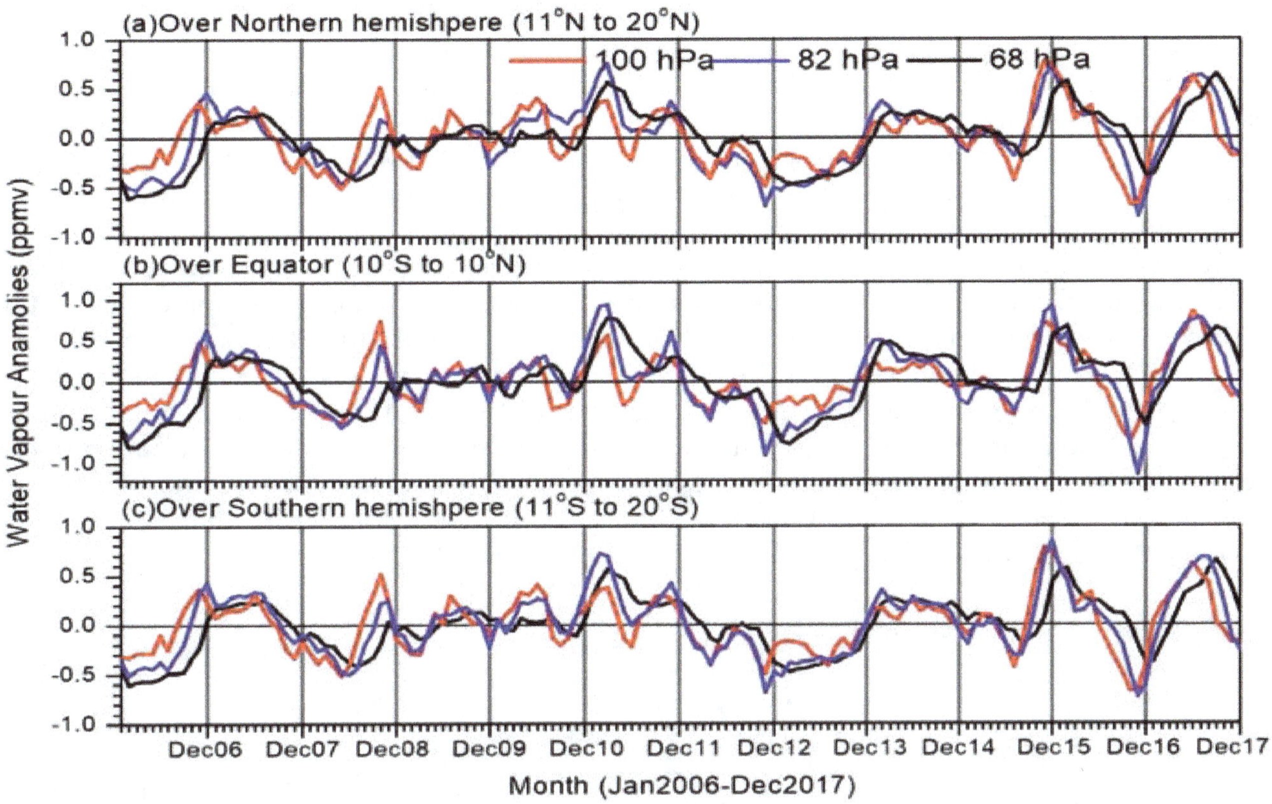

Figure 10. Zonal mean water vapour anomalies observed (**a**) over northern hemisphere (averaged over 11°N–20°N) (**b**) over equator (averaged over 10°N–10°S) and (**c**) over southern hemisphere (averaged over11°S–20°S). Different colours indicate different pressure levels.

4. Summary and Conclusions

During an El Niño event, warm waters over the Western Pacific (WP) and Indonesian region shift towards the central to EP regions accompanied by shifting of the convection towards central and EP regions. With a shift in the convection pattern and changes in the Walker circulation, the El Niño events strongly alter the precipitation pattern, which leads to strong regional moisture variability, drought conditions and forest fires along with biomass burning especially over Indonesia region [12]. The effects of El Niño events are found to be strong over the tropical WP and most severe over Indonesia, leading to large scale changes in the atmospheric chemical composition [13–16]. Cooling of the tropical LS, strengthening of the BDC, negative O_3 and temperature anomalies in tropical LS are observed changes in the El Niño events [10,32]. These changes due to El Niño have a significant impact on the distribution of WV, O_3 and other trace gases in the UTLS region over entire tropical region. Recent 2015–16 El Niño was one of the strongest and long lasted events in the 21st century. Lots of unusual changes happened in the atmosphere in 2015–16 El Niño event and were well reported by several studies [7–9,11,25,26].

In the present study, we quantified the observed changes in the trace gases, O_3, CO and WV in the UTLS region (146, 100 and 82/68 hPa) along with CPT over the tropics (20°N–20°S) from July to December 2015 using Aura MLS/AIRS/COSMIC GPS RO satellite measurements. The background climatology was calculated from 2006 to 2014 period. Before reaching its peak intensity during winter 2015–16, a remarkable change in the UTLS trace gases (O_3 and CO) concentrations over the tropics took place. Due to the fact that strong 2015–16 El Niño induced biomass burning and forest fires, huge amount of carbon emissions were released into the atmosphere in September and October [18–20]. Due to large carbon emissions, the CO was released into the atmosphere and transported to the UTLS region and recorded very high values in October and November 2015. The high resolution GPS RO observations clearly show the strong positive cold point tropopause temperatures over the SEA and WP regions in November and December. The variability of these trace gases shows some delay in the time period between them as depicted in Figure 11. The percentage change in the trace gases concentrations over equatorial region with respect to the background climatology clearly shows strong increase/decrease in trace gases concentrations in 2015–16 El Nino event. In October 2015, the CO shows 40% increase at 215 hPa whereas similar increase in CO shows at 146 hPa in November 2015. At 100 hPa, the increase of CO is ~25% in November 2015. Interestingly, the CO change at the 68 hPa shows continues increase from November 2015 to November 2016. This clearly shows the tropical tape recorder signal in CO. In O_3, the change is insignificant at 261 hPa and even at 146 hPa also. However, a significant decrease of O_3 at 100 hPa and 82 hPa is clearly noticed. Compared to the 82 hPa, the decrease in O_3 is quite high at 100 hPa in July and August 2015. At 100 hPa, maximum decrease in O_3 is observed in July 2015 whereas at 82 hPa the maximum decrease is observed in August 2015. WV also shows the maximum change (increase) in December 2015 at 82 hPa. However, the change in the WV was started from the October and reached maximum in December at 82 hPa. The strong decrease of ozone in the LS and at the tropopause level is observed in July and August 2015 and this loss in the O_3 is the record amount during the MLS data period. The observed strong decrease in LS O_3 is mainly due to the 2015–16 El Niño induced changes in the BDC in the LS.

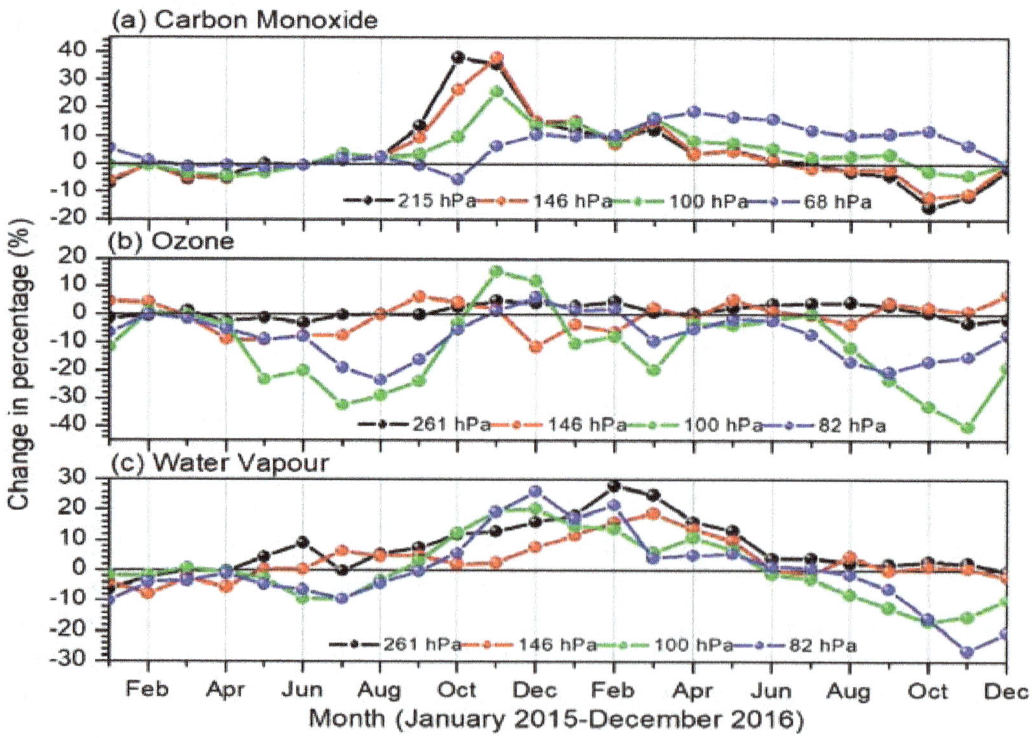

Figure 11. Percentage change with respect to the background climatology (2006–2014) in (**a**) carbon monoxide, (**b**) ozone and (**c**) water vapour observed over equatorial region (averaged over 10°S–10°N). Different colours represent different pressure levels.

The major findings from the present study are summarized below:

(1) A 32% (23%) decrease in the zonal mean equatorial O_3 is observed at 100 hPa (82 hPa) in July (August) 2015 and decrease in the O_3 is recorded maximum in the recent decade.
(2) A 38% (25%) increase of the zonal mean equatorial CO is observed at 146 hPa (100 hPa) in November 2015. The observed increased changes in the CO concentrations are recorded maximum in the MLS data period. The carbon emissions observed over SEA and WP regions play a crucial role on the increased high zonal mean CO in 2015–16 El Nino period.
(3) Large anomalies of cold point tropopause temperatures (5 K) are noticed from the COSMIC GPS RO observations over SEA and WP regions in December 2015.
(4) A 26% (20%) increase of the zonal mean equatorial WV is found at 82 hPa (100 hPa) in December 2015 whereas a ~30% WV decrease is observed in 2016.

Author Contributions: S.R. conceived the project, conducted research, performed initial analyses and wrote the first manuscript draft. M.V.R. and G.B. provided helpful discussions during conception of the project. N.N.R. assisted in processing and analyses of data. Y.-A.L. edited the first manuscript and finalized it for the first communication with the journal.

Acknowledgments: We thank the MLS and AIRS team for providing data which is used in the present study through their ftp sites. We also thank the NOAA for providing the MEI index data and COSMIC Data Analysis and Archive Centre (CDAAC) for providing GPS RO data used in the present study through their FTP site (http://cdaac-www.cosmic.ucar.edu/cdaac/products.html). Data used in the present study can be obtained freely from respective websites. We also thank Masatomo Fujiwara for his suggestions to improve the manuscript. Thanks are also given to the Ministry of Science and Technology (MOST) of Taiwan for financial support through grants No. 105-2221-E-008-056-MY3 and 107-2111-M-008-036.

References

1. Shepherd, T.G. Transport in the middle atmosphere. *Meteorol. Soc. Jpn. II* **2007**, *85B*, 165–191. [CrossRef]
2. Forster, P.M.; Shine, K.P. Stratospheric water vapour changes as a possible contributor to observed stratospheric cooling. *Geophys. Res. Lett.* **1999**, *26*, 3309–3312. [CrossRef]
3. Forster, P.; Ramaswamy, V.; Artaxo, P.; Berntsen, T.; Betts, R.; Fahey, D.W.; Haywood, J.; Lean, J.; Lowe, D.C.; Myhre, G.; et al. Radiative Forcing of Climate Change. In *Climate Change 2007: The Physical Science Basis. Contribution of Working Group I to the Fourth Assessment Report of the Intergovernmental Panel on Climate Change*; Cambridge University Press: Cambridge, UK, 2007; pp. 129–234.
4. Solomon, S.; Rosenlof, K.H.; Portmann, R.W.; Daniel, J.S.; Davis, S.M.; Sanford, T.; Plattner, G.-K. Contributions of Stratospheric Water Vapor to Decadal Changes in the Rate of Global Warming. *Science* **2010**, *327*, 1219–1223. [CrossRef]
5. Maycock, A.C.; Shine, K.P.; Joshi, M.M. The temperature response to stratospheric water vapour changes. *Q. J. R. Meteorol. Soc.* **2011**, *137*, 1070–1082. [CrossRef]
6. Riese, M.; Ploeger, F.; Rap, A.; Vogel, B.; Konopka, P.; Dameris, M.; Forster, P. Impact of uncertainties in atmospheric mixing on simulated UTLS composition and related radiative effects. *J. Geophys. Res.* **2012**, *117*, D16305. [CrossRef]
7. Avery, M.A.; Davis, S.M.; Rosenlof, K.H.; Ye, H.; Dessler, A.E. Large anomalies in lower stratospheric water vapour and ice during the 2015–2016 El Niño. *Nat. Geosci.* **2017**. [CrossRef]
8. Domeisen, D.I.; Garfinkel, C.I.; Butler, A.H. The teleconnection of El Niño Southern Oscillation to the stratosphere. *Rev. Geophys.* **2019**, 57. [CrossRef]
9. Diallo, M.; Riese, M.; Birner, T.; Konopka, P.; Müller, R.; Hegglin, M.I.; Santee, M.L.; Baldwin, M.; Legras, B.; Ploeger, F. Response of stratospheric water vapor and ozone to the unusual timing of El Niño and the QBO disruption in 2015–2016. *Atmos. Chem. Phys.* **2018**, *18*, 13055–13073. [CrossRef]
10. Randel, W.J.; Garcia, R.R.; Calvo, N.; Marsh, D. ENSO influence on zonal mean temperature and ozone in the tropical lower stratosphere. *Geophys. Res. Lett.* **2009**, 39. [CrossRef]
11. Diallo, M.; Konopka, P.; Santee, M.L.; Müller, R.; Tao, M.; Walker, K.A.; Legras, B.; Riese, M.; Ern, M.;

Ploeger, F. Structural changes in the shallow and transition branch of the Brewer–Dobson circulation induced by El Niño. *Atmos. Chem. Phys.* **2109**, *19*, 425–446. [CrossRef]

12. Nassar, R.; Logan, J.A.; Megretskaia, I.A.; Murray, L.T.; Zhang, L.; Jones, D.B.A. Analysis of tropical tropospheric ozone, carbon monoxide and water vapor during the 2006 El Niño using TES observations and the GEOS-Chem model. *J. Geophys. Res.* **2009**, *114*, D17304. [CrossRef]
13. Logan, J.A.; Megretskaia, I.; Nassar, R.; Murray, L.T.; Zhang, L.; Bowman, K.W.; Worden, H.M.; Luo, M. Effects of the 2006 El Niño on tropospheric composition as revealed by data from the Tropospheric Emission Spectrometer (TES). *Geophys. Res. Lett.* **2008**, *35*, L03816. [CrossRef]
14. Chandra, S.; Ziemke, J.R.; Duncan, B.N.; Diehl, T.L.; Livesey, N.J.; Froidevaux, L. Effects of the 2006 El Niño on tropospheric ozone and carbon monoxide: Implications for dynamics and biomass burning. *Atmos. Chem. Phys.* **2009**, *9*, 4239–4249. [CrossRef]
15. Chandra, S.; Ziemke, J.R.; Schoeberl, M.R.; Froidevaux, L.; Read, W.G.; Levelt, P.F.; Bhartia, P.K. Effects of the 2004 El Niño on tropospheric ozone and water vapour. *Geophys. Res. Lett.* **2007**, *34*, L06802. [CrossRef]
16. Doherty, R.M.; Stevenson, D.S.; Johnson, C.E.; Collins, W.J.; Sanderson, M.G. Tropospheric ozone and El Niño–Southern Oscillation: Influence of atmospheric dynamics, biomass burning emissions and future climate change. *J. Geophys. Res.* **2006**, *111*, D19304. [CrossRef]
17. Tweedy, O.V.; Waugh, D.W.; Randel, W.J.; Abalos, M.; Oman, L.D.; Kinnison, D.E. The impact of boreal summer ENSO events on tropical lower stratospheric ozone. *J. Geophys. Res. Atmos.* **2018**, *123*, 9843–9857. [CrossRef]
18. Huijnen, V.; Wooster, M.; Kaiser, J.; Gaveau, D.; Flemming, J.; Parrington, M.; Inness, A.; Murdiyarso, D.; Main, B.; van Weele, M. Fire Carbon Emissions Over Maritime Southeast Asia in 2015 Largest Since 1997. *Sci. Rep.* **2016**, *6*, 26886. [CrossRef]
19. Field, R.D.; van der Werf, G.R.; Fanin, T.; Fetzer, E.J.; Fuller, R.; Jethva, H.; Levy, R.; Livesey, N.J.; Luo, M.; Torres, O.; et al. Indonesian fire activity and smoke pollution in 2015 show persistent nonlinear sensitivity to El Niño induced drought. *Proc. Natl. Acad. Sci. USA* **2016**, *113*, 9204–9209. [CrossRef]
20. Heymann, J.; Reuter, M.; Buchwitz, M.; Schneising, O.; Bovensmann, H.; Burrows, J.P.; Massart, S.; Kaiser, J.W.; Crisp, D. CO_2 emission of Indonesian fires in 2015 estimated from satellite-derived atmospheric CO_2 concentrations. *Geophys. Res. Lett.* **2017**. [CrossRef]
21. Parker, R.J.; Boesch, H.; Wooster, M.J.; Moore, D.P.; Webb, A.J.; Gaveau, D.; Murdiyarso, D. Atmospheric CH_4 and CO_2 enhancements and biomass burning emission ratios derived from satellite observations of the 2015 Indonesian fire plumes. *Atmos. Chem. Phys.* **2016**, *16*, 10111–10131. [CrossRef]
22. Whitburn, S.; Van Damme, M.; Clarisse, L.; Turquety, S.; Clerbaux, C.; Coheur, P.F. Doubling of Annual Ammonia Emissions from the Peat Fires in Indonesia During the 2015 El Niño. *Geophys. Res. Lett.* **2016**, *43*, 11007–11014. [CrossRef]
23. Osprey, S.M.; Butchart, N.; Knight, J.R.; Scaife, A.A.; Hamilton, K.; Anstey, J.A.; Schenzinger, V.; Zhang, C. An unexpected disruption of the atmospheric quasi-biennial oscillation. *Science* **2016**, *353*, 1424–1427. [CrossRef]
24. Coy, L.; Newman, P.; Pawson, S.; Lait, L.R. Dynamics of the disrupted 2015–16 quasi-biennial oscillation. *J. Clim.* **2017**, *30*, 5661–5674. [CrossRef]
25. Tweedy, O.V.; Kramarova, N.A.; Strahan, S.E.; Newman, P.A.; Coy, L.; Randel, W.J.; Park, M.; Waugh, D.W.; Frith, S.M. Response of trace gases to the disrupted 2015–2016 quasi-biennial oscillation. *Atmos. Chem. Phys.* **2017**, *17*, 6813–6823. [CrossRef]
26. Garfinkel, C.I.; Gordon, A.; Oman, L.D.; Li, F.; Davis, S.; Pawson, S. Nonlinear response of tropical lower-stratospheric temperature and water vapor to ENSO. *Atmos. Chem. Phys.* **2018**, *18*, 4597–4615. [CrossRef]
27. Garfinkel, C.I.; Hurwitz, M.M.; Oman, L.D.; Waugh, D.W. Contrasting effects of Central Pacific and Eastern Pacific El Niño on stratospheric water vapor. *Geophys. Res. Lett.* **2013**, *40*, 4115–4120. [CrossRef]
28. Fueglistaler, S.; Dessler, A.E.; Dunkerton, T.J.; Folkins, I.; Fu, Q.; Mote, P.W. Tropical Tropopause Layer. *Rev. Geophys.* **2009**, *47*, G1004. [CrossRef]
29. Randel, W.J.; Jensen, E.J. Physical processes in the tropical tropopause layer and their roles in a changing climate. *Nat. Geosci.* **2013**, *6*, 169–176. [CrossRef]
30. Ravindra Babu, S.; Venkat Ratnam, M.; Basha, G.; Krishnamurthy, B.V. Indian summer monsoon onset signatures on the tropical tropopause layer. *Atmos. Sci. Lett.* **2019**, e884. [CrossRef]

31. Gettelman, A.; Forster, P.; Fujiwara, M.; Fu, Q.; Vomel, H.; Gohar, L.; Johason, C.; Ammerman, M. The radiation balance of the tropical tropopause layer. *J. Geophys. Res.* **2004**, *109*, D07103. [CrossRef]
32. Oman, L.D.; Douglass, A.R.; Ziemke, J.R.; Rodriguez, J.M.; Waugh, D.W.; Nielsen, J.E. The ozone response to ENSO in Aura satellite measurements and a chemistry climate model. *J. Geophys. Res.* **2013**, *118*, 965976. [CrossRef]
33. Lelieveld, J.; Gromov, S.; Pozzer, A.; Taraborrelli, D. Global tropospheric hydroxyl distribution, budget and reactivity. *Atmos. Chem. Phys.* **2016**, *16*, 12477–12493. [CrossRef]
34. Wolter, K.; Timlin, M.S. Measuring the strength of ENSO events—How does 1997/98 rank? *Weather* **1998**, *53*, 315–324. [CrossRef]
35. Wolter, K.; Timlin, M.S. El Nino/Southern Oscillation behavior since 1871 as diagnosed in an extended multivariate ENSO index (MEI. ext). *Int. J. Climatol* **2011**, *31*, 1074–1087. [CrossRef]
36. Livesey, N.J.; Read, W.G.; Wagner, P.A.; Froidevaux, L.; Lambert, A.; Manney, G.L.; Pumphrey, H.C.; Santee, M.L.; Schwartz, M.J.; Wang, S.; et al. *Aura Microwave Limb Sounder (MLS) Version 4.2× Level 2 Data Quality and Description Document*; JPL D-33509 Rev.D, Tech. Rep.; NASA Jet Propulsion Laboratory, California Institute of Technology: Pasadena, CA, USA, 2018.
37. Anthes, R.A.; Bernhardt, P.A.; Chen, Y.; Cucurull, L.; Dymond, K.F.; Ector, D.; Healy, S.B.; Ho, S.P.; Hunt, D.C.; Kuo, Y.H.; et al. The COSMIC/Formosat/3 mission: Early results. *Bull. Am. Meteorol. Soc.* **2008**, *89*, 313–333. [CrossRef]
38. Rao, D.N.; Ratnam, M.V.; Mehta, S.; Nath, D.; Ghouse Basha, S.; Jagannadha Rao, V.V.M.; Krishna Murthy, B.V.; Tsuda, T.; Nakamura, K. Validation of the COSMIC radio occultation data over Gadanki (13.48 N, 79.2 E): A tropical region. *Terr. Atmos. Ocean. Sci* **2009**, *20*, 59–70. [CrossRef]
39. Kishore, P.; Basha, G.; Ratnam, M.V.; Ouarda, T.B.M.J.; Velicogna, I. Evaluation of CMIP5 upper troposphere and lower stratosphere temperature with COSMIC, 21st century projections and trends. *Clim. Dyn.* **2016**. [CrossRef]
40. Kishore, P.; Ratnam, M.V.; Namboothiri, S.P.; Velicogna, I.; Basha, G.; Jiang, J.H.; Igarashi, K.; Rao, S.V.B.; Sivakumar, V. Global distribution of water vapor observed by COSMIC GPS RO: Comparison with GPS radiosonde, NCEP, ERA-Interim and JRA-25 reanalysis data sets. *J. Atmos. Sol.-Terr. Phys.* **2013**. [CrossRef]
41. Liou, Y.A.; Pavelyev, A.G.; Matyugov, S.S.; Yakovlev, O.I.; Wickert, J. *Radio Occultation Method for Remote Sensing of the Atmosphere and Ionosphere*; INTECH: Vukovar, Croatia, 2010; 170p.
42. Fong, C.J.; Shiau, W.T.; Lin, C.T.; Kuo, T.C.; Chu, C.H.; Yang, S.K.; Yen, N.; Chen, S.S.; Kuo, Y.H.; Liou, Y.A.; et al. Constellation deployment for FORMOSAT-3/COSMIC mission. *IEEE Trans. Geosci. Remote Sens.* **2008**, *46*, 3367–3379. [CrossRef]
43. Fong, C.J.; Yang, S.K.; Chu, C.H.; Huang, C.Y.; Yeh, J.J.; Lin, C.T.; Kuo, T.C.; Liu, T.Y.; Yen, N.; Chen, S.S.; et al. FORMOSAT-3/COSMIC constellation spacecraft system performance: After one year in orbit. *IEEE Trans. Geosci. Remote Sens.* **2008**, *46*, 3380–3394. [CrossRef]
44. Fong, C.J.; Yen, N.; Chu, C.H.; Hsiao, C.C.; Liou, Y.A.; Chi, S. Space-based global weather monitoring System: FORMOSAT-3/COSMIC constellation and its follow-on mission. *J. Spacecr. Rocket.* **2009**, *46*, 883–891. [CrossRef]
45. Fong, C.J.; Yen, N.; Chu, V.; Yang, E.; Shiau, A.; Huang, C.Y.; Chi, S.; Chen, S.S.; Liou, Y.A.; Kuo, Y.H. FORMOSAT-3/COSMIC spacecraft constellation system, mission results and prospect for follow-on mission. *Terr. Atmos. Ocean. Sci.* **2009**, *20*, 1–19. [CrossRef]
46. Liou, Y.A.; Pavelyev, A.G.; Huang, C.-Y.; Igarashi, K.; Hocke, K.; Yan, S.-K. Analytic method for observation of the gravity waves using radio occultation data. *Geophys. Res. Lett.* **2003**, *30*, 2021. [CrossRef]
47. Liou, Y.A.; Pavelyev, A.G.; Wickert, J. Observation of the gravity waves from GPS/MET radio occultation data. *J. Atmos. Sol.-Terr. Phys.* **2005**, *67*, 219–228. [CrossRef]
48. Liou, Y.A.; Pavelyev, A.G.; Wickert, J.; Huang, C.Y.; Yan, S.K.; Liu, S.-F. Response of GPS occultation signals to atmospheric gravity waves and retrieval of gravity wave parameters. *GPS Solut.* **2004**, *8*, 103–111. [CrossRef]
49. Liou, Y.A.; Pavelyev, A.G.; Wicker, J.; Liu, A.A.; Schmidt, T.; Igarashi, K. Application of GPS radio occultation method for observation of the internal waves in the atmosphere. *J. Geophys. Res.* **2006**, *111*, D06104. [CrossRef]
50. Chane Ming, F.; Ibrahim, C.; Barthe, C.; Jolivet, S.; Keckhut, P.; Liou, Y.A.; Kuleshov, K. Observation and a numerical study of gravity waves during tropical cyclone Ivan (2008). *Atmos. Chem. Phys.* **2014**, *14*, 641–658. [CrossRef]

51. Ravindra Babu, S.; Venkata Ratnam, M.; Basha, G.; Krishnamurthy, B.V.; Venkateswararao, B. Effect of tropical cyclones on tropical tropopause parameters observed using COSMIC GPS RO data. *Atmos. Chem. Phys.* **2015**. [CrossRef]
52. Venkata Ratnam, M.; Ravindra Babu, S.; Das, S.S.; Basha, G.; Krishnamurthy, B.V.; Venkateswararao, B. Effect of tropical cyclones on the stratosphere–troposphere exchange observed using satellite observations over the north Indian Ocean. *Atmos. Chem. Phys.* **2016**, *16*, 8581–8591. [CrossRef]
53. Biondi, R.; Steiner, A.K.; Kirchengast, G.; Rieckh, T. Characterization of thermal structure and conditions for overshooting of tropical and extratropical cyclones with GPS radio occultation. *Atmos. Chem. Phys.* **2015**, *15*, 5181–5193. [CrossRef]
54. Blunden, J.; Arndt, D.S. State of the Climate in 2015. *Bull. Am. Meteorol. Soc.* **2016**, 97. [CrossRef]
55. Shindel, D.T. Climate and ozone response to increased stratospheric water vapour. *Geophys. Res. Lett.* **2001**, *28*, 1551–1554. [CrossRef]
56. Blunden, J.; Arndt, D.S. State of the Climate in 2016. *Bull. Am. Meteorol. Soc.* **2017**, 98. [CrossRef]

7

Temporal and Spatial Characteristics of EVI and its Response to Climatic Factors in Recent 16 years based on Grey Relational Analysis in Inner Mongolia Autonomous Region, China

Dong He [1,2], Guihua Yi [3,*], Tingbin Zhang [1,2,*], Jiaqing Miao [2,4], Jingji Li [5,6] and Xiaojuan Bie [1]

1 College of Earth Science, Chengdu University of Technology, Chengdu 610059, China; 17828068016@163.com (D.H.); biexiaojuan06@cdut.cn (X.B.)
2 The Engineering & Technical College of Chengdu University of Technology, Leshan 614000, China; jmiao@mtech.edu
3 College of Management Science, Chengdu University of Technology, Chengdu 610059, China
4 Department of Geophysical Engineering, Montana Tech of the University of Montana, Butte, MT 59801, USA
5 Chengdu University of Technology, College of Environmental and Civil Engineering Institute, Chengdu 610059, China; jijingli.ok@163.com
6 Chengdu University of Technology, Institute of Ecological Resource and Landscape, Chengdu 610059, China
* Correspondence: yigh@cdut.edu.cn (G.Y.); zhangtb@cdut.edu.cn (T.Z.)

Abstract: The Inner Mongolia Autonomous Region (IMAR) is a major source of rivers, catchment areas, and ecological barriers in the northeast of China, related to the nation's ecological security and improvement of the ecological environment. Therefore, studying the response of vegetation to climate change has become an important part of current global change research. Since existing studies lack detailed descriptions of the response of vegetation to different climatic factors using the method of grey correlation analysis based on pixel, the temporal and spatial patterns and trends of enhanced vegetation index (EVI) are analyzed in the growing season in IMAR from 2000 to 2015 based on moderate resolution imaging spectroradiometer (MODIS) EVI data. Combined with the data of air temperature, relative humidity, and precipitation in the study area, the grey relational analysis (GRA) method is used to study the time lag of EVI to climate change, and the study area is finally zoned into different parts according to the driving climatic factors for EVI on the basis of lag analysis. The driving zones quantitatively show the characteristics of temporal and spatial differences in response to different climatic factors for EVI. The results show that: (1) The value of EVI generally features in spatial distribution, increasing from the west to the east and the south to the north. The rate of change is 0.22/10°E from the west to the east, 0.28/10°N from the south to the north; (2) During 2000–2015, the EVI in IMAR showed a slightly upward trend with a growth rate of 0.021/10a. Among them, the areas with slight and significant improvement accounted for 21.1% and 7.5% of the total area respectively, ones with slight and significant degradation being 24.6% and 4.3%; (3) The time lag analysis of climatic factors for EVI indicates that vegetation growth in the study area lags behind air temperature by 1–2 months, relative humidity by 1–2 months, and precipitation by one month respectively; (4) During the growing season, the EVI of precipitation driving zone (21.8%) in IMAR is much larger than that in the air temperature driving zone (8%) and the relative humidity driving zone (11.6%). The growth of vegetation in IMAR generally has the closest relationship with precipitation. The growth of vegetation does not depend on the change of a single climatic factor. Instead, it is the result of the combined action of multiple climatic factors and human activities.

Keywords: EVI; climatic factor; driving force; grey relational analysis (GRA); Inner Mongolia Autonomous Region (IMAR)

1. Introduction

Vegetation is a key component of ecosystems, and any change in terrestrial ecosystems is bound to result in fluctuations in vegetation types, quantity, or quality [1–3]. As a comprehensive indicator, vegetation reflects the changes of the ecological environment, and studying its response to climate change has become one of the main contents of the current global change research [4–6]. The enhanced vegetation index (EVI) is an important quantitative index, reflecting the growth status of the surface vegetation and is also one of the most important basic data in ecosystems research. The change of EVI plays an important role in indicating the changes of regional ecosystems and environment [7–9]. The Inner Mongolia Autonomous Region (IMAR) is located in the transitional zone from arid and semi-arid climates to humid and semi-humid monsoon climates of the southeast coast [10–12]. It is a water conservation area of Songhua River and functions as an important ecological barrier of northern China [11,12]. The ecological environment in IMAR is characterized by distinctive geographical differences, fragile ecological conditions, and complex ecological types [12–14], where vegetation types show a northeast-southwest pattern of surface cover of forests, steppes and deserts [12,15]. Quantitative assessment of the dynamic changes and driving forces of the vegetation ecosystem in Inner Mongolia will help people to understand the feedback between the global climate change and vegetation ecosystems [12,14–16], which is of great theoretical and practical significance for evaluating the environmental quality of terrestrial ecosystems and regulating ecological processes [1,16,17].

A lot of work has been done in studying the relationship between terrestrial vegetation and climatic factors. Related studies have found that vegetation changes in the Sahel [18], the tropical Africa [19], the Central Plains of the United States [20], and the Eurasian continent [21] are closely related to the amount of precipitation, while air temperature is the main factor that influences the growth of terrestrial vegetation in the northern Finno-Candea [22], the Arctic [23], and North America [24]. China is located in the monsoon region of eastern Asia, with complex climate types and rich vegetation types. There are also distinct differences in the correlation between vegetation changes and hydrothermal factors in different regions [25–27], as well as the time lag of response to climatic factors [13,28,29].

Most of the previous studies on the relationship between vegetation remote sensing quantitative factors and climate change have been made using the method of correlation analysis [11,13,19], but the correlation analysis method usually requires that each variable should follow the joint normal distribution; thus, the extreme values of the factors in the analysis process would have a great impact on the results of correlation analysis [30–33]. Some scholars use the grey relational analysis (GRA) to study the relationship between vegetation index and climatic factors [30,32,34], but related studies are mostly conducted on the overall statistical value or limited sample sites of the study area. The results of those studies lack detailed descriptions of how different vegetation types and vegetation growing in different geomorphological characteristics of the study area respond to different climatic factors. In order to quantitatively display and evaluate the temporal and spatial features of the response of EVI to different climatic factors, the relationship between EVI and climatic factors are discussed using GRA method on the basis of time lag analysis.

2. Materials and Methods

2.1. Study Area

IMAR is located in the northern part of China (Figure 1a), stretching over the northwestern, northern, and northeastern parts of China, a narrow band from the west to the east. The total area of IMAR is 1.183 million square kilometers, accounting for 12.3% of the country's total area. With rich resource reserves, it is known as "Forests in the East, Mines in the West, Agriculture in the South and Animal Husbandry in the North" and it ranks first in terms of the steppe, forest, and per capita arable land in the country, also the largest prairie pastoral area in the country [10,35,36]. The average elevation of the study area is about 1000 m. The climate of IMAR belongs to the transitional zone of arid and semi-arid monsoon climates to humid and semi-humid climates. The hydrothermal conditions show a northeast-southwest zonal distribution [11,12]. Due to differences in natural factors, such as hydrothermal conditions and landforms, the vegetation in IMAR has obvious east-west zonal distribution features (Figure 1b). The types of vegetation cover from the east to the west are mainly coniferous forests, broadleaf forests, steppes, desert steppes, deserts, and so on.

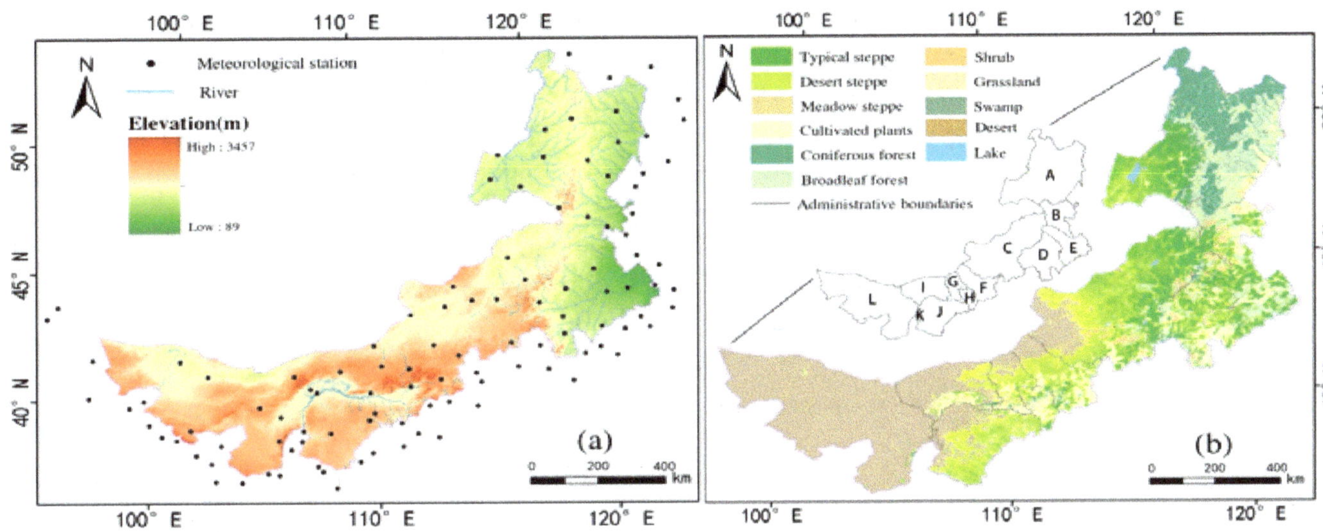

Figure 1. The Digital Elevation Model (DEM) and meteorological station distribution (**a**) and the vegetation types (**b**) in IMAR. (A—Hulunbuir; B—Hinggan League; C—Xilingol League; D—Chifeng; E—Tongliao; F—Ulanqab League; G—Baotou; H—Hohhot; I—Bayannur League; J—Erdos; K—Wuhai; L—Alxa League).

2.2. Data Sources

The EVI data came from the MODIS C6 MOD13Q1 dataset published by the National Aeronautics and Space Administration (NASA) of the United States. Compared with C5 data, C6 data further solves the problem of data attenuation and distortion due to the aging of satellite sensors, resulting in a significant increase in data quality. IMAR involves seven scenes of h27v04, h26v05, h26v04, h26v03, h25v05, h25v04, and h25v03 with a spatial resolution of 250 m and a time resolution of 16d.

The meteorological data were selected from the monthly mean air temperature, monthly mean relative humidity, and monthly cumulative precipitation data from 110 meteorological stations in IMAR and its neighboring areas from February to September during year 2000–2015. The data were provided by China Meteorological Science Data Sharing Service Website (http://data.cma.gov.cn).

The vegetation type data were taken from the 1:1,000,000 national vegetation type dataset published by the Data Center of Chinese Academy of Resources and Environmental Sciences (http://www.resdc.cn).

Digital Elevation Model (DEM) data were taken from Advanced Spaceborne Theemal Emission and Reflection Radiometer (ASTER) global digital elevation model (ASTER GDEM) V2 digital elevation data from geospatial data cloud platform (http://www.gscloud.cn) with a spatial resolution of 30 m.

2.3. Data Processing

2.3.1. EVI Data Processing

EVI data was affected by factors such as aerosol, ice and snow, solar illumination angle, and sensor observation angle in the process of collection and processing. There may be anomalous or missing data, which caused the seasonal trend of EVI curve to be insignificant. Therefore, in order to make EVI time series data reflect seasonal variation of vegetation correctly, it was necessary to conduct a filtering process [9,37,38]. Firstly, the invalid pixels in the quality control file attached to MODIS13Q1 with a value of 65,535 were removed. With full consideration of the growth periodicity of vegetation, the 16d EVI data of IMAR were smoothed using harmonic analysis of time series (HANTS) (Figure 2). During processing, the valid range was −3000~10,000, the period was 23, and the frequency was 2 (11, 23). The time series data after reconstruction can reflect the periodic variation of the EVI curve [9,38]. Then, using the MODIS reprojection tools (MRT), the MOD13Q1 data were pre-georeferenced to the UTM zone 48 North projection WGS-84 datum resampled with the resolution of 500 m. Next, the monthly EVI data for vegetation in the growing season (May–September) were calculated using the method of maximum value composite (MVC). Subsequently, the effects of bare soil and sparsely vegetated areas were eliminated according to the following rules: (1) The annual mean value of EVI in the growing season was greater than 0.07; (2) The annual maximum value of EVI was greater than 0.10; (3) The annual maximum value of EVI should appear in July–September. Finally, the EVI pixels from May to September of the year 2000–2015, which met the requirements above, were used as the mean annual value of EVI in the growing season.

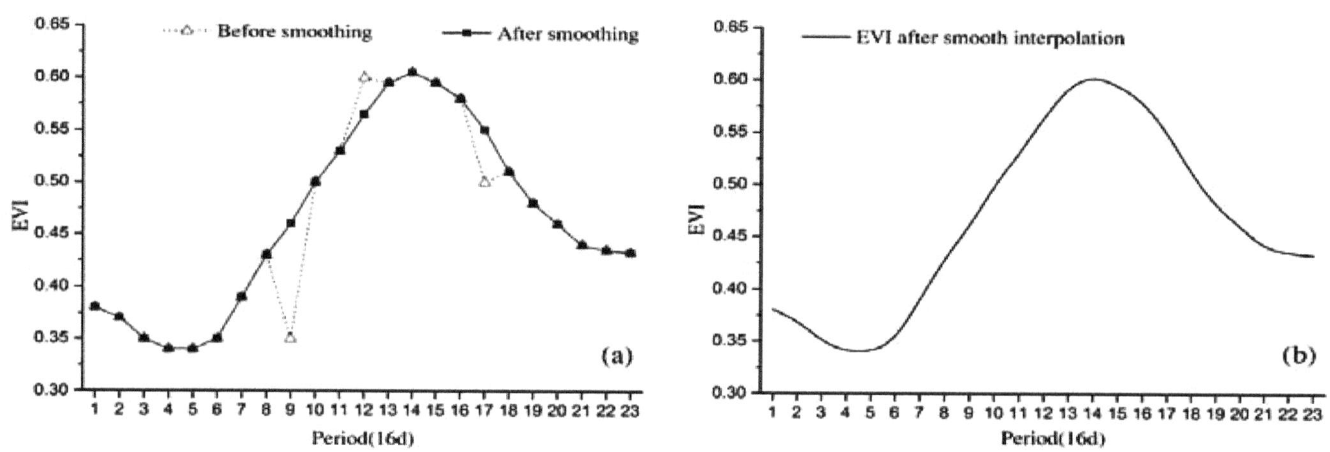

Figure 2. The EVI time series data before (**a**) and after (**b**) HANTS processing.

2.3.2. Spatial Interpolation of Meteorological Data

Changes in meteorological sites, observation instruments and observation methods, or environmental changes in the surroundings will lead to non-uniformity of collected data. It was necessary to conduct homogeneity tests on the data of meteorological sites [39] to eliminate invalid site data. The existing research on meteorological factor interpolation adopts different methods—Kriging and IDW (Inverse Distance Weighted) [11,40] for relative humidity, Kriging and IDW [11,32] for precipitation., and TPS (Thin Plate Spline) and Kriging [40,41] for air temperature. Considering that the spatial distribution was significantly different in relative humidity, precipitation, and air temperature, and air temperature had a certain degree of altitude sensitivity [40–42], on the basis of the previous study method, after being tested by comparing meteorological site data and other methods, IDW was selected to interpolate the relative humidity, and Kriging to interpolate the precipitation. TPS was selected and DEM acted as the covariate to assist spatial interpolation of air temperature data. The spatial resolutions of meteorological element interpolation were unified to 500 m.

2.4. Methods

2.4.1. Trend Analytical Method

The linear trend analysis of EVI in the growing season from 2000 to 2015 was carried out with the trend analytical method. The formula is as follows:

$$\text{Slope} = \frac{n \times \sum_{i=1}^{n} (i \times EVI_i) - \sum_{i=1}^{n} i \sum_{i=1}^{n} EVI_i}{n \times \sum_{i=1}^{n} i^2 - (\sum_{i=1}^{n} i)^2}, \tag{1}$$

where n stands for the time series (2000–2015), i.e., n = 16; EVI_i the mean EVI of the year i; Slope the inter-annual change slope of certain EVI pixel. When the value of Slope is positive, it indicates that EVI shows a trend of increase as time goes by, and vice versa. The formula is highly reliable and has been widely used [11,43,44].

2.4.2. Grey Relational Analysis

Grey relation refers to the uncertain correlation between things and system factors, or factors and the main behavior. GRA is based on the geometrical approximation of the sequence of behavioral factors and the main behavioral sequence, and it is used to analyze and determine the degree of influence between factors or the contribution measure of factors to the main behavior [45,46]. Not restricted by the type of samples and the distribution of probability distribution, GRA is a new method to study uncertain issues with limited data and information [30].

The reference factor sequence can be expressed as:

$$X_0 = [X_0(1), X_0(2), \ldots, X_0(n)] \tag{2}$$

The comparative factor sequence can be expressed as:

$$X_i = [X_i(1), X_i(2), \ldots, X_i(n)], i = 1, 2, \ldots, m. \tag{3}$$

Considering the relevant factor sequence, the point relational coefficient is defined:

$$r(x_0(k), x_i(k)) = \frac{x(\min) + \xi x(\max)}{\Delta_{0i}(k) + \xi x(\max)}, \tag{4}$$

And

$$\Delta_{0i}(k) = |x_0(k) - x_i(k)|, \tag{5}$$

$$x(\min) = \min_i \min_k \Delta_{0i}(k), \tag{6}$$

$$x(\max) = \max_i \max_k \Delta_{0i}(k), \tag{7}$$

the Grey Relational Grade (GRG) of $\gamma(X_0, X_i)$ between $X_i(i = 1, 2, \cdots, m)$ and X_0:

$$\gamma(X_0, X_i) = \frac{1}{n} \sum_{k=1}^{n} r(x_0(k), x_i(k)), \tag{8}$$

where $\gamma(x_0(k), x_i(k))$ is the relational coefficient between x_i and x_0 when it meets the condition of comparative factor k and the grey resolution coefficient ξ. The GRG is a measure of the influence of the comparative factor sequence on the reference factor sequence. When the value is closer to 1, the effect of the comparative factor sequence on the reference factor sequence becomes more significant.

The value of the GRG can be used as an indicator that reflects the influence of comparative factor sequence on reference factor sequence [47,48].

In this study, EVI in the growing season of 2000–2015 was used as the reference factor sequence (X_0), and the comparative factor sequence (X_i) was made up of three climatic factors of air temperature, relative humidity, and precipitation during 2000–2015.

2.4.3. Time Lag Analysis

The EVI sequence (May–September) and the monthly mean air temperature in the growing season of the study area from 2000 to 2015 were used as two sets of variables to calculate GRG between EVI sequence and the monthly mean air temperature. Similarly, the GRG between EVI sequence in the growing season and the monthly mean air temperature (April–August, March–July and February–June) could be calculated, and that between the monthly mean relative humidity and the monthly cumulative precipitation (May–September, April–August, and March–July, February–June) could also be calculated respectively. By comparing the GRG between EVI in the growing season and different comparative factor sequence, the authors further discuss the time lag effect of EVI for climatic factors.

3. Results

3.1. Inter-Annual Change of EVI in the Growing Season and Spatial Distribution Pattern

3.1.1. Inter-Annual Change of EVI in the Growing Season

According to the mean EVI in the growing season of IMAR from 2000 to 2015, the inter-annual change of EVI during the study period was obtained (Figure 3). In general, the mean value of EVI in IMAR showed a rising trend with a growth rate of 0.021/10a and a multi-year mean value of 0.274. Among them, the mean values of EVI in 2000, 2001, 2003–2007, and 2009–2011 were lower than that of the multi-year mean level. The mean values of EVI in 2002, 2008, and 2012–2015 were higher than the multi-year mean value. The minimum EVI (0.251) appeared in 2001, and the maximum EVI (0.301) appeared in 2012.

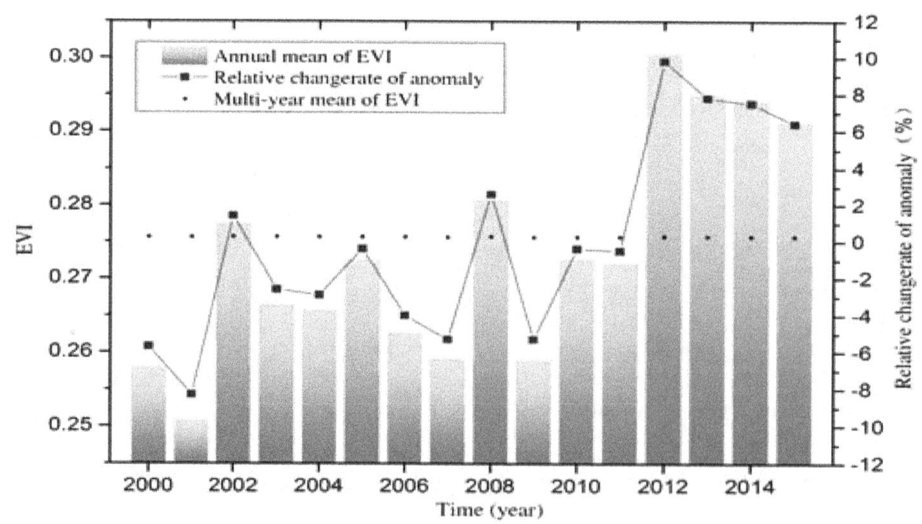

Figure 3. Inter-annual change of EVI in the study area from 2000 to 2015.

3.1.2. Features of Spatial Distribution of Multi-Year Mean EVI

The elevation of the study area had a characteristic of a lower height in the east and the north, and a higher height in the west and the south. The sensitivity of the vegetation to the climate gradient was strong, resulting in an uneven spatial distribution of the ecological pattern. On the whole, the EVI in the growth season showed a distribution pattern decreasing from the north to the south with a change rate of 0.28/10°N and from the east to the west with a rate of 0.22/10°E (Figure 4). The statistical results showed that areas with high EVI value in the growing season were mainly located in the areas of Hulunbuir, Hinggan League, Tongliao, and Chifeng, with an elevation below 1250 m, between 44°–50°N and 115°–125°E (Figure 4). On the other hand, areas with low EVI value are mainly located in Xilingol League, Ulanqab League, Baotou, Bayannur League, and Erdos. The overall EVI was lower in areas where the elevation was above 2300 m, and the vegetation types in these areas were mainly meadow steppe and desert steppe.

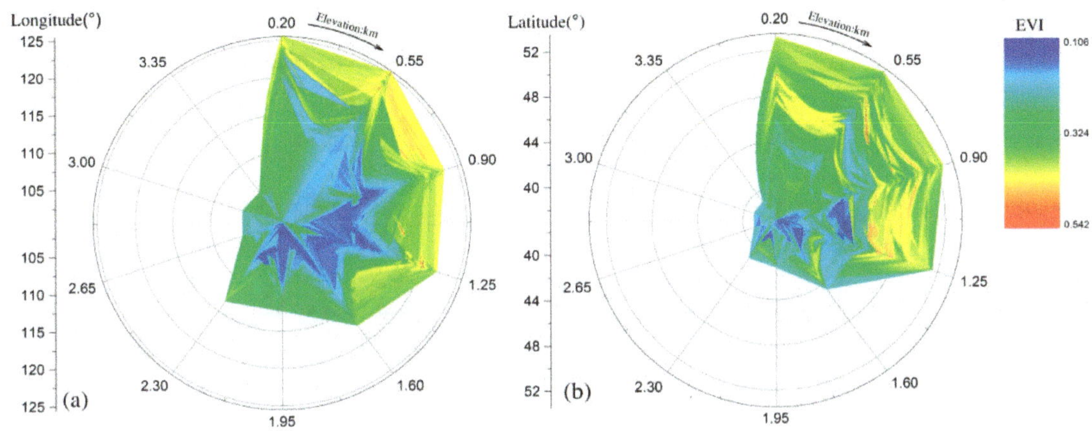

Figure 4. Spatial distribution of multi-year mean of EVI in the growing season in the study area from 2000 to 2015. (**a**) Longitude and altitude statistics; (**b**) Latitude and altitude statistics.

3.2. Temporal and Spatial Variation of EVI in the Growing Season

With the trend analytical method, the trend of EVI in the growing season of the study area for 16 years was drawn (Figure 5a). With the standard deviation classification method, the change trend of EVI in the study area can be divided into seven types, significant degradation, moderate degradation, slight degradation, basically unchanged, slight improvement, moderate improvement, and significant improvement (Figure 5b). On the whole, the EVI in the growing season in IMAR was basically unchanged, slightly improved, and slightly degraded. Among them, the basically unchanged areas were mainly in northern, central, and southeastern part of the study area, accounting for 42.5% of the area; and the vegetation types were mainly evergreen coniferous forests, deciduous coniferous forests, deciduous broadleaf forests, coniferous and broadleaf mixed forests, typical steppe and swamps. The areas with slight improvement were mainly in northeastern, west-central part of the study area, accounting for 21.1% of the total area; and the vegetation types were typical steppe and desert steppe. The areas where it displayed a slight degradation were mainly in the northern and southeastern part of IMAR, accounting for 24.6% of the total area; and the vegetation types were mainly typical steppe, desert steppe, deciduous coniferous forests and deciduous broadleaf forests. The areas with significant improvement mainly included Hinggan League, Hohhot, northern Xilingol League, western Hulunbuir, and southern Tongliao, accounting for 1.6% of the total area; and the vegetation types are shrubs, meadow steppe, deciduous broadleaf forests, and mixed coniferous and broadleaf forests. The areas with moderate improvement were mainly in Chifeng, northeastern Hulunbuir, central Xilingol, and southern Erdos, accounting for 5.9% of the total area; and the vegetation types included typical steppe, desert steppe, shrubs, and meadow grasslands. The areas with moderate degradation were mainly in Baotou, Ulanqab League, southwestern Xilingol League and eastern Hulunbuir, accounting for 3.5% of the total area; and the types of vegetation were swamps and typical steppe. The areas where there was a significant degradation were mainly located in central Hulunbuir, southeast of Xilingol League, and southeast of Ulanqab League, only accounting for 0.8%, a small part of the total area, and the vegetation types are shrubs and typical steppe. The vegetation types that showed a trend of improvement in the study area were mainly the typical steppe in the north-central part and the desert steppe in the southern part. The vegetation types that showed a trend of degradation were mainly desert steppe in the south and coniferous and broadleaf mixed forest in the north.

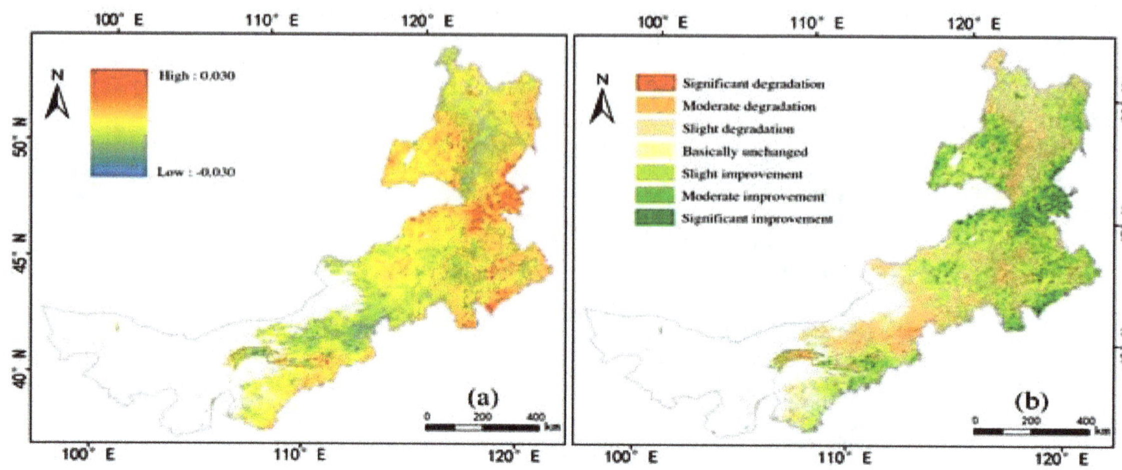

Figure 5. EVI trend (**a**) and EVI change (**b**) in the study area from 2000 to 2015.

3.3. Time Lag Analysis of EVI in the Growing Season to Climatic Factors

Previous studies have shown that the response of vegetation index to climatic factors show significant differences in both time and space [49,50]. And different climatic factors have different impact on vegetation with certain degree of time lag effects [32,51,52]. The GRGs are computed using the EVI in the growing season (from May to September) with different climatic factors (mean air temperature, mean relative humidity, and accumulated precipitation) in different time series (from March to July, from April to August and from May to September), resulting in the largest GRGs between EVI and climatic factors being air temperature (from April to August), relative humidity, (from March to July), and precipitation (from April to August), with GRG mean values of 0.649, 0.609, and 0.623 respectively (Figure 6). It showed that the response of growing season EVI to air temperature and relative humidity lagged by 1–2 months, and the response of growing season EVI to precipitation lagged by one month. In other words, the vegetation in the growing season of the study area responded more quickly to precipitation, and more slowly to air temperature and relative humidity.

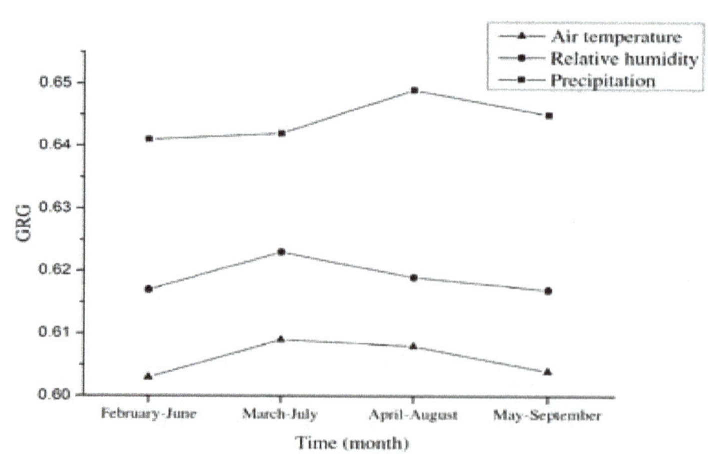

Figure 6. The GRGs of growing season EVI to air temperature, relative humidity and precipitation in February–June, March–July, April–August, and May–September.

3.4. GRAs between EVI and Climatic Factors

3.4.1. Features of Inter-Annual Change of Climatic Factors

According to the inter-annual trends of the mean air temperature, mean relative humidity, and cumulative precipitation in the growing season of the study area from 2000 to 2015 (Figure 7), the mean air temperature reached a maximum of 11.47 °C in 2008, then gradually decreased, and reached its minimum of 9.69 °C in 2010. The multi-year mean air temperature was 10.56 °C at a decline rate of 0.37 °C/10a and the linear trend was not obvious. The mean relative humidity reached a minimum of 46.77% in 2006 and then gradually increased, reaching a maximum of 53.8% in 2013, and the multi-year mean relative humidity was 49.94%. It generally showed an upward trend with a rate of 0.91%/10a, and the linear trend was not obvious. The minimum of cumulative precipitation (210.69 mm) occurred in 2007 after which there is a rising trend, reaching a maximum of 378.87 mm in 2013, and it showed a downward trend after 2013, which generally showed an upward trend. The multi-year mean cumulative precipitation was 267.43 mm, at a growth rate of 5.833 mm/a.

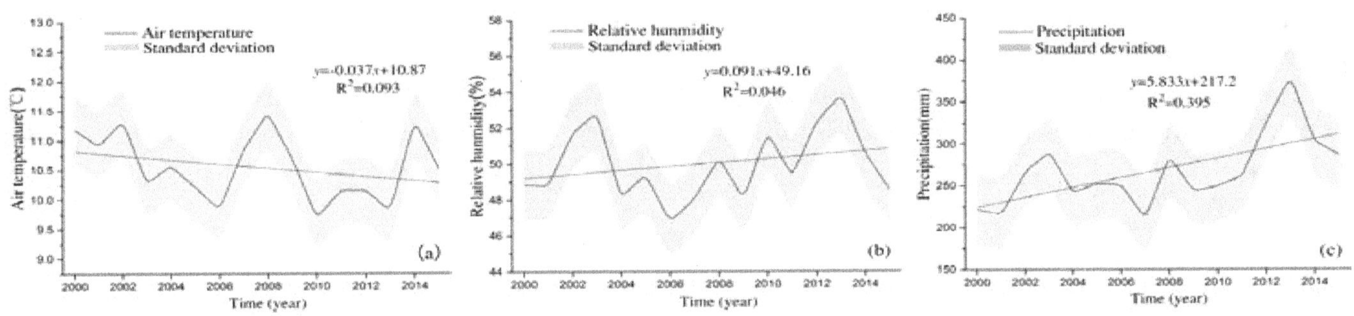

Figure 7. Inter-annual change of air temperature (**a**); relative humidity (**b**); and precipitation (**c**) in IMAR from 2000 to 2015.

3.4.2. GRAs between the Growing Season EVI and the Climatic Factors

The GRA results of the EVI with climatic factors (Figure 8a) show that the GRG between EVI and air temperature was 0.388–0.880 with a mean value of 0.609. The areas with high GRG were mainly located in Hulunbuir, western Xilingol League, and eastern Bayannur League, the elevation in the areas below 2000 m. However, the EVI value of the high GRG area was low, and the vegetation types were mainly swamps, meadow steppe, typical steppe, and desert steppe. While the low GRG areas were mainly distributed in the south-central part of Chifeng and southeastern Erdos, the elevation below 2000 m, the EVI value of the area was high, and the vegetation types were mainly evergreen broadleaf forests, deciduous broadleaf forests, swamps, and typical steppe. It indicated that air temperature had a greater influence on the vegetation in the low-EVI areas (steppe and meadow steppe, etc.) than that of the high EVI areas (coniferous forest, broadleaf forest, etc.).

The GRA results between EVI and relative humidity in the study area (Figure 8b) display that the GRG between EVI and relative humidity was 0.385–0.902, with a mean value of 0.623. In northern Hulunbuir and western Xilingol League with the elevation of 200–1000 m, the GRG was relatively high and the vegetation types were mainly evergreen coniferous forests, deciduous coniferous forests, deciduous broadleaf forests, coniferous and broadleaf mixed forests. In the southern Chifeng, eastern Tongliao and southern Hohhot, the elevation from 100 to 2500 m, the GRG was relatively low and the vegetation types were mainly typical steppe, desert steppe, and shrubs. It can be seen that relative humidity had a much greater impact on forest ecological areas than on steppe ecological areas.

According to Figure 8c, the GRA results between EVI and cumulative precipitation in the study area can be found that the GRG between EVI and cumulative precipitation was 0.398–0.893, and the mean value was 0.649. High GRG areas were mainly located in the northwestern Hulunbuir and southeastern Hinggan League, the elevation below 2000 m. Evergreen coniferous forests, deciduous coniferous forests, deciduous broadleaf forests, meadow steppe and swamps were the main vegetation types in the high GRG areas. Low GRG areas were mainly distributed in the southeastern Hulunbuir, northern Hinggan League and southern Chifeng with elevation of 300–1500 m. The vegetation types of the low GRG area were mainly swamps, typical steppe, desert steppe, and coniferous and broadleaf mixed forests. The results show that although the impact of precipitation on the forest ecological areas was slightly higher than that on the steppe ecological areas, precipitation and EVI showed a high correlation in the two major areas.

According to the GRA results, there was a distinct spatial difference in the degree of vegetation response to climatic factors in the study area. In forest ecological areas, the response of vegetation to precipitation and relative humidity was generally higher than that of air temperature, the relational correlations between vegetation and precipitation and between vegetation and relative humidity showed an obvious difference. In the steppe ecological areas, the response of vegetation to precipitation and air temperature was generally higher than that of relative humidity, but the GRGs of EVI and precipitation and EVI and air temperature were very close. Figure 8 shows that at certain air temperatures, the water from precipitation directly affected the surface of the vegetation, but the water from relative humidity existed in the air. Therefore, in the typical steppe and forest areas with moderate air humidity in the northern part of the study area, precipitation had a closer relationship with vegetation growth. On the contrary in the desert steppe areas with less precipitation in the western and central parts of the study area, relative humidity played a more dominant role in vegetation growth than precipitation.

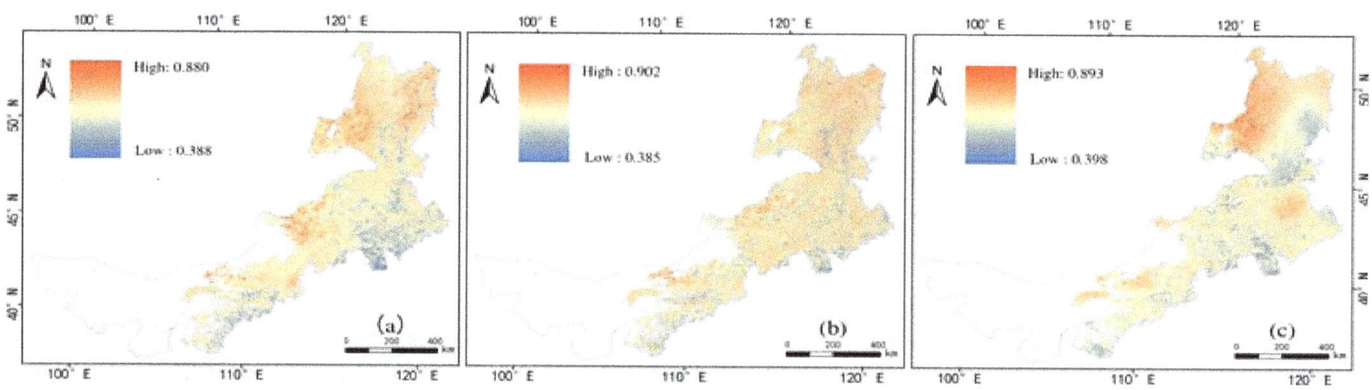

Figure 8. GRG between EVI and air temperature (**a**); relative humidity (**b**); and precipitation (**c**).

3.5. Zoning of EVI for Climatic Driving Forces

The growth of vegetation was closely related to the climate and the environment. Climate conditions, especial changes in air temperature, relative humidity, and precipitation, would have a great influence on the distribution and growth of vegetation [53]. In this paper, air temperature, relative humidity, and precipitation were chosen to zone different driving forces for EVI (Table 1).

The zoning results of driving force for EVI change in the study area (Figure 9) show that the zones where EVI change was driven by air temperature were mainly located in the north and scattered in the central and southern parts, including northeastern Hulunbuir and western Xilingol League with an elevation of 600–1400 m. The air temperature driving zones account for 8% of the total area and the vegetation types were mainly typical steppe; and including deciduous broadleaf forests, deciduous coniferous forests, and desert steppe. The zones where relative humidity was the main driving force for EVI change were mainly distributed in the northern part of the study area, and scattered in the central and southern parts, involving Xilingol League, Chifeng, Tongliao, Baotou, and southeastern Hulunbuir, with an elevation from 200 to 1800 m, being 11.6% of the total area. The vegetation types mainly included evergreen coniferous forests, deciduous coniferous forests, deciduous broadleaf forests, and coniferous and broadleaf mixed forests. In the northeastern, central and western part of IMAR, precipitation driving areas scattered in the northwest of Hulunbuir and southern part of Hinggan League and the northern part of Tongliao, where the elevation was 300–1500 m, accounting for 21.8% of the area. EVI there was driven by precipitation, and the vegetation types were mainly evergreen coniferous forests, deciduous coniferous forest and typical steppe. Others include typical steppe, swamps, and desert steppe.

Table 1. The zoning rules of driving forces for EVI change.

Driving Factors for EVI Change		Rules of Zoning		
		g_T [1]	g_P [2]	g_R [3]
Climatic Factors	T [4]	$g_T > 0.7$		
	P [5]		$g_P > 0.7$	
	R [6]			$g_R > 0.7$
	[T + P]+ [7]	$g_T > 0.7$	$g_P > 0.7$	
	[T + R]+ [8]	$g_T > 0.7$		$g_R > 0.7$
	[P + R]+ [9]		$g_P > 0.7$	$g_R > 0.7$
	[T + R + P]+ [10]	$g_T > 0.7$	$g_P > 0.7$	$g_R > 0.7$

[1] The GRG between EVI and air temperature; [2] The GRG between EVI and precipitation; [3] The GRG between EVI and relative humidity; [4] EVI change driven by air temperature; [5] EVI change driven by precipitation; [6] EVI change driven by relative humidity; [7] EVI change driven by both air temperature and precipitation; [8] EVI change driven by both air temperature and relative humidity; [9] EVI change driven by both precipitation and relative humidity; [10] EVI change driven by air temperature, relative humidity and precipitation.

The zones where EVI change was driven by both air temperature and precipitation were mainly distributed in the southeastern Hulunbuir with an elevation of 400–1000 m, accounting for 2.7% of the total area. The vegetation types were mainly meadow steppe, desert steppe, and typical steppe. The zones where air temperature and relative humidity co-driving EVI change were mainly located in the eastern Hulunbuir, the elevation from 500 to 1000 m, accounting for 3.1% of the total area. The vegetation types were mainly deciduous coniferous forests and deciduous broadleaf forests. The zones where precipitation and relative humidity co-driving EVI change were mainly located in the northern Hulunbuir, the elevation from 600 to 1100 m, accounting for 3.8% of the total area. The vegetation types mainly include evergreen coniferous forests, deciduous broadleaf forests. and meadow steppe. The zones where EVI change was driven by air temperature, precipitation and relative humidity were mainly distributed in the northeastern Hulunbuir, the elevation from 300 to 800 m, accounting for 0.93% of the total area. The vegetation types were mainly evergreen coniferous forests, deciduous broadleaf forests, and typical steppe. The weak driving force of climatic factors made little contribution to the study, so the authors will not further analyze it.

The vegetation in different zones relied on water, heat, or both to different degrees. The typical steppe, coniferous forests, and cultivated plants in east-central and northwestern parts of the study area clearly showed that water was a dominant condition for vegetation growth. However, in the typical steppe area where there was a remarkable joint drive of air temperature and precipitation, it indicated that the growth of typical steppe was more dependent on the collective effect of water and heat. In the swamps, broadleaf forests, and desert steppe of the west-central part, as well as southeastern region of the northern part of the study area, air temperature and relative humidity were the main driving forces, indicating that these zones were more dependent on air temperature and humidity than on water. In the coniferous and broadleaf mixed forests of the northeastern part of the study area, where EVI was driven by air temperature, relative humidity and precipitation, it indicated that the vegetation had greater demand for air temperature, relative humidity, and precipitation.

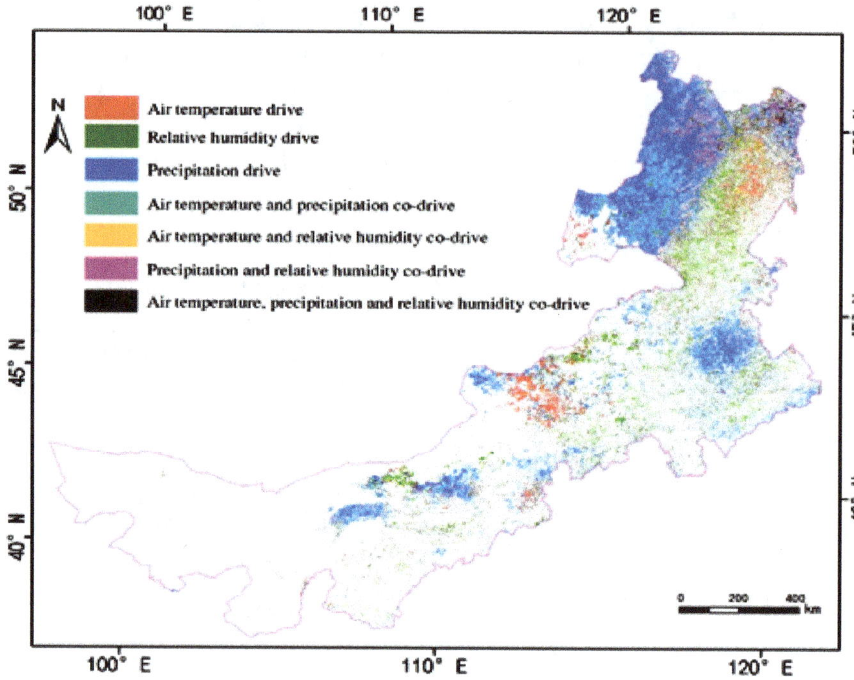

Figure 9. Areas of precipitation drive, air temperature drive, and relative humidity drive in the study area.

4. Discussion

Long et al. [10] used the normalized difference vegetation index (NDVI) to study the response of vegetation to climatic factors in IMAR from 1982 to 2006 and found that different vegetation types had different responses to climatic factors and different time lag, but only air temperature and precipitation were taken into consideration. Based on the MODIS data, Mu, et al. [11] studied the response of vegetation cover to air temperature and precipitation in IMAR from 2001 to 2010. It was suggested that the relationship between vegetation cover and precipitation in IMAR was closer, and there was a certain lag in the response of vegetation growth and climatic factors but the exact time lag was not found. The NDVI data of vegetation was used to study the lagged response of precipitation in Xilingol League of IMAR from 2001 to 2007. Liu et al. [52] believed that the vegetation growth of Xilingol League in the growing season has an obvious lag to precipitation, and the time lag was about 50–60 days. Different types of steppe have different time lag. Li et al. [32] used the field sampling data and climate data from the Inner Mongolia Steppe Ecosystem Research Station of the Chinese Academy of Sciences to study the response of Carex korshinskyi to climatic factors in 2014 and found that the biomass of Carex korshinskyi lagged behind air temperature by 16 days and the relative humidity by 15 days.

The zoning of driving force in this paper shows that precipitation driving areas (21.8%) in IMAR were much larger than the air temperature driving areas (8%) and relative humidity driving areas (11.6%), which was basically consistent with the view of Mu et al. [11]. Since this paper was based on the result of time lag analysis, the response of the forest ecological areas to precipitation and relative humidity was generally higher than that of air temperature in the spatial distribution of different zones, and the response of the steppe ecological areas to precipitation and air temperature was generally higher than that of relative humidity. Regarding the time lag of EVI in the growing season, we draw a conclusion that the time lag of EVI in IMAR to different climatic factors was about 1–2 months for air temperature, 1–2 months for relative humidity, and one month for precipitation. On one hand, data sources (especially time resolution), the selection of analysis index, and experimental errors may cause differences in results. On the other hand, the spatial resolution of remote sensing data, the size of interpolation grid for climatic factors, and the changes in ecological complexity may also lead to large differences in the response and time lag of the EVI to climatic factors in IMAR [11,32,52].

This paper mainly discusses the relationship between changes of IMAR vegetation EVI and climatic factors (air temperature, relative humidity, and precipitation). In addition, there was also a certain correlation between vegetation EVI and extreme weather [15]. In the desert biota of IMAR, other climate factors like wind speed also had a great impact on local biomass [54]. Changes in vegetation EVI were not only closely related to climate factors but also related to human activities and other factors [15,55,56]. Before the 21st century, IMAR had a declining trend in overall vegetation growth under the combined effects of climate change, social economy, and local policy changes [57,58]. Since the beginning of the 21st century, China has implemented a series of ecological restoration projects, such as harnessing wind and sand in the Beijing-Tianjin region, returning farmland to forests and grasslands, returning grazing land to grassland, enclosing and transferring, and IMAR is one of the key implementation areas for these projects [11,59]. At the beginning of the study period, IMAR has begun implementing state-owned afforestation, encouraging artificial afforestation, returning farmland to forests and grasslands, closing mountains for afforestation, and new closures of non-forested land and open forest land, which has resulted in effective protection and improvement of forest lands and grasslands, especially forest lands [11,26,52]. In semi-arid areas, human management of ecological zones has become a major driving force of vegetation change [54]. Similar to many studies, this study find that the vegetation in some dusty land and desert steppe (such as SonidLeft Banner and Etuoke Banner) in IMAR remained a good growing trend from 2000 to 2015, and the implementation of ecological restoration measures (such as planting tree and grass, prohibiting grazing, preventing and governing the sand) played an important role in this process [10,11,59]. According to the IMAR Statistical Yearbook [60], from 2000 to 2002, the crop area affected by natural disasters in the IMAR reduced from 48,000 km^2 to 32,000 km^2, the total number of grazing livestock decreased from 73.01 million to 63.27 million, the area of arable land reduced from 73,000 km^2 to 69,000 km^2, the area of artificial afforestation increased from 5900 km^2 to 9100 km^2, and the population decreased from 23.77 million to 22.79 million. The EVI index increased significantly from 0.251 in 2001 to 0.278 in 2002 under the combined effect of reduced intensity in natural disasters, grazing and human activity, in addition to greater intensity in returning farmland to forests and grasslands, and the impact of climate change. Subsequently, from 2002 to 2007, the area of artificial afforestation decreased from 9100 km^2 to 5900 km^2, the total population increased from 22.79 million to 24.05 million, the total number of livestock increased from 63.27 million to a maximum of 100.04 million, resulting in an increase of grazing intensity. The superposition effect resulted in a slow decline in the EVI index between 2002 and 2007, with the lowest mean value of 0.259. From 2007 to 2008, the total number of livestock decreased by 5% and the area of artificial afforestation increased to 7200 km^2, with EVI reaching a peak mean value of 0.281 in 2008. From 2008 to 2011, the total population, the crop area affected by natural disasters and the total number of livestock increased slowly, while the area of artificial afforestation remains basically unchanged, therefore, the EVI of IMAR vegetation slowly declined from 2008 to 2011. In 2012, the area of artificial afforestation increased to 7800 km^2, and the crop area affected by natural disasters fell to 26,000 km^2. Therefore, the EVI of IMAR increased significantly by 10.44%, reaching the highest mean value of 0.301 during the study period. After 2012, the area of artificial afforestation remained basically unchanged, but the total number of livestock increased by 20%. The crop area affected by natural disasters increased significantly from 26,000 km^2 in 2012 to 101,000 km^2 in 2015. This, together with the impact of climate change, led to a declining trend of EVI in IMAR from 2012 to 2015. In general, from 2000 to 2015, the total population increased by 6.3%, the total number of livestock increased by 65.7%, so the increase in the intensity of human activities year by year will cause the growth rate of vegetation in IMAR to slow down. Human disturbance at different time periods has different effects on vegetation growth, and such positive or negative influences will increase the inter-annualecological complexity of IMAR, and change the response sensitivity of vegetation EVI to climate factors. Eventually, it led to differences in the response degree and response speed of EVI to air temperature, relative humidity, and precipitation [11].

5. Conclusions

Based on the MOD13Q1 EVI remote sensing data, vegetation type data, topographical data, and three climatic data (air temperature, relative humidity, and precipitation) of different time series, the temporal and spatial distribution patterns and changing trends of the EVI in growing season were analyzed in IMAR from 2000 to 2015. Using the method of GRA, we studied the time lag of the EVI responding to climatic factors during the 16 years and finally zone the EVI driving areas according to the different climatic factors on the basis of time lag analysis. The main conclusions were as follows:

(1) The mean EVI value in the growing season in IMAR from 2000 to 2015 was 0.274. The spatial distribution was significantly different. The EVI value generally showed a spatial distribution of increase from west to east and from south to north. The rate of change from west to east was 0.22/10°E, and that from south to north is 0.28/10°N.

(2) During 2000–2015, the overall EVI in the growing season in IMAR showed a slight increasing trend, with a growth rate of 0.021/10a. The areas with slight and significant improvement during the study period accounted for 21.1% and 7.5% of the total study area. The areas with slight and significant degradation accounted for 24.6% and 4.3% of the total study area.

(3) The results of time lag analysis show that the response time of EVI in IMAR to the three climatic factors (air temperature, relative humidity and precipitation) was different. The EVI lagged behind air temperature by 1–2 months, relative humidity by 1–2 months, and precipitation by one month.

(4) The precipitation driving areas (21.8%) in IMAR were much larger than air temperature driving ones (8%) and the relative humidity driving ones (11.6%). The EVI in the study area had the closest relationship with precipitation, followed by the relative humidity, and then air temperature. However, the growth of vegetation did not depend on the change of a single climate factor, but was the result of the collective effect of multiple climatic factors and human activities.

Author Contributions: D.H., G.Y., and T.Z. jointly designed the study and wrote the thesis. J.M., J.L., and X.B. all participated in the process of discussion, editing, and revision.

Acknowledgments: The author thanks anonymous reviewers for providing invaluable comments on the original manuscript.

References

1. Guo, L.; Wu, S.; Zhao, D.; Yin, Y.; Leng, G.; Zhang, Q. NDVI-Based vegetation change in Inner Mongolia from 1982 to 2006 and its relationship to climate at the biome scale. *Adv. Meteorol.* **2014**, *4*, 79–92. [CrossRef]
2. Xu, G.; Zhang, H.; Chen, B.; Zhang, H.; Innes, J.; Wang, G.; Yan, J.; Zheng, Y.; Zhu, Z.; Mvneni, R. Changes in Vegetation Growth Dynamics and Relations with Climate over China's Landmass from 1982 to 2011. *Remote Sens.* **2014**, *6*, 3263–3283. [CrossRef]
3. Hantson, S.; Knorr, W.; Schurgers, G.; Pugh, T.A.M.; Arneth, A. Global isoprene and monoterpene emissions under changing climate, vegetation, CO_2, and land use. *Atmos. Environ.* **2017**, *155*, 35–45. [CrossRef]
4. Xin, Z.; Xu, J.; Zheng, W. Spatiotemporal variations of vegetation cover on the Chinese Loess Plateau (1981–2006): Impacts of climate changes and human activities. *Sci. China Ser. D* **2008**, *51*, 67–78. [CrossRef]
5. Stralberg, D.; Bayne, E.M.; Cumming, S.G.; Solymos, P.; Song, S.J.; Schmiegelow, F.K.A. Conservation of future boreal forest bird communities considering lags in vegetation response to climate change: A modified refugia approach. *Divers. Distrib.* **2015**, *21*, 1112–1128. [CrossRef]
6. Keersmaecker, W.D.; Lhermitte, S.; Hill, M.; Tits, L.; Coppin, P.; Somers, B. Assessment of regional vegetation

response to climate anomalies: A case study for australia using GIMMS NDVI time series between 1982 and 2006. *Remote Sens.* **2017**, *9*, 34. [CrossRef]

7. Sarkar, S.; Kafatos, M. Interannual variability of vegetation over the Indian sub-continent and its relation to the different meteorological parameters. *Remote Sens. Environ.* **2004**, *90*, 268–280. [CrossRef]
8. Peters, A.J. A spatial regression procedure for evaluating the relationship between AVHRR-NDVI and climate in the northern Great Plains. *Int. J. Remote Sens.* **2004**, *25*, 297–311. [CrossRef]
9. Cao, R.; Chen, J.; Shen, M.; Tang, Y. An improved logistic method for detecting spring vegetation phenology in grasslands from MODIS EVI time-series data. *Agric. For. Meteorol.* **2015**, *200*, 9–20. [CrossRef]
10. Long, H.; Li, X.; Bao, Y.; Huang, L.; Li, Z. Time lag analysis between vegetation and climate change in Inner Mongolia. In Proceedings of the IEEE International Geoscience and Remote Sensing Symposium, Honolulu, HI, USA, 25–30 July 2010; pp. 1513–1516.
11. Mu, S.; Li, J.; Chen, Y.; Gang, C.; Zhou, W.; Ju, W. Spatial differences of variations of vegetation coverage in Inner Mongolia during 2001–2010. *Acta Geogr. Sin.* **2012**, *67*, 1255–1268. [CrossRef]
12. Ren, S.; Yi, S.; Peichl, M.; Wang, X. Diverse responses of vegetation phenology to climate change in different grasslands in Inner Mongolia during 2000–2016. *Remote Sens.* **2017**, *10*, 17. [CrossRef]
13. Piao, S.; Fang, J.; He, J. Variations in vegetation net primary production in the Qinghai-Xizang Plateau, China, from 1982 to 1999. *Clim. Chang.* **2006**, *74*, 253–267. [CrossRef]
14. Tong, S.; Zhang, J.; Si, H.; Lai, Q.; Ma, Q. Dynamics of Fractional Vegetation Coverage and Its Relationship with Climate and Human Activities in Inner Mongolia, China. *Remote Sens.* **2016**, *8*, 776. [CrossRef]
15. Li, C.L.; Wang, J.; Hu, R.C.; Yin, S.; Bao, Y.H.; Ayal, D.Y. Relationship between vegetation change and extreme climate indices on the Inner Mongolia Plateau, China, from 1982–2013. *Ecol. Indic.* **2018**, *89*, 101–109. [CrossRef]
16. Chen, L.Y.; Li, H.; Zhang, P.J.; Zhao, X.; Zhou, L.H.; Liu, T.Y.; Hu, H.F.; Bai, Y.F.; Shen, H.H.; Fang, J.Y. Climate and native grassland vegetation as drivers of the community structures of shrub-encroached grasslands in Inner Mongolia, China. *Landsc. Ecol.* **2015**, *30*, 1627–1641. [CrossRef]
17. Bai, Y.; Wu, J.; Xing, Q.; Pan, Q.; Huang, J.; Yang, D.; Han, X. Primary production and rain use efficiency across a precipitation gradient on the Mongolia Plateau. *Ecology* **2008**, *89*, 2140–2153. [CrossRef] [PubMed]
18. Anyamba, A.; Tucker, C.J. Analysis of Sahelian vegetation dynamics using NOAA-AVHRR NDVI data from 1981–2003. *J. Arid Environ.* **2005**, *63*, 596–614. [CrossRef]
19. Camberlin, P.; Martiny, N.; Philippon, N.; Richard, Y. Determinants of the interannual relationships between remote sensed photosynthetic activity and rainfall in tropical Africa. *Remote Sens. Environ.* **2007**, *106*, 199–216. [CrossRef]
20. Wardlow, B.D.; Egbert, S.L. Large-area crop mapping using time-series MODIS 250m NDVI data: An assessment for the U.S. Central Great Plains. *Remote Sens. Environ.* **2008**, *112*, 1096–1116. [CrossRef]
21. Piao, S.; Wang, X.; Ciais, P.; Zhu, B.; Wang, T.; Liu, J. Changes in satellite-derived vegetation growth trend in temperate and boreal Eurasia from 1982 to 2006. *Glob. Chang. Biol.* **2011**, *17*, 3228–3239. [CrossRef]
22. Karlsen, S.R.; Tolvanen, A.; Kubin, E.; Poikolainen, J.; Hogda, K.A.; Johansen, B.; Danks, F.S.; Aspholm, P.; Wielgolaski, F.E.; Makarova, O. MODIS-NDVI-based mapping of the length of the growing season in northern Fennoscandia. *Int. J. Appl. Earth Obs.* **2008**, *10*, 253–266. [CrossRef]
23. Raynolds, M.K.; Comiso, J.C.; Waiker, D.A.; Verbyla, D. Relationship between satellite-derived land surface temperatures, arctic vegetation types, and NDVI. *Remote Sens. Environ.* **2008**, *112*, 1884–1894. [CrossRef]
24. Wang, X.; Piao, S.; Ciais, P.; Li, J.; Friedlingstein, P.; Koven, C.; Chen, A. Spring temperature change and its implication in the change of vegetation growth in North America from 1982 to 2006. *Proc. Natl. Acad. Sci. USA* **2011**, *108*, 1240–1245. [CrossRef] [PubMed]
25. Zhao, X.; Tan, K.; Zhao, S.; Fang, J. Changing climate affects vegetation growth in the arid region of the northwestern China. *J. Arid Environ.* **2011**, *75*, 946–952. [CrossRef]
26. Dan, S.; Li, H.; Ping, L.; De, X. Effects of Climate Change on Vegetation in Desert Steppe Inner Mongolia. *Nat. Resour.* **2013**, *4*, 319–322. [CrossRef]
27. Qu, H.; Wang, C.J.; Zhang, Z.X. Planning priority conservation areas under climate change for six plant species with extremely small populations in China. *Nat. Conserv.* **2018**, *25*, 89–106. [CrossRef]
28. Hu, M.Q.; Mao, F.; Sun, H.; Hou, Y.Y. Study of normalized difference vegetation index variation and its

correlation with climate factors in the three-river-source region. *Int. J. Appl. Earth Obs.* **2011**, *13*, 24–33. [CrossRef]

29. Lin, Y.; Xin, X.; Zhang, H.; Wang, X. The implications of serial correlation and time-lag effects for the impact study of climate change on vegetation dynamics—A case study with Hulunber meadow steppe, Inner Mongolia. *Int. J. Remote Sens.* **2015**, *36*, 5031–5044. [CrossRef]
30. Zhang, Y.; Zhou, W. Correlation analysis between vegetation fraction and vegetation indices in reclaimed forest: A case study in Pingshuo mining area. In Proceedings of the IEEE International Workshop on Earth Observation and Remote Sensing Applications, Guangzhou, China, 4–6 July 2016; pp. 122–126.
31. Ballantyne, M.; Treby, D.L.; Quarmby, J.; Pickering, C.M. Comparing the impacts of different types of recreational trails on grey box grassy-woodland vegetation: Lessons for conservation and management. *Aust. J. Bot.* **2016**, *64*, 246–259. [CrossRef]
32. Li, J.; Zhang, C.Q.; Zhang, J. The Lag Response of the Growth Dynamics of Dominant Grasses to Meteorological Factors in Typical Steppe of Inner Mongolia. *Acta Agrestia Sin.* **2017**, *25*, 267–272. [CrossRef]
33. Klinge, M.; Dulamsuren, C.; Erasmi, S.; Karger, D.N.; Hauck, M. Climate effects on vegetation vitality at the treeline of boreal forests of Mongolia. *Biogeosciences* **2018**, *15*, 1–25. [CrossRef]
34. Jin, X.; Xu, X. Rmote sensing of leaf water content for winter wheat using grey relational analysis (GRA), stepwise regression method (SRM) and partial least squares (PLS). In Proceedings of the IEEE First International Conference on Agro-Geoinformatics, Shanghai, China, 2–4 August 2012; pp. 1–5.
35. Liu, D.; Yue, L.; Wang, T.; Peylin, P.; Macbean, N.; Ciais, P.; Jia, G.S.; Ma, M.G.; Ma, Y.M.; Shen, M.G.; et al. Contrasting responses of grassland water and carbon exchanges to climate change between Tibetan Plateau and Inner Mongolia. *Agric. For. Meteorol.* **2018**, *249*, 163–175. [CrossRef]
36. Dong, Z.Q.; Pan, Z.H.; He, Q.J.; Wang, J.L.; Huang, L.; Pan, Y.Y.; Han, G.L.; Xue, X.P.; Chen, Y.C. Vulnerability assessment of spring wheat production to climate change in the Inner Mongolia region of China. *Ecol. Indic.* **2018**, *85*, 67–78. [CrossRef]
37. Malingreau, J.P. Global vegetation dynamics: Satellite observations over Asia. *Int. J. Remote Sens.* **1986**, *7*, 1121–1146. [CrossRef]
38. Roerink, G.J.; Menenti, M.; Verhoef, W. Reconstructing cloudfree NDVI composites using Fourier analysis of time series. *Int. J. Remote Sens.* **2000**, *21*, 1911–1917. [CrossRef]
39. Xu, Z.X.; Wang, D.J.; Gao, J. Homogeneity Test on Temperature Series in Liaoning Province. *J. Anhui Agric. Sci.* **2010**, *27*, 15149–15151. [CrossRef]
40. Zhang, X.; Shao, J.; Luo, H. Spatial interpolation of air temperature with ANUSPLIN in Three Gorges Reservoir Area. In Proceedings of the IEEE International Conference on Remote Sensing, Environment and Transportation Engineering, Nanjing, China, 24–26 June 2011; pp. 3465–3468.
41. Yan, L.; Zhou, G.S.; Wang, Y.H.; Hu, T.Y.; Sui, X.H. The spatial and temporal dynamics of carbon budget in the alpine grasslands on the Qinghai-Tibetan Plateau using the Terrestrial Ecosystem Model. *J. Clean Prod.* **2015**, *107*, 195–201. [CrossRef]
42. Way, R.G.; Lewkowicz, A.G.; Bonnaventure, P.P. Development of moderate-resolution gridded monthly air temperature and degree-day maps for the Labrador-Ungava region of northern Canada. *Int. J. Climatol.* **2017**, *37*, 493–508. [CrossRef]
43. Xu, Y.; Yang, J.; Chen, Y. NDVI-based vegetation responses to climate change in an arid area of China. *Theor. Appl. Climatol.* **2016**, *126*, 213–222. [CrossRef]
44. Testa, S.; Soudani, K.; Boschetti, L.; Mondino, E.B. MODIS-derived EVI, NDVI and WDRVI time series to estimate phenological metrics in French deciduous forests. *Int. J. Appl. Earth Obs.* **2018**, *64*, 132–144. [CrossRef]
45. Chan, J.W.K. Product end-of-life options selection: Grey relational analysis approach. *Int. J. Prod. Res.* **2008**, *46*, 2889–2912. [CrossRef]
46. Shen, C.; Wang, Y.; Wei, Y.; Yu, L. A lag analysis of R&D investment driving economic growth using grey relational model. In Proceedings of the IEEE International Conference on Environmental Science and Information Application Technology, Wuhan, China, 17–18 July 2010; pp. 61–64.
47. Yi, G.H.; Deng, W.; Li, A.N.; Zhang, T.B. Response of Lakes to Climate Change in Xainza Basin Tibetan Plateau Using Multi-Mission Satellite Data from 1976 to 2008. *J. Mt. Sci. Engl.* **2015**, *12*, 604–613. [CrossRef]
48. Yi, G.H.; Zhang, T.B. Delayed Response of Lake Area Change to Climate Change in Siling Co Lake, Tibetan Plateau, from 2003 to 2013. *Int. J. Environ. Res. Public Health* **2015**, *12*, 13886–13900. [CrossRef] [PubMed]

49. Lei, J.; Peters, A.J. Assessing vegetation response to drought in the northern Great Plains using vegetation and drought indices. *Remote Sens. Environ.* **2003**, *87*, 85–98. [CrossRef]
50. Nezlin, N.P.; Kostianoy, A.G.; Li, B.L. Inter-annual variability and interaction of remote-sensed vegetation index and atmospheric precipitation in the Aral Sea region. *J. Arid Environ.* **2005**, *62*, 677–700. [CrossRef]
51. Goward, S.N.; Prince, S.D. Transient effects of climate on vegetation dynamics: Satellite observations. *J. Biogeogr.* **1995**, *22*, 549–564. [CrossRef]
52. Liu, C.L.; Fan, R.H.; Wu, J.J.; Yan, F. Temporal lag of grassland vegetation growth reponse to precipitation in xilinguolemeng. *Arid Land Geogr.* **2009**, *32*, 512–518. [CrossRef]
53. Pyankov, V.I.; Black, C.C. C4 plants in the vegetation of Mongolia: Their natural occurrence and geographical distribution in relation to climate. *Oecologia* **2000**, *123*, 15–31. [CrossRef] [PubMed]
54. Xu, L.; Tu, Z.; Zhou, Y.; Yu, G. Profiling Human-Induced Vegetation Change in the Horqin Sandy Land of China Using Time Series Datasets. *Sustainability* **2018**, *10*, 1068. [CrossRef]
55. Greene, R.S.B.; Nettleton, W.D.; Chartres, C.J.; Leys, J. Runoff and micromorphological properties of a grazed haplargid, near Cobar, NSW, Australia. *Soil Res.* **1998**, *36*, 87–108. [CrossRef]
56. Schuman, G.E.; Reeder, J.D.; Manley, J.T.; Hart, R.H.; Manley, W.A. Impact of grazing management on the carbon and nitrogen balance of a mixed-grass rangeland. *Ecol Appl.* **1999**, *9*, 65–71. [CrossRef]
57. John, R.; Chen, J.; Lu, N.; Wilske, B. Land cover/land use change in semi-arid Inner Mongolia: 1992–2004. *Environ. Res. Lett.* **2009**, *4*, 4–13. [CrossRef]
58. Yin, H. Understanding Land Use And Land Cover Change In Inner Mongolia Using Remote Sensing Time Series. Ph.D. Thesis, Humboldt-Universitat zu Berlin, Berlin, Germany, 2014.
59. Yin, H.; Pflugmacher, D.; Li, A.; Li, Z.; Hostert, P. Land use and land cover change in Inner Mongolia-understanding the effects of China's re-vegetation programs. *Remote Sens. Environ.* **2018**, *204* 918–930. [CrossRef]
60. Bureau of stati of Inner Mongolia Autonomous Region. Available online: http://www.nmgtj.gov.cn/ (accessed on 15 June 2018).

8

Quantifying Snow Albedo Radiative Forcing and its Feedback during 2003–2016

Lin Xiao [1,2], Tao Che [1,3,*], Linling Chen [4], Hongjie Xie [5] and Liyun Dai [1,6]

1 Key Laboratory of Remote Sensing of Gansu Province, Heihe Remote Sensing Experimental Research Station, Northwest Institute of Eco-Environment and Resources, Chinese Academy of Sciences, Lanzhou 730000, China; xiaolin@lzb.ac.cn (L.X.); dailiyun@lzb.ac.cn (L.D.)
2 University of Chinese Academy of Sciences, Beijing 100049, China
3 Center for Excellence in Tibetan Plateau Earth Sciences, Chinese Academy of Sciences, Beijing 100101, China
4 Nansen Environmental and Remote Sensing Center, N-5006 Bergen, Norway; linling.chen@nersc.no
5 Laboratory for Remote Sensing and Geoinformatics, Department of Geological Sciences, The University of Texas at San Antonio, One UTSA Circle, San Antonio, TX 78249, USA; Hongjie.Xie@utsa.edu
6 Jiangsu Center of Collaborative Innovation in Geographical Information Resource Development and Application, Nanjing 210023, China
* Correspondence: chetao@lzb.ac.cn

Academic Editors: Yuei-An Liou, Jean-Pierre Barriot, Chung-Ru Ho, Yuriy Kuleshov, Chyi-Tyi Lee, Richard Müller and Prasad S. Thenkabail

Abstract: Snow albedo feedback is one of the most crucial feedback processes that control equilibrium climate sensitivity, which is a central parameter for better prediction of future climate change. However, persistent large discrepancies and uncertainties are found in snow albedo feedback estimations. Remotely sensed snow cover products, atmospheric reanalysis data and radiative kernel data are used in this study to quantify snow albedo radiative forcing and its feedback on both hemispheric and global scales during 2003–2016. The strongest snow albedo radiative forcing is located north of 30°N, apart from Antarctica. In general, it has large monthly variation and peaks in spring. Snow albedo feedback is estimated to be 0.18 ± 0.08 W·m^{-2}·°C^{-1} and 0.04 ± 0.02 W·m^{-2}·°C^{-1} on hemispheric and global scales, respectively. Compared to previous studies, this paper focuses specifically on quantifying snow albedo feedback and demonstrates three improvements: (1) used high spatial and temporal resolution satellite-based snow cover data to determine the areas of snow albedo radiative forcing and feedback; (2) provided detailed information for model parameterization by using the results from (1), together with accurate description of snow cover change and constrained snow albedo and snow-free albedo data; and (3) effectively reduced the uncertainty of snow albedo feedback and increased its confidence level through the block bootstrap test. Our results of snow albedo feedback agreed well with other partially observation-based studies and indicate that the 25 Coupled Model Intercomparison Project Phase 5 (CMIP5) models might have overestimated the snow albedo feedback, largely due to the overestimation of surface albedo change between snow-covered and snow-free surface in these models.

Keywords: snow albedo radiative forcing; snow albedo feedback; radiative kernel; remote sensing

1. Introduction

The globally averaged surface temperature has increased by 0.85 °C over the period of 1880–2012 [1], with particularly strong warming signals appearing at high northern latitudes [2–4]. This is known as the Arctic amplification [5–7]. One of the main contributions to the amplified warming is perhaps the positive surface albedo feedback [8–10], primarily snow and ice albedo feedback. The ice, specifically, refers to bare ice, melting ice (mainly includes sea ice, glaciers and ice caps), snow-covered ice, open water, etc. [1]. In the warming climate, the decreasing snow cover (ice) extent and snow (ice) depth are leading to a less reflective planet that absorbs more solar radiation, and thus warming the earth further [11,12].

Climate feedback variables are valuable indicators of climate change due to their sensitivity to temperature. Specifically, a central task of climate change research is to quantify the Equilibrium Climate Sensitivity (ECS, [13–15]), which refers to the equilibrium change in annual mean surface temperature with a doubling of the atmospheric equivalent carbon dioxide concentration [16–18]. However, no agreement has been reached on the magnitude of the ECS [19], and the large uncertainty in the ECS is primarily attributed to the inaccurate estimation of individual feedbacks [20–22].

Models are useful tools for climate feedback study for their capability of long-term simulation. Energy balance models were first used mainly for the recognition of feedback mechanisms and dynamic processes [23–25]. Later on, General Circulation Models (GCMs) were widely used to investigate whether variables act as positive or negative feedbacks and the magnitude of each feedback [26–28]. In recent development, the Atmosphere–Ocean General Circulation Models (AOGCMs) make it possible for more detailed and comprehensive estimation of individual feedbacks [17,29,30]. However, feedback estimations from model simulations are still associated with large uncertainties, i.e., fivefold intermodel difference on surface albedo feedback was reported in the Fourth Assessment Report (AR4) of Intergovernmental Panel on Climate Change (IPCC) models [31]. Such uncertainty remains in IPCC AR5 [1,32].

Observation-based research has also made its contribution to feedback assessment, with more realistic representation of the climate from observations. For example, Flanner et al. found that the Coupled Model Intercomparison Project Phase 3 (CMIP3) models underestimated the snow and ice albedo feedback in Northern Hemisphere, as compared with results based on satellite observations of the Extended Advanced Very High Resolution Radiometer Polar Pathfinder (APP) product and Moderate Resolution Imaging Spectroradiometer (MODIS) data [33]. Fletcher et al. showed that the modeled average snow albedo feedback in the CMIP3 models was slightly larger than satellite observations [34]. Dessler reported an agreement on both global average and spatial pattern between model results and individual feedbacks that were calculated based on reanalysis data of ERA-Interim [35] and Modern-Era Retrospective Analysis for Research and Applications (MERRA, [36]) [37]. However, problems also exist in observation-based results. For example, the coarse resolution of observational data would either temporally or spatially smooth the feedbacks, especially in snowmelt seasons [38,39].

Due to the fact that large differences and uncertainties remain in snow albedo feedback assessments, an effective way to improve its accuracy is to use observation-based results to constrain model assessments [1]. Nonetheless, most snow albedo feedback studies were based on model simulations, and only few on observations [40–42]. Moreover, most of these studies calculate the combining effects of snow albedo feedback and ice albedo feedback, namely surface albedo feedback [13,37,43]. As a result, it is impossible to separate the contribution of snow or ice albedo feedback, as well as their uncertainties.

Being motivated by these scientific findings and limitations, this study specifically focuses on the quantification of snow albedo feedback by remotely sensed snow cover products of MODIS (MOD10C1 and MYD10C1), atmospheric reanalysis data and radiative kernel data. The purpose is to examine the source of the differences between our result and partially observation-based results, as well as Coupled Model Intercomparison Project Phase 5 (CMIP5) model-based results, and to provide precise information of snow cover change and snow albedo radiative forcing for model parameterization.

Data and method are described in Section 2. Spatial and temporal variability of snow cover, snow albedo radiative forcing and its feedback are calculated and analyzed in Section 3. Comparison and discrepancy analysis with both partially observational-based and model-based studies, strengths and uncertainties of this work are discussed in Section 4. Conclusion follows in Section 5.

2. Materials and Methods

2.1. Data

2.1.1. Remote Sensing Data

MODIS is a key imaging instrument onboard two complementary satellites, Terra and Aqua, and provides two series of datasets with different passing times each day. With relatively high spatial, temporal and spectral resolutions, MODIS offers timely monitoring for land, ocean, and atmosphere research [44]. In this paper, daily L3 Climate Modeling Grid (CMG) of fractional snow cover products from Terra (MOD10C1) and Aqua (MYD10C1) were selected.

The two products are with the same spatial resolution of 0.05°. MOD10C1 products were available since February 2000, and MYD10C1 products from July 2002. Both products are still being updated. Therefore, data duration of this study is chosen from 1 September 2002 to 31 August 2016. During the study period, if either image of Terra (MOD10C1) or Aqua (MYD10C1) were missing, the alternative image would represent the data of the day. If both images are missing (occurs only once during the study period), the mean value of the previous day and the latter day is considered as the missing day's snow cover. We define every snow year from 1 September of the previous year to 31 August of the current year. Seasons are defined as autumn (September–October–November), winter (December–January–February), spring (March–April–May) and summer (June–July–August). Note that the spatial resolution of snow cover data were resampled to 0.25° when calculating snow albedo radiative forcing and snow albedo feedback, while it remained 0.05° in snow cover change analysis.

A general disadvantage of snow cover optical remote sensing products, e.g., MODIS snow cover products, is that clouds shadow the information of their underlying surface, usually resulting in an underestimate of snow cover. In addition, cloud changes in position and extent at the different passing time with different viewing angle of Terra and Aqua, thus the two sensors possibly detect different snow cover information. Therefore, daily combination method was used to get the potentially maximum snow cover against the block of cloud, which could be a more realistic representation of the actual snow cover amount [45]. The combination method combines snow cover information of Terra and Aqua at daily scale, and contains two circumstances. Firstly, if either image of Terra or Aqua in the same day considers the pixel to be snow, the pixel is considered as snow. Secondly, if both images of Terra and Aqua consider the pixel to be snow, but with different fractional snow cover value, the larger one is used as the fractional snow cover of the day.

2.1.2. Atmospheric Reanalysis Data

Atmospheric reanalysis data were extracted from ERA-Interim [35], which is the latest product of European Centre for Medium-Range Weather Forecasts (ECMWF). ERA-Interim is a global atmospheric reanalysis product that started in 1979, and is continuously updated in near real time [46].

Snow albedo and albedo data were extracted on global scale, with spatial and temporal resolutions of 0.25° and 6 h, respectively. Specially, the albedo data in ERA-Interim refer to the background snow-free surface albedo. Monthly mean air temperature data at 2 m height were also acquired to estimate temperature change during the study period, with the same spatial resolution of 0.25°. All these datasets were obtained for the period 2003–2016. Snow albedo and snow-free albedo datasets (four per day) were separately averaged to daily resolution.

2.1.3. Radiative Kernel Data

Radiative kernel data [47,48] were obtained from Community Atmosphere Model version 3 (CAM3, [49]) of National Center for Atmospheric Research (NCAR) (Boulder, CO, USA). Surface albedo radiative kernel under all sky (K_α) was obtained, describing the response of net shortwave radiation at top of the atmosphere (TOA) to a 1% additive increase in surface albedo [47]. K_α is a function of latitude, longitude, and month of year. The spatial resolution of K_α is approximately 2.8° on average, and varies a little with latitude. The data were then bilinear interpolated to 0.25° for spatial consistency with other products.

Spatial distribution of monthly surface albedo radiative kernel is illustrated in Figure 1. Monthly variability of K_α is strongly affected by the movement of the sun. High values mostly appear in low latitudes where there is large incident solar radiation. K_α decreases as latitude increases, and diminishes to 0 in areas with polar darkness. Moreover, as clouds act to mask the effects of changes in the albedo of the underlying surface, K_α has the lower values in persistent cloud-covered regions, (e.g., mid-latitude storm track area), and relatively higher values in cloud-free areas (e.g., low-latitude desert area) [47].

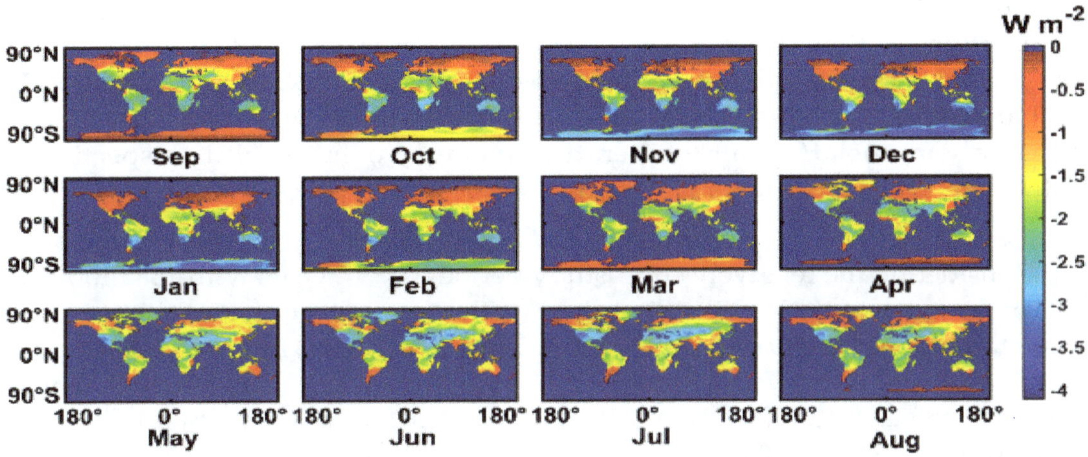

Figure 1. Spatial distribution of monthly surface albedo radiative kernel under all sky (2.8° on average resampled to 0.25° cell size) (these data are based on NCAR's monthly surface albedo radiative kernel [47]).

2.2. Method

For the advantage of less computationally expensive and easier comparison with other results, radiative kernel method is widely used in snow albedo radiative forcing and feedback estimations [50–52].

Feedback estimations using radiative kernel method usually decompose feedback into two factors, radiative kernel and climate response pattern. The former depends only on the radiative algorithm and base climate. The latter describes the change in mean climatology of the feedback variable between the two climate states [47]. Thus, the magnitude of feedback is simply the product of the two factors.

However, on purpose of getting more process information of feedback, snow albedo radiative forcing was calculated first in our study, using Equation (1):

$$G_s(t,R)\ ,\ \begin{cases} 0, & ,\ S(t,r)=0 \\ \frac{\int_R \frac{\partial\alpha}{\partial S}(t,r)\times\frac{\partial H}{\partial\alpha}(t,r)\times dA(r)}{A(R)} & ,\ S(t,r)>0 \end{cases} \tag{1}$$

Here, $G_s(t,R)$ is snow albedo radiative forcing that describes the instantaneous influence of snow cover on TOA energy budget, with unit of $W\cdot m^{-2}$ [33]. $G_s(t,R)$ is a function of time t and study area R, and R has a total area of A and grid size r. $S(t,r)$ represents snow cover at time t and pixel r. Being the prerequisite of snow albedo radiative forcing, snow cover of each pixel is checked before calculation. According to Equation (1), if pixel r is snow-free at time t, i.e., $S(t,r) = 0$, no snow albedo radiative forcing occurs, i.e., $G_s(t,R) = 0$. Only pixels with a fractional snow cover larger than 0 contribute to the snow albedo radiative forcing. The magnitude of $G_s(t,R)$ depends on the albedo contrast between snow-covered and snow-free circumstances, as well as the radiative kernel data. Specifically, $\frac{\partial\alpha}{\partial S}(t,r)$ is surface albedo change induced by snow cover change, which can be simplified as albedo contrast between snow-covered and snow-free circumstances ([33,53]). $\frac{\partial H}{\partial \alpha}(t,r)$ is the radiative change against the coincident change of albedo, namely the surface albedo radiative kernel. Note that $\frac{\partial \alpha}{\partial S}(t,r)$ and $S(t,r)$, $\frac{\partial H}{\partial \alpha}(t,r)$ and $\alpha(t,r)$ are assumed independent from each other, and to avoid the inconvenience of negative numbers, the absolute value of K_α is used in the calculation.

Then, snow albedo feedback can be quantified by the amount of additional net shortwave radiation at TOA as surface albedo decreases in association with a 1 °C temperature increase [11]:

$$\lambda = \frac{Gs(t,R)}{T(t,R)} \tag{2}$$

where λ is the feedback parameter, here considered as snow albedo feedback, with unit of $W\cdot m^{-2}\cdot{}^{\circ}C^{-1}$. $\Delta G_s(t,R)$ and $\Delta T(t,R)$ are monthly anomalies of $G_s(t,R)$ and $T(t,R)$ during the study period, respectively, and both are functions of time t and study area R. In this study, t is set to be monthly, and when calculating the regional mean, R is defined as landmasses north of 30°N, namely the North Hemisphere Extratropical Land (NEL). Accordingly, $G_s(t,R)$ and $T(t,R)$ are the monthly snow albedo radiative forcing and monthly mean surface air temperature averaged over the NEL, respectively. Note that $G_s(t,R)$ and $T(t,R)$ were area weighted [34] during the calculation process.

The coefficient of least square fit of $\Delta G_s(t,R)$ and $\Delta T(t,R)$ represents the magnitude of snow albedo feedback [54]. Considering the relatively short study period, i.e., 14 years with 168 monthly anomalies, and the small magnitude of both $\Delta G_s(t,R)$ and $\Delta T(t,R)$, the result may be subjected to substantial uncertainty due to random variations of $G_s(t,R)$ and $T(t,R)$. Therefore, in order to get a precisely snow albedo feedback and confidence level, the block bootstrap test was used. Specifically, the method considers data of each year (12 monthly $\Delta G_s(t,R)$ and $\Delta T(t,R)$) as a block, thus 14 blocks of data are contained in the original dataset. The process includes three steps:

Step 1: First, randomly pick one block of data from the 14 blocks, pick another from the 14 blocks of data (there is possibility that the same block is chosen again), pick a third block, etc. Repeat this process until all the 14 blocks are included in the newly picked dataset, in which, some blocks of data may appear more than once.

Step 2: Evaluate snow albedo feedback of the newly picked dataset based on least square fit.

Step 3: Repeat Step 1 and Step 2 for a large number of times, 10,000 times in our case. Estimate the mean snow albedo feedback and its confidence level according to the 10,000 snow albedo feedback results.

3. Results

3.1. Spatial and Temporal Variability of Snow Cover

Through the daily combination method, annual mean fractional snow cover from 2003 to 2016 increased 3.94% and 4.69% from the original MOD10C1 and MYD10C1 datasets, respectively. Being the original dataset of MOD10C1, MOD10A1 has an absolute overall accuracy of ~93% [55]. Nevertheless, accuracy decreases when cloud is taken into consideration. Due to the large uncertainties in accuracy assessment of MOD10C1 (MYD10C1), i.e., accuracy difference between MOD10A1 and MOD10C1, cloud variabilities, as well as the uncertainties in contribution of snow cover under cloud to snow albedo radiative forcing and snow albedo feedback, the contribution of the accuracy improvement through daily combination method is not discussed.

Global mean fractional snow cover during 2003–2016 (Figure 2) is calculated as the ratio between the sum of daily fractional snow cover and total days of the study period. With large spatial variability, snow cover mainly distributes in landmasses north of 30°N, excluding Antarctica. Greenland is the most densely snow-covered area over the NEL where most of the land is covered by snow and ice all year round. The Tibetan Plateau should also be mentioned for its abundant snow cover in such low latitude. Furthermore, the whole Antarctica continent is deliberately mapped as snow in MODIS snow cover products, thus it is not discussed in the paper. Apart from Antarctica, snow cover can be seen in limited areas of the Southern Hemisphere, i.e., the Andes Mountains, southeast corner of Australia, and a few parts of New Zealand.

As MODIS is an optical sensor, snow-covered areas inside of the Arctic Circle during polar night cannot be detected. This would certainly result in an underestimate of the total snow amount and maybe some discrepancies on trend analysis. However, as solar radiation is the precondition of snow albedo radiative forcing and feedback, areas in darkness receive no insolation, consequently exert no radiative forcing and feedback to the climate system. In other words, the "missing data" have little influence on radiative forcing and feedback assessment, thus the detected snow cover can be regarded as effective snow cover in our study.

Time series of monthly and annual mean fractional snow cover over the NEL during 2003–2016 are presented in Figure 3a. Here, the NEL rather than the globe was chosen for analysis, because snow cover over the NEL makes up for about 98.45% of the total snow amount during the study period, apart from Antarctica. Monthly variations are large in a sense that snow cover from winter and spring months accounts for over 75.28% of the total snow amount on average during 2003–2016, while summer months contributes only about 6.98%. Specifically, March has the largest fractional snow cover through the year. February, April and January follow and the four months have a much larger fractional snow cover value above the average. On the contrary, apart from summer months, September has the smallest snow cover amount, and its fractional snow cover amount is similar to that of June. Annual mean fractional snow cover over the NEL shows small interannual variability, while there is a slight, but insignificant (at the $p = 0.05$ level) decline in total. Months with large fractional snow cover (e.g., March, February, April, and January) are also the main contributors to interannual variability. Inconsistency tendencies of interannual variabilities are found in months, e.g., January and November experience an increasing trend of fractional snow cover in the recent three years while a decreasing trend are found in most of the rest months.

The climatological monthly mean fractional snow cover over the NEL during 2003–2016 is displayed in Figure 3b. The amount of the effective snow cover presents a single-peaked shape (peaks in March) through the year. Effective snow cover increases continuously from September to March, indicating a successive process of accumulation, during which winter months are the strongest and fastest accumulation period. Accordingly, snow cover decreases from March to July (the August value being slightly higher than the July value), and the fastest ablation occurs in spring from March to May.

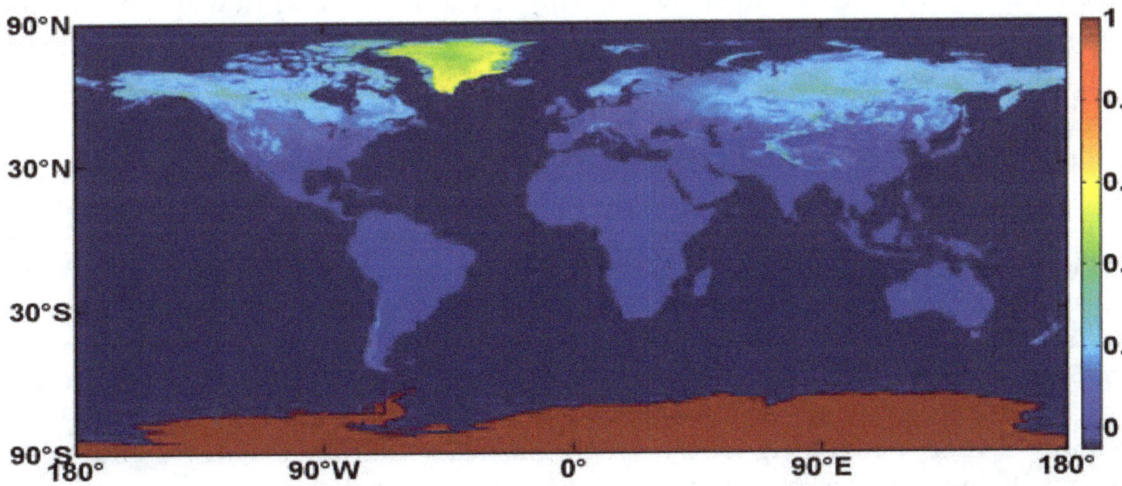

Figure 2. Spatial distribution of mean fractional snow cover during 2003–2016.

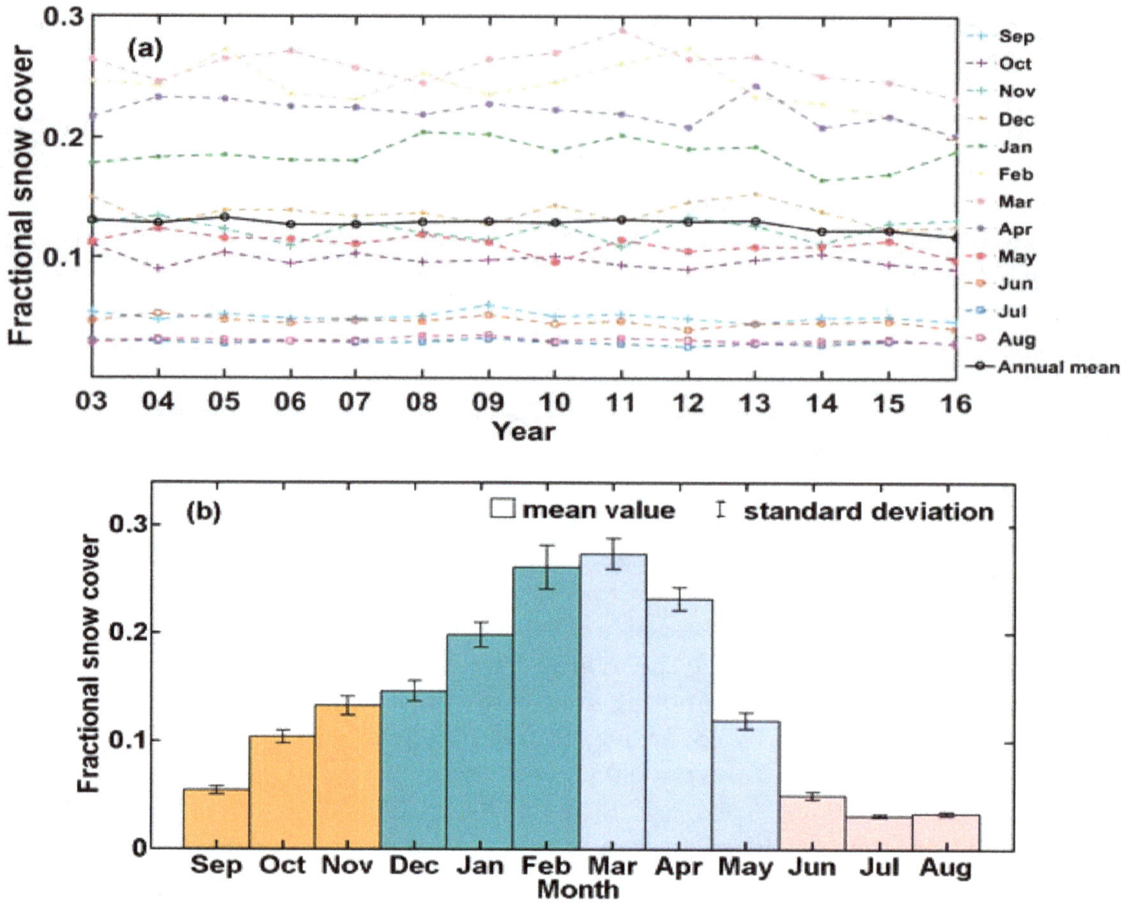

Figure 3. Fractional snow cover over the North Hemisphere Extratropical Land during 2003–2016: (**a**) monthly and annual mean; and (**b**) climatological monthly mean.

3.2. Snow Albedo Radiative Forcing

Annual mean snow albedo radiative forcing during 2003–2016 is shown in Figure 4. Areas with strong snow albedo radiative forcing also appear over the NEL, excluding Antarctica. However, areas with large fractional snow cover do not always represent strong snow albedo radiative forcing, and the peak value appears in the Tibetan Plateau, rather than Greenland. This is partly because the Tibetan Plateau is located at the relatively low latitude with much more insolation than the higher latitudes, and snow there can persist until relatively late spring. In addition, areas in western part of the U.S., northern part of Canada, mid-high latitudes of Russia, and the Andes Mountains along the west coast of South America also exhibit strong snow albedo radiative forcing.

Climatological monthly mean snow albedo radiative forcing over the NEL during 2003–2016 is displayed in Figure 5a. Standard deviations are shown as whiskers, which represent interannual variability of each month during the study period. Autumn months from September to November exhibit small snow albedo radiative forcing. Because snow cover over the NEL starts to accumulate in autumn (Figure 3b), there is neither much snow cover nor strong insolation. Winter experiences the largest expansion of snow cover extent and the smallest insolation among the four seasons. Snow albedo radiative forcing is much larger than that in autumn, and it increases from December to February. Snow albedo radiative forcing is the largest in spring months and peaks in April, as a result of both large snow cover extent and strong solar radiation. Spring also experiences the largest decrease of snow cover extent throughout the year. This leaves summer with very little snow cover amount. As a result, in spite of its largest insolation, snow albedo radiative forcing in summer is much smaller than that in spring.

April exhibits the largest snow albedo radiative forcing throughout the year, and its interannual variability is displayed in Figure 5b. There is an overall insignificant (at $p = 0.05$ level) decline in April snow albedo radiative forcing over the NEL during 2003–2016. Large interannual variability with a continuous decline during the last 4 years are experienced during the study period. In general, April snow albedo radiative forcing decreased about 6.10% during 2003–2016.

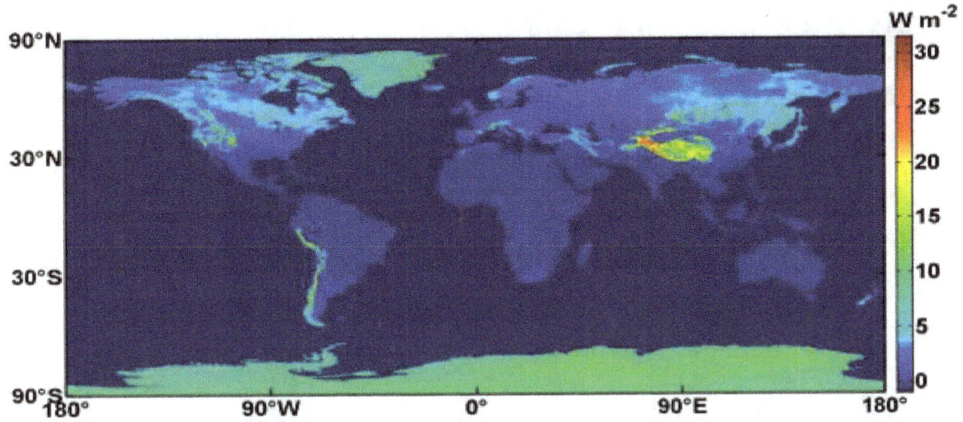

Figure 4. Spatial distribution of annual mean snow albedo radiative forcing during 2003–2016.

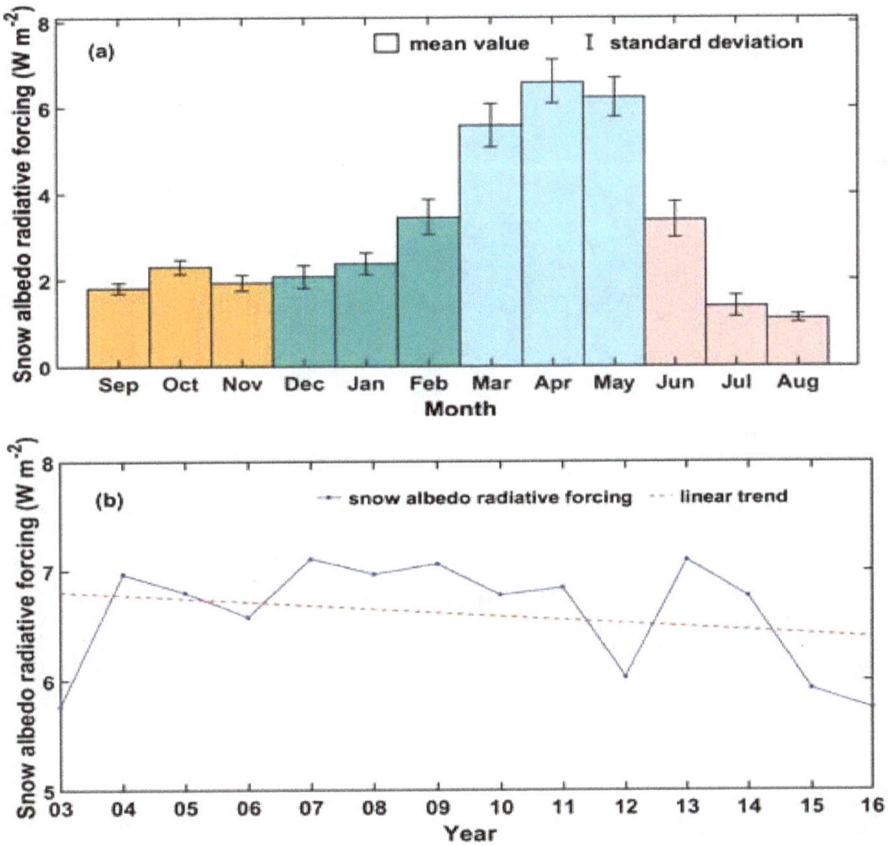

Figure 5. Snow albedo radiative forcing over the North Hemisphere Extratropical Land during 2003–2016: (**a**) climatological monthly mean; and (**b**) April interannual variability.

3.3. *Snow Albedo Feedback*

Snow albedo feedback estimation based on the block bootstrap test is explained in detail with an example trial, according to the method introduced previously (Section 2.2). Specifically, following

Step 1, it takes 27 times of random choice before all the 14 blocks of data are picked. Details of each pick are displayed in Table 1. Specifically, Pick ID represents the count of each pick, while Block ID is the block of data of the corresponding year. During the experiment, it is not until the occurrence of Block 15 (the 27th pick) that all 14 blocks of the original dataset are included. Thus, 27 blocks with 324 data are contained in the newly generated dataset.

Following Step 2, snow albedo feedback can be quantified based on the newly picked dataset. Scatterplot of the 324 $\Delta G_S(t,R)$ and $\Delta T(t,R)$ data are shown in Figure 6a. Colors of the dots represent the frequency of their occurrence (corresponding to Table 2). The dash line is the least square fit of the 324 data, i.e., snow albedo feedback strength of this trial.

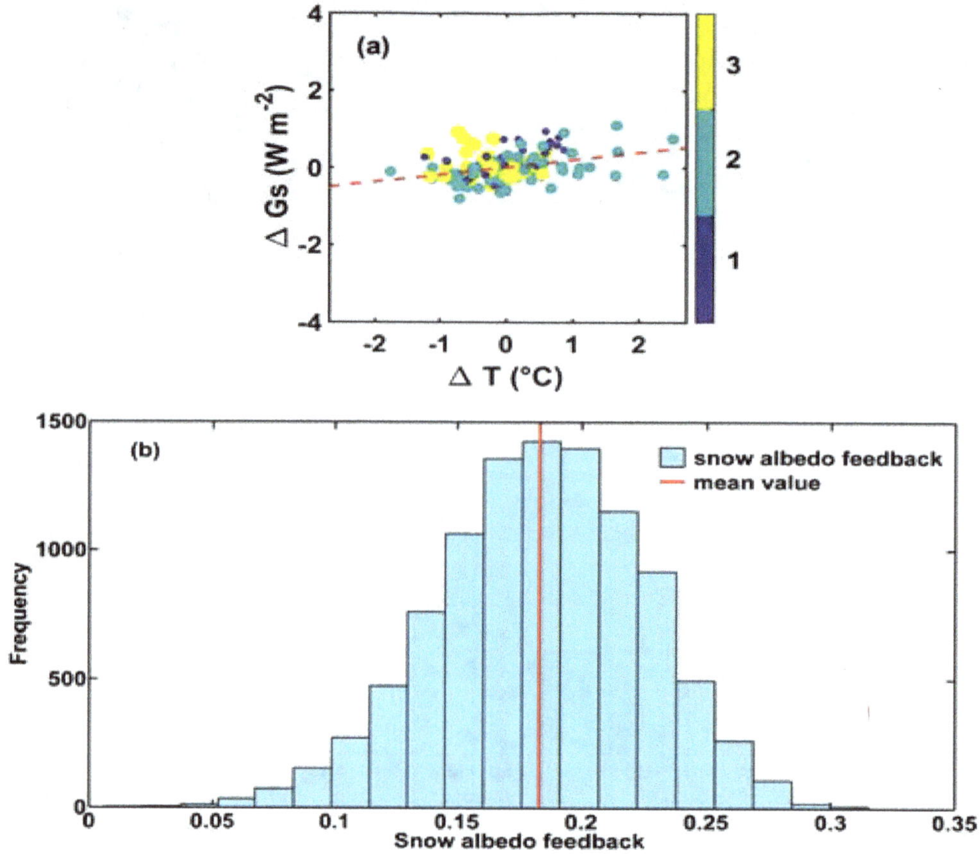

Figure 6. Snow albedo feedback estimations over the North Extratropical Land during 2003–2016 from the block bootstrap test: (**a**) an example of one test (Scatterplot of monthly snow albedo radiative forcing anomalies ($\Delta G_S(t,R)$) versus surface air temperature anomalies ($\Delta T(t,R)$), dots with different colors represent their frequency, and the least square fit coefficient suggests the magnitude of snow albedo feedback); and (**b**) probability density function of snow albedo feedback estimations from the 10,000 tests.

Following Step 3, the mean snow albedo feedback and the 95% confidence level can be estimated after repeating the experiment for 10,000 times. The histogram of the 10,000 snow albedo feedback is shown in Figure 6b. The value of the red line refers to the mean snow albedo feedback, in our case, 0.18 ± 0.08 W·m^{-2}·°C^{-1} over the NEL during 2003–2016. This implies that as surface albedo decreases in association with 1 °C temperature increase, snow albedo feedback would cause an increase of 0.18 ± 0.08 W·m^{-2} in the net shortwave radiation at TOA, averaged over the NEL.

By rescaling the NEL result with product of two factors: the ratio of the NEL area to the global area and the ratio of annual mean surface air temperature change of the NEL to global mean change [56], global snow albedo feedback can be scaled as 0.04 ± 0.02 W·m^{-2}·°C^{-1}. However, due to the limited sample amount, the result may be subjected to substantial uncertainty.

Table 1. Every Pick of the example test.

Pick ID	**1**	**2**	**3**	**4**	**5**	**6**	**7**	**8**	**9**	**10**	**11**	**12**	**13**	**14**
Block ID	07	13	06	04	09	06	10	08	08	04	13	16	11	16
Pick ID	**15**	**16**	**17**	**18**	**19**	**20**	**21**	**22**	**23**	**24**	**25**	**26**	**27**	
Block ID	03	05	12	03	14	10	14	03	10	14	07	11	15	

Table 2. Frequency of each block of data based on the example test.

Block ID	**03**	**04**	**05**	**06**	**07**	**08**	**09**	**10**	**11**	**12**	**13**	**14**	**15**	**16**
Frequency	3	2	1	2	2	2	1	3	2	1	2	3	1	2

4. Discussion

4.1. Comparison with Partially Observation-Based Studies

There are extremely limited observation-based snow albedo feedback studies, so only three of the compatible ones (with similar study area of the North Hemisphere) are chosen for further comparison: snow and ice albedo radiative forcing estimation by Flanner et al. for 1979–2008 [33], both snow albedo radiative forcing and feedback estimations by Chen et al. for 1982–2013 [57] and snow albedo feedback estimation by Peng et al. for 1980–2006 [58]. It should be noted that, although feedback was also quantified in Flanner and coworkers' study, the regional mean value of the combining snow and ice albedo feedback makes it incomparable with our snow albedo feedback results.

The magnitude of snow albedo feedback is examined first. By using a linear fit between snow albedo radiative forcing change and temperature increase during the study period, Chen et al. calculated snow albedo feedback over the North Hemisphere to be 0.17 ± 0.008 W·m^{-2}·K^{-1} [57]. By multiplying the sensitivity of end date of snow cover to temperature by difference in black surface albedo before and after snowmelt, the result is 0.19 ± 0.17 W·m^{-2}·K^{-1}, according to Peng et al. [58]. With the method of block bootstrap test, snow albedo feedback of this study is 0.18 ± 0.08 W·m^{-2}·°C^{-1}. Despite the different methods and datasets applied in each work, results of the three studies agree well.

Snow albedo radiative forcing is also compared, as it's a key parameter of feedback and offers valuable spatial information. Both Flanner and coworkers' work and our work show similar snow albedo radiative forcing pattern in terms of spatial distribution and temporal variation [33]. Large snow albedo radiative forcing is found in the northern part of Eurasia, the mid-high latitude of North America and the Tibetan Plateau in both works, though the Tibetan Plateau exhibits even larger values in our work. In addition, seasonal cycle of snow albedo radiative forcing in both works show a single peak (peaks in April), with the largest value in spring months. Chen and coworkers' work, however, shows a relatively weaker consistency with us [57]: smaller value in the Tibetan Plateau as compared with Flanner et al. and our work, and a clear tendency of snow albedo radiative forcing with latitude, i.e., the higher latitudes exhibit larger radiative forcing.

While different surface albedo kernels used in these studies have proven to vary only a little [33,56], differences among the results of snow albedo radiative forcing are considered mainly due to albedo change caused by snow cover change ($\frac{\partial \alpha}{\partial S}(t,r)$). In Chen and coworkers' work, surface albedo instead of snow albedo data was used, thus the influence of vegetation change was imported to snow albedo change [57]. In addition, according to the conclusions of Singh et al., the coarse resolution of snow data (1° and monthly [33]) is likely to result in an overestimate of snow albedo radiative forcing [53].

4.2. Comparison with Model-Based Studies

Based on the agreement among observation-based studies, snow albedo feedbacks of our work are compared with those from the 25 models that participated in CMIP5 (Qu and Hall [56], Table 1 of their paper).

Snow albedo feedback over the NEL estimated from the 25 models ranges from 0.18 to 0.78 W·m^{-2}·K^{-1}, with the ensemble mean of 0.42 ± 0.15 W·m^{-2}·K^{-1}. Our result is 0.18 ± 0.08 W·m^{-2}·K^{-1}. The global snow albedo feedback of the 25 models ranges from 0.03 to 0.16 W·m^{-2}·K^{-1}, with ensemble mean of 0.08 ± 0.03 W·m^{-2}·K^{-1}, and, in our case, it is 0.04 ± 0.02 W·m^{-2}·K^{-1}. In general, results of our work fall near the lower bound of the 25 models. The ensemble means of both hemispheric and global snow albedo feedbacks estimated by Qu and Hall [56] are larger than our results and the other partially observation-based results mentioned above. This might indicate an overall overestimation of snow albedo feedback by most of the 25 models. For the purpose of offering detailed information for model optimism, the possible source of discrepancies between this study and the 25 models is discussed below.

According to Qu and Hall, snow albedo feedback is mainly determined by two terms: one represents the variations in planetary albedo with surface albedo ($\frac{\partial \alpha_P}{\partial \alpha_S}$), and the other is the change in surface albedo associated with a 1 °C increase in T_s ($\frac{\alpha_S}{T_S}$) [56]. The spread of $\frac{\partial \alpha_P}{\partial \alpha_S}$ is mainly determined by the surface albedo kernel, because this term is calculated as the ratio of surface albedo kernel to the TOA shortwave radiation. As the kernels used in models are also used in this study, we consider this coefficient is of little contribution to the difference. As a result, the major source of the discrepancy would be originated from $\frac{\alpha_S}{T_S}$. Specifically, increase in the area-averaged surface air temperature (*Ts*) is relatively consistent both among the models and between model simulations and observations, thus surface albedo change (α_S) could be the largest source of discrepancy between our work and the 25 models. Surface albedo change, which is mainly determined by albedo contrast between snow-covered and snow-free surface, is also considered to be most uncertain (threefold spread in CMIP3 and persists to be large in CMIP5) among models. However, as the spread of surface albedo change between snow-covered and snow-free surface is not directly analyzed in Qu and Hall's work [56], further investigations are still required.

Despite the fact that there are other factors contributing to the discrepancies, surface albedo change is considered as the predominate difference and possibly being overestimated by most of the 25 models. Therefore, model parameterization should specifically focus on this factor.

Moreover, according to Hall and Qu, the surface albedo decrease associated with loss of snow cover, rather than the reduction in snow albedo due to snow metamorphosis is more important in the determination of snow albedo feedback [41]. Thus, in turn, the significance of albedo constraint by snow cover data of high spatial and temporal resolutions in this study is strengthened. The information of snow cover change, as well as the constrained albedo data, can be guidance for the model parameterizations.

4.3. Strengths and Limitations

The strengths of this study compared with the previous can be summarized as the following:

1. Instead of a combination assessment of snow and ice albedo feedback, snow albedo feedback is examined exclusively in this study, thus the contribution of snow albedo feedback and its uncertainty to the surface albedo feedback can be independently achieved.
2. Satellite-based MODIS snow cover data of high spatial resolution (0.05°) is used to constrain and determine the areas of snow albedo radiative forcing and feedback. In this study, when snow cover data were included in the calculation, the snow albedo radiative forcing decreased as much as 27.63% (not shown), compared to when only albedo contrast data and surface albedo kernel data were used for calculation. Meanwhile, it offers accurate information on constraining surface albedo decrease associated with loss of snow cover of models on spatial aspect.
3. High temporal resolution (daily) of snow cover and albedo data offers detailed information of snow cover change, which is relevant because changing snow cover occurs rapidly. In particular, the snow melt in mid-latitudes generally lasts for less than one month, especially in spring. For example, the Tibetan Plateau, one of the most intensive and important snow albedo feedback areas, has the largest inconsistency in snow albedo radiative forcing according to the comparisons

previously (Section 4.1), since rapid snow accumulation and ablation processes in spring always last for less than one week. Thus, the monthly mean data, which are commonly used in other studies [35,43,59], would easily smooth the snow and feedback processes. In addition, surface albedo decrease associated with loss of snow cover is estimated precisely on temporal scale, which offers guidance for model optimization.

4. The block bootstrap test is used effectively to reduce the uncertainty of snow albedo feedback. Considering the fact that most observation-based studies are short in time duration, a simple linear regression [28,35,43] to compute snow albedo feedback and its confidence limits would probably give misleading results due to the random variations of variables. By enlarging the sample amount and enhancing the number of tests (10,000 times in our case), the block bootstrap test should have obtained more reliable results.

Our study has some limitations that could lead to the uncertainties of the results:

1. Compared to model simulations, the available observational data of only 14 years would be potentially biased by internal climate variation.
2. Data from multiple sources with different spatial and temporal resolutions are applied to our work. Therefore, the processes of unifying their spatial and temporal resolutions, i.e., interpolation and resampling, would add errors to the spatial distribution of the results. Meanwhile, different temporal resolutions between daily datasets (fractional snow cover and albedo contrast) and monthly dataset (the radiative kernel) would also add uncertainty in temporal variation of the results.
3. The import of snow cover data is intended to constrain albedo data, quantifying a more precise area of snow albedo radiative forcing and feedback. However, due to different data sources and different computation methods, discrepancies are imported as well, and would probably be the major source of uncertainty.
4. Long-term observational data are not available for feedback study at present, and, at the same time, intermodel spread cannot be constrained effectively. Therefore, the best effort we can make is probably observing short-term variations and comparing the results with those from climate models [22,60]. Even though there has not been substantial progress in using observation results to constrain model simulations directly, we believe that, by using improved observational datasets and methods, observation-based results would help in better understanding the origin of intermodel differences, as well as the assessment of reliability of different model simulations. Finally, the goal is to get better description of feedback processes and finer estimation of feedbacks, more accurate ECS, and better projections of future climate [2,16].

5. Conclusions

We quantified snow albedo feedback and calculated snow albedo radiative forcing during the period of 2003–2016, with remote sensing, atmospheric reanalysis and radiative kernel data. The results were then compared with those from partially observation-based calculations and model-based estimations.

The results suggest that, excluding Antarctica, both snow cover and snow albedo radiative forcing are the largest in landmasses north of 30° N. Snow albedo radiative forcing peaks in spring, due to strong insolation and large snow cover extent. Snow albedo feedback over the NEL is estimated to be $0.18 \pm 0.08\ \mathrm{W{\cdot}m^{-2}{\cdot}{}^{\circ}C^{-1}}$ and the global mean is $0.04 \pm 0.02\ \mathrm{W{\cdot}m^{-2}{\cdot}{}^{\circ}C^{-1}}$. Results were compared with partially observation-based studies first. The regional mean snow albedo feedbacks were consistent, while the spatial pattern of snow albedo radiative forcing was different. The differences probably originated from snow albedo data and albedo data, as well as resolutions of data. Compared to the 25 models that participated in CMIP5, a general overestimation of models was found, mainly due to the overestimation of surface albedo variation between snow-covered and snow-free surface in the models. Surface albedo change is also with the largest spread among snow albedo feedback determinants.

Therefore, model parameterization should specifically focus on the constraint of this factor. Meanwhile, remotely sensed snow cover data with high spatial and temporal resolutions, the constrained albedo contrast data between snow-covered and snow-free surface, as well as the resulted high resolution of snow albedo radiative forcing in this study, can offer valuable information for model parameterization.

Acknowledgments: The authors would like to thank Karen M. Shell and Brian Soden for sharing radiative kernels of CAM3 from NCAR, which is freely available online at http://people.oregonstate.edu/~shellk/kernel.html. We would also thank the anonymous reviewers for the constructive suggestions and comments. This study was supported by the Chinese National Natural Science Foundation under Grant numbers 41690143, 91547210 and 41271356 and the Chinese Academy of Sciences Project under Grant KJZD-EW-G03.

Author Contributions: Lin Xiao and Tao Che designed the study and wrote the paper, Linling Chen, Hongjie Xie and Liyun Dai all contributed to the discussions, edits, and revisions.

References

1. Intergovernmental Panel on Climate Change (IPCC). *Climate Change 2013: The Physical Science Basis*; Contribution of Working Group I to the Fifth Assessment Report of the Intergovernmental Panel on Climate Change; Stocker, T.F., Qin, D., Plattner, G.K., Tignor, M., Allen, S.K., Boschung, J., Nauels, A., Xia, Y., Bex, V., Midgley, P.M., Eds.; Cambridge University Press: Cambridge, UK, 2013.
2. Cubasch, U.; Meehl, G.A.; Boer, G.J.; Stouffer, R.J.; Dix, M.; Noda, A.; Senior, C.A.; Raper, S.; Yap, K.S. *Climate Change 2001: The Scientific Basis*; Projections of Future Climate Change; Kim, J.W., Stone, J., Eds.; Cambridge University Press: Cambridge, UK, 2001; pp. 525–582.
3. Trenberth, K.E.; Philip, D.J. Observations: Surface and atmospheric climate change. In *Climate Change 2007: The Physical Science Basis*; Contribution of Working Group I to the Fourth Assessment Report of the Intergovernmental Panel on Climate Change; Solomon, S., Qin, D., Manning, M., Chen, Z., Marquis, M., Averyt, K.B., Tignor, M., Miller, H.L., Eds.; Cambridge University Press: Cambridge, UK, 2007; pp. 237–336.
4. Screen, J.; Simmonds, I. The central role of diminishing sea ice in recent Arctic temperature amplification. *Nature* **2010**, *464*, 1334–1337. [CrossRef] [PubMed]
5. Clark, P.U.; Alley, R.B.; Pollard, D. Northern Hemisphere ice-sheet influences on global climate change. *Science* **1999**, *286*, 1104–1111. [CrossRef]
6. Serreze, M.C.; Francis, J.A. The arctic amplification debate. *Clim. Chang.* **2006**, *76*, 241–264. [CrossRef]
7. Euskirchen, E.S.; McGuire, A.D.; Chapin, F.S. Energy feedbacks of northern high-latitude ecosystems to the climate system due to reduced snow cover during 20th century warming. *Glob. Chang. Biol.* **2007**, *13*, 2425–2438. [CrossRef]
8. Graversen, R.; Wang, M. Polar amplification in a coupled climate model with locked albedo. *Clim. Dyn.* **2009**, *33*, 629–643. [CrossRef]
9. Kumar, A.; Perlwitz, J.; Eischeid, J.; Quan, X.; Xu, T.; Zhang, T.; Hoerling, M.; Jha, B.; Wang, W. Contribution of sea ice loss to Arctic amplification. *Geophys. Res. Lett.* **2010**, *37*, 21. [CrossRef]
10. Kay, J.; Holland, M.; Jahn, A. Inter-annual to multi-decadal Arctic sea ice extent trends in a warming world. *Geophys. Res. Lett.* **2011**, *38*, L15708. [CrossRef]
11. Hall, A. The role of surface albedo feedback in climate. *J. Clim.* **2004**, *17*, 1550–1568. [CrossRef]
12. Box, J.E.; Fettweis, X.; Stroeve, J.C.; Tedesco, M.; Hall, D.K.; Steffen, K. Greenland ice sheet albedo feedback: Thermodynamics and atmospheric drivers. *Cryosphere* **2012**, *6*, 821–839. [CrossRef]
13. Soden, B.J.; Held, I.M. An assessment of climate feedbacks in coupled ocean-atmosphere models. *J. Clim.* **2006**, *19*, 3354–3360. [CrossRef]
14. Roe, G.H.; Baker, M.B. Why is climate sensitivity so unpredictable? *Science* **2007**, *318*, 629–632. [CrossRef] [PubMed]
15. Bony, S.; Bellon, G.; Klocke, D.; Sherwood, S.; Fermepin, S.; Denvil, S. Robust direct effect of carbon dioxide on tropical circulation and regional precipitation. *Nat. Geosci.* **2013**, *6*, 447–451. [CrossRef]
16. Bony, S.; Colman, R.; Kattsov, V.M.; Allan, R.P.; Bretherton, C.S.; Dufresne, J.L.; Hall, A.; Soden, B.J.; Tselioudis, G.; Webb, M.J. How well do we understand and evaluate climate change feedback processes? *J. Clim.* **2006**, *191*, 3445–3482. [CrossRef]

17. Randall, D.A.; Wood, R.A.; Bony, S.; Colman, R.; Fichefet, T.; Fyfe, J.; Kattsov, V.; Pitman, A.; Shukla, J.; Srinivasan, J.; et al. Climate models and their evaluation. In *Climate Change 2007: The Physical Science Basis*; Contribution of Working Group I to the Fourth Assessment Report of the Intergovernmental Panel on Climate Change; Solomon, S., Qin, D., Manning, M., Chen, Z., Marquis, M., Averyt, K.B., Tignor, M., Miller, H.L., Eds.; Cambridge University Press: Cambridge, UK; New York, NY, USA, 2007.
18. Vial, J.; Dufresne, J.L.; Bony, S. On the interpretation of inter-model spread in CMIP5 climate sensitivity estimates. *Clim. Dyn.* **2013**, *41*, 3339–3362. [CrossRef]
19. Andrews, T.; Gregory, J.M.; Webb, M.J.; Taylor, K.E. Forcing, feedbacks and climate sensitivity in CMIP5 coupled atmosphere-ocean climate models. *Geophys. Res. Lett.* **2012**, *39*. [CrossRef]
20. Hansen, J.; Lacis, A.; Rind, D.; Russell, G.; Stone, P.; Fung, I.; Ruedy, R.; Lerner, J. Climate sensitivity: Analysis of feedback mechanisms. *Clim. Process. Clim. Sensit.* **1984**, *5*, 130–163.
21. Colman, R.A.; McAvaney, B. Climate feedbacks under a very broad range of forcing. *Geophys. Res. Lett.* **2009**, *36*, L01702. [CrossRef]
22. Dalton, M.M.; Shell, K.M. Comparison of short-term and long-term radiative feedbacks and variability in twentieth-century global climate model simulations. *J. Clim.* **2013**, *26*, 10051–10070. [CrossRef]
23. Budyko, M.I. The effect of solar radiation variations on the climate of the earth. *Tellus* **1969**, *21*, 611–619. [CrossRef]
24. Sellers, W.D. A global climatic model based on the energy balance of the earth-atmosphere system. *J. Appl. Meteorol.* **1969**, *8*, 392–400. [CrossRef]
25. Chen, D.; Gerdes, R.; Lohmann, G.A. 1-D atmospheric energy balance model developed for ocean modeling. *Theor. Appl. Climatol.* **1995**, *51*, 25–38. [CrossRef]
26. Colman, R.A.; McAvaney, B.J.; Fraser, J.R.; Rikus, L.J.; Dahni, R.R. Snow and cloud feedbacks modelled by an atmospheric general circulation model. *Clim. Dyn.* **1994**, *9*, 253–265. [CrossRef]
27. Watterson, I.G.; Dix, M.R.; Colman, R.A. A comparison of present and doubled CO_2 climates and feedbacks simulated by three GCMs. *J. Geophys. Res.* **1999**, *104*, 1943–1956. [CrossRef]
28. Colman, R.A. Surface albedo feedbacks from climate variability and change. *J. Geophys. Res.* **2013**, *118*, 2827–2834. [CrossRef]
29. Gregory, J.M.; Ingram, W.J.; Palmer, M.A.; Jones, G.S.; Stott, P.A.; Thorpe, R.B.; Lowe, J.A.; Johns, T.C.; Williams, K.D. A new method for diagnosing radiative forcing and climate sensitivity. *Geophys. Res. Lett.* **2004**, *31*, L03205. [CrossRef]
30. Sanderson, B.M.; Shell, K.M.; Ingram, W. Climate feedbacks determined using radiative kernels in a multi-thousand member ensemble of AOGCMs. *Clim. Dyn.* **2010**, *35*, 1219–1236. [CrossRef]
31. Intergovernmental Panel on Climate Change (IPCC). *The Physical Science Basis*; Contribution of Working Group I to the Fourth Assessment Report of the Intergovernmental Panel on Climate Change; Solomon, S., Qin, D., Manning, M., Chen, Z., Marquis, M., Tignor, K.B., Averyt, M., Miller, H.L., Eds.; Cambridge University Press: Cambridge, UK; New York, NY, USA, 2007; p. 996.
32. Webb, M.J.; Lambert, F.H.; Gregory, J.M. Origins of differences in climate sensitivity, forcing and feedback in climate models. *Clim. Dyn.* **2013**, *40*, 677–707. [CrossRef]
33. Flanner, M.G.; Shell, K.M.; Barlage, M.; Perovich, D.K.; Tschudi, M.A. Radiative forcing and albedo feedback from the Northern Hemisphere cryosphere between 1979 and 2008. *Nat. Geosci.* **2011**, *4*, 151–155. [CrossRef]
34. Fletcher, C.G.; Zhao, H.; Kushner, P.J.; Fernandes, R. Using models and satellite observations to evaluate the strength of snow albedo feedback. *J. Geophys. Res.* **2012**, *117*, D11117. [CrossRef]
35. Dee, D.P.; Uppala, S.M.; Simmons, A.J.; Berrisford, P.; Poli, P.; Kobayashi, S.; Andrae, U.; Balmaseda, M.A.; Balsamo, G.; Bauer, P.; et al. The ERA-Interim reanalysis: Configuration and performance of the data assimilation system. *Q. J. R. Meteorol. Soc.* **2011**, *137*, 553–597. [CrossRef]
36. Rienecker, M.M.; Suarez, M.J.; Gelaro, R.; Todling, R.; Bacmeister, J.; Liu, E.; Bosilovich, M.G.; Schubert, S.D.; Takacs, L.; Kim, G.K.; et al. MERRA: NASA's modern-era retrospective analysis for research and applications. *J. Clim.* **2011**, *24*, 3624–3648. [CrossRef]
37. Dessler, A.E. Observations of climate feedbacks over 2000–10 and comparisons to climate models. *J. Clim.* **2013**, *26*, 333–342. [CrossRef]
38. Loth, B.; Graf, H.; Oberhuber, J.M. Snow cover model for global climate simulations. *J. Geophys. Res.* **1993**, *98*, 10451–10464. [CrossRef]

39. Song, X.; Zhang, G.J.; Cai, M. Characterizing the climate feedback pattern in the NCAR CCSM3-SOM using hourly data. *J. Clim.* **2014**, *27*, 2912–2930. [CrossRef]
40. Déry, S.J.; Brown, R.D. Recent Northern Hemisphere snow cover extent trends and implications for the snow-albedo feedback. *Geophys. Res. Lett.* **2007**, *34*, L22504. [CrossRef]
41. Qu, X.; Hall, A. What controls the strength of snow-albedo feedback? *J. Clim.* **2007**, *20*, 3971–3981. [CrossRef]
42. Loranty, M.M.; Berner, L.T.; Goetz, S.J.; Jin, Y.; Randerson, J.T. Vegetation controls on northern high latitude snow-albedo feedback: Observations and CMIP5 model simulations. *Glob. Chang. Biol.* **2014**, *20*, 594–606. [CrossRef] [PubMed]
43. Crook, J.A.; Forster, P.M. Comparison of surface albedo feedback in climate models and observations. *Geophys. Res. Lett.* **2014**, *41*, 1717–1723. [CrossRef]
44. Hall, D.K.; Salomonson, V.V.; Riggs, G.A. *MODIS/Terra Snow Cover Daily L3 Global 0.05 Deg CMG*; Version 5; Digital Media; National Snow and Ice Data Center: Boulder, CO, USA. Available online: http://nsidc.org/data/mod10c1 (accessed on 26 June 2017).
45. Xia, Q.; Gao, X.; Chu, W.; Sorooshian, S. Estimation of daily cloud-free, snow-covered areas from MODIS based on variational interpolation. *Water Resour. Res.* **2012**, *48*. [CrossRef]
46. Berrisford, P.; Dee, D.P.; Fielding, K.F.; Fuentes, K.M.; Kallberg, P.; Kobayashi, S.; Uppala, S. *The ERA-Interim Archive*; ERA Report Series; ECMWF: Reading, UK, 2009; pp. 1–16.
47. Soden, B.J.; Held, I.M.; Colman, R.; Shell, K.M.; Kiehl, J.T.; Shields, C.A. Quantifying climate feedbacks using radiative kernels. *J. Clim.* **2008**, *21*, 3504–3520. [CrossRef]
48. Shell, K.M.; Kiehl, J.; Shields, C.A. Using the radiative kernel technique to calculate climate feedbacks in NCAR's community atmospheric model. *J. Clim.* **2008**, *21*, 2269–2282. [CrossRef]
49. Collins, W.D.; Rasch, P.J.; Boville, B.A.; Hack, J.J.; McCaa, J.R.; Williamson, D.L.; Briegleb, B.P.; Bitz, C.M.; Lin, S.; Zhang, M. The formulation and atmospheric simulation of the community atmosphere model version 3 (CAM3). *J. Clim.* **2006**, *19*, 2144–2161. [CrossRef]
50. Sanderson, B.M.; Shell, K.M. Model-specific radiative kernels for calculating cloud and noncloud climate feedbacks. *J. Clim.* **2012**, *25*, 7607–7624. [CrossRef]
51. Jonko, A.K.; Shell, K.M.; Sanderson, B.M.; Danabasoglu, G. Climate feedbacks in CCSM3 under changing CO_2 forcing. Part I: Adapting the linear radiative kernel technique to feedback calculations for a broad range of forcings. *J. Clim.* **2012**, *25*, 5260–5272. [CrossRef]
52. Chen, X.; Long, D.; Hong, Y.; Liang, S.; Hou, A. Observed radiative cooling over the Tibetan Plateau for the past three decades driven by snow-cover-induced surface albedo anomaly. *J. Geophys. Res.* **2017**, *122*, 6170–6185. [CrossRef]
53. Singh, D.; Flanner, M.G.; Perket, J. The global land shortwave cryosphere radiative effect during the MODIS era. *Cryosphere* **2015**, *9*, 2057–2070. [CrossRef]
54. Dessler, A.E. A determination of the cloud feedback from climate variations over the past decade. *Science* **2010**, *330*, 1523–1527. [CrossRef] [PubMed]
55. Hall, D.K.; Riggs, G.A. Accuracy assessment of the MODIS snow products. *Hydrol. Process.* **2010**, *21*, 1534–1547. [CrossRef]
56. Qu, X.; Hall, A. On the persistent spread in snow-albedo feedback. *Clim. Dyn.* **2014**, *42*, 69–81. [CrossRef]
57. Chen, X.; Liang, S.; Cao, Y. Satellite observed changes in the Northern Hemisphere snow cover phenology and the associated radiative forcing and feedback between 1982 and 2013. *Environ. Res. Lett.* **2016**, *11*, 084002. [CrossRef]
58. Peng, S.; Piao, S.; Ciais, P.; Friedlingstein, P.; Zhou, L.; Wang, T. Change in snow phenology and its potential feedback to temperature in the Northern Hemisphere over the last three decades. *Environ. Res. Lett.* **2013**, *8*, 014008. [CrossRef]
59. Fernandes, R.; Zhao, H.; Wang, X.; Key, J.; Qu, X.; Hall, A. Controls on Northern Hemisphere snow albedo feedback quantified using satellite Earth observations. *Geophys. Res. Lett.* **2009**, *36*, L21702. [CrossRef]
60. Qu, X.; Hall, A. Assessing snow albedo feedback in simulated climate change. *J. Clim.* **2006**, *19*, 2617–2630. [CrossRef]

9

Improving Estimation of Gross Primary Production in Dryland Ecosystems by a Model-Data Fusion Approach

Haibo Wang [1], Xin Li [2,3,*], Mingguo Ma [4] and Liying Geng [1]

1. Key Laboratory of Remote Sensing of Gansu Province, Heihe Remote Sensing Experimental Research Station, Northwest Institute of Eco-Environment and Resources, Chinese Academy of Sciences, 320 Donggang West Road, Lanzhou 730000, China; whb@lzb.ac.cn (H.W.); gengly@lzb.ac.cn (L.G.)
2. Institute of Tibetan Plateau Research, Chinese Academy of Sciences, Beijing 100101, China
3. CAS Center for Excellence in Tibetan Plateau Earth Sciences, Chinese Academy of Sciences, Beijing 100101, China
4. Chongqing Engineering Research Center for Remote Sensing Big Data Application, School of Geographical Sciences, Southwest University, Beibei, Chongqing 400715, China; mmg@lzb.ac.cn

* Correspondence: lixin@lzb.ac.cn

Abstract: Accurate and continuous monitoring of the production of arid ecosystems is of great importance for global and regional carbon cycle estimation. However, the magnitude of carbon sequestration in arid regions and its contribution to the global carbon cycle is poorly understood due to the worldwide paucity of measurements of carbon exchange in arid ecosystems. The Moderate Resolution Imaging Spectroradiometer (MODIS) gross primary productivity (GPP) product provides worldwide high-frequency monitoring of terrestrial GPP. While there have been a large number of studies to validate the MODIS GPP product with ground-based measurements over a range of biome types. Few studies have comprehensively validated the performance of MODIS estimates in arid and semi-arid ecosystems, especially for the newly released Collection 6 GPP products, whose resolution have been improved from 1000 m to 500 m. Thus, this study examined the performance of MODIS-derived GPP by compared with eddy covariance (EC)-observed GPP at different timescales for the main ecosystems in arid and semi-arid regions of China. Meanwhile, we also improved the estimation of MODIS GPP by using in situ meteorological forcing data and optimization of biome-specific parameters with the Bayesian approach. Our results revealed that the current MOD17A2H GPP algorithm could, on the whole, capture the broad trends of GPP at eight-day time scales for the most investigated sites. However, GPP was underestimated in some ecosystems in the arid region, especially for the irrigated cropland and forest ecosystems (with $R^2 = 0.80$, RMSE = 2.66 $gC/m^2/day$ and $R^2 = 0.53$, RMSE = 2.12 $gC/m^2/day$, respectively). At the eight-day time scale, the slope of the original MOD17A2H GPP relative to the EC-based GPP was only 0.49, which showed significant underestimation compared with tower-based GPP. However, after using in situ meteorological data to optimize the biome-based parameters of MODIS GPP algorithm, the model could explain 91% of the EC-observed GPP of the sites. Our study revealed that the current MODIS GPP model works well after improving the maximum light-use efficiency (ε_{max} or LUE_{max}), as well as the temperature and water-constrained parameters of the main ecosystems in the arid region. Nevertheless, there are still large uncertainties surrounding GPP modelling in dryland ecosystems, especially for desert ecosystems. Further improvements in GPP simulation in dryland ecosystems are needed in future studies, for example, improvements of remote sensing products and the GPP estimation algorithm, implementation of data-driven methods, or physiology models.

Keywords: terrestrial ecosystem; MODIS GPP product; calibration; arid region; oasis-desert ecosystem

1. Introduction

Drylands, including arid and semi-arid ecosystems, cover 30%–45% of the Earth's land surface [1,2], and play an important role in the global carbon cycle and future carbon sequestration [3,4]. Accurate and continuous monitoring of terrestrial ecosystem production in arid and semi-arid regions is of great importance to improve the understanding of the role of arid terrestrial ecosystems in the global carbon cycle. However, the worldwide paucity of measurement of carbon exchange in arid ecosystems has hindered the full understanding of the magnitude of carbon sequestration and the accurate prediction of the carbon cycle [5,6].

Terrestrial gross primary production (GPP) is the largest component of the global carbon cycle and is essential to understand and quantify the contribution of terrestrial ecosystems to the global carbon cycle [7]. Satellite remote sensing provides continuous and temporally repetitive observation of land surfaces and has advanced tremendously over the past few decades that has become a useful tool in estimating the terrestrial ecosystem production across broad temporal and spatial scales. Production efficiency models (PEMs), developed for predicting global GPP with remote sensing, have been widely used to quantify the spatial and temporal variation of terrestrial ecosystem productivity [8–10]. In the absence of widespread ground observations, remote sensing models are also commonly used to estimate dryland CO_2 exchange [4,11]. Previous data and remote sensing models comparisons have only included a few dryland sites [12]. Thus, there is a need to understand how well commonly used remote sensing models capture the magnitude and inter-annual variability of measured CO_2 exchange [13].

Since 2000, satellite-based GPP estimation have increasingly used data from the Moderate Resolution Imaging Spectroradiometer (MODIS) due to its continuous worldwide availability [8]. The MODIS GPP algorithm (i.e., MOD17) is a type of PEM, which provides high frequency worldwide observations of GPP [14,15]. To date, MODIS has issued multiple versions of GPP [14,16]. Currently, the MOD17 product has been updated to Collection 6 (C6), which has improved the algorithm parameters and forcing data of previous collections [15,17], as the spatial resolution has increased from 1000 m to 500 m. A large number of studies have validated the capacity of MODIS GPP products with eddy covariance (EC) measurements across multiple biomes, such as forests [18,19], shrublands [20], grasslands [21,22], savanna [23], croplands [24], and across biomes [12,25–27]. However, most of these studies validated previous versions of MODIS GPP products (i.e., Collection 4 and 5). Comprehensive evaluation of the performance of MODIS GPP C6 products in arid regions of China remains limited to this date [1].

Previous studies showed no consistent results in the validation of Collection 4 and 5 of MODIS GPP products. MODIS GPP may underestimate at some sites, such as at cropland sites [24], overestimate at some low productivity sites [25,28], or agree well [26] with tower-based GPP. Meanwhile, the MODIS GPP Collection 6 products (i.e., MOD17A2H) also tend to overestimate GPP in alpine meadows of the Tibetan Plateau [22] and underestimate flux-derived GPP at most sites across the globe [27]. However, because of inadequate observations in arid regions compared with other regions, it remains uncertain whether these biases also exist in other ecosystems in arid regions for the improved Collection 6 GPP products. Therefore, it is necessary to validate the performance of the latest version of MODIS products in arid regions.

The overall uncertainty of carbon flux modelling includes uncertainty of input variables, model structure, and model parameters [29], which can significantly impact carbon flux at regional scales. Several attempts have been made to address the uncertainties of the PEM algorithm [26,27,30,31]. For the MOD17 products, inaccuracies in the parameterization of model parameters (such as maximum light-use efficiency (ε_{max} or LUE_{max})) were found to be one of the most important factors attributed to the bias of MODIS GPP [12,20]. The current MOD17 algorithm uses the constant maximum LUE and other parameters for one ecosystem [18], which is not suitable for variability of climate conditions and ecosystems. Previous studies found that the LUE parameter in the MOD17 algorithm was underestimated [26]. Several attempts have been made to calibrate the maximum LUE parameters and to improve the performance of MODIS GPP estimation [24,26,32]. However, most of these studies overlooked the potential impacts of other model parameters' uncertainty on the estimation of GPP, e.g., water-limited factors, which are important factors for GPP estimation, especially for ecosystems in the arid region.

Research community have established that by adjusting the key parameters of the model can improve GPP estimation using MODIS GPP algorithm, which can compensate for the errors introduced by the model structures [23]. A model–data fusion approach provides powerful tools for optimizing the model parameters and quantifying the influence of uncertainties, and is being increasingly used to estimate the parameters of ecological models [33–38]. Model–data fusion approaches include Bayesian and non-Bayesian approaches. Non-Bayesian approaches, such as global optimization algorithms, can efficiently determine the optimal parameter solutions by minimizing (or maximizing) objective functions [36], but cannot quantify uncertainty. In contrast, the Bayesian approach can be employed to update the parameter distributions when new information becomes available [37], and produce reliable estimates of parameter and predictive uncertainty [38]. Some past studies have strengthened the importance of parameters estimation in carbon cycle models [32,39], but have mainly focused on single site to constrain the parameters of a given plant functional type (PFT), in addition, few studies have assessed the variability of parameters within a PFT [40]. For MODIS GPP validation, since the PFT parameters in the MOD17 algorithm are obtained from flux towers worldwide, they are not appropriate for specific regions such as the arid regions of China.

Thus, this study aims to examine the performance of newly released MODIS GPP C6 products and MOD17 algorithms in predicting GPP in a typical arid region of China. The overall goals of this study are to: (1) Evaluate the model performance of the MODIS GPP Collection 6 products at eight-day to annual time scale across various ecosystem types in a typical arid region of China; (2) analyze the uncertainty of remote sensing models in simulating GPP in typical arid regions; and (3) quantify the parameter uncertainties in GPP estimation for the main ecosystem types in arid regions of China by using a Bayesian approach with calibration of maximum LUE and water and temperature-limited factors. This research will contribute to the development and improvement of GPP estimates in arid regions.

2. Materials and Methods

2.1. In Situ Meteorological Observations and Carbon Flux Data

The fluxes and meteorological data used in this study are mainly based on a flux observation network located in a typical inland river basin: The Heihe River Basin (HRB) in the arid region of Northwest China. The HRB (37.7°–42.7° N, 97.1°–102.0° E), second largest inland river basin in China, is located in the middle part of the Hexi corridor and covers an area of approximately 1,432,000 km^2 [41]. The HRB is a unique region in China and can be viewed as an epitome of the arid region of western China for its varied distributed landscapes of alpine meadow, wetland–oasis–desert and natural oasis–desert ecosystems from upstream to downstream [42,43]. We constructed a comprehensive flux observation network in the whole river basin to investigate the complexity of hydrological and ecological processes in the arid region (Figure 1). In this study, we compiled 12 EC flux sites covering 3 grassland sites, 3 desert grassland sites, 3 cropland sites (including a wetland site), and 3 forest sites, which almost covered the major plant function types (PFTs) and typical ecosystem types in the arid region of an inland river basin. Figure 2 shows the meteorological observations of all the flux tower sites over HRB including precipitation, air temperature (T), and vapor pressure deficit (VPD). A large variability of climate conditions exist within and across the species. The specific locations and related information of the sites are shown in Table 1.

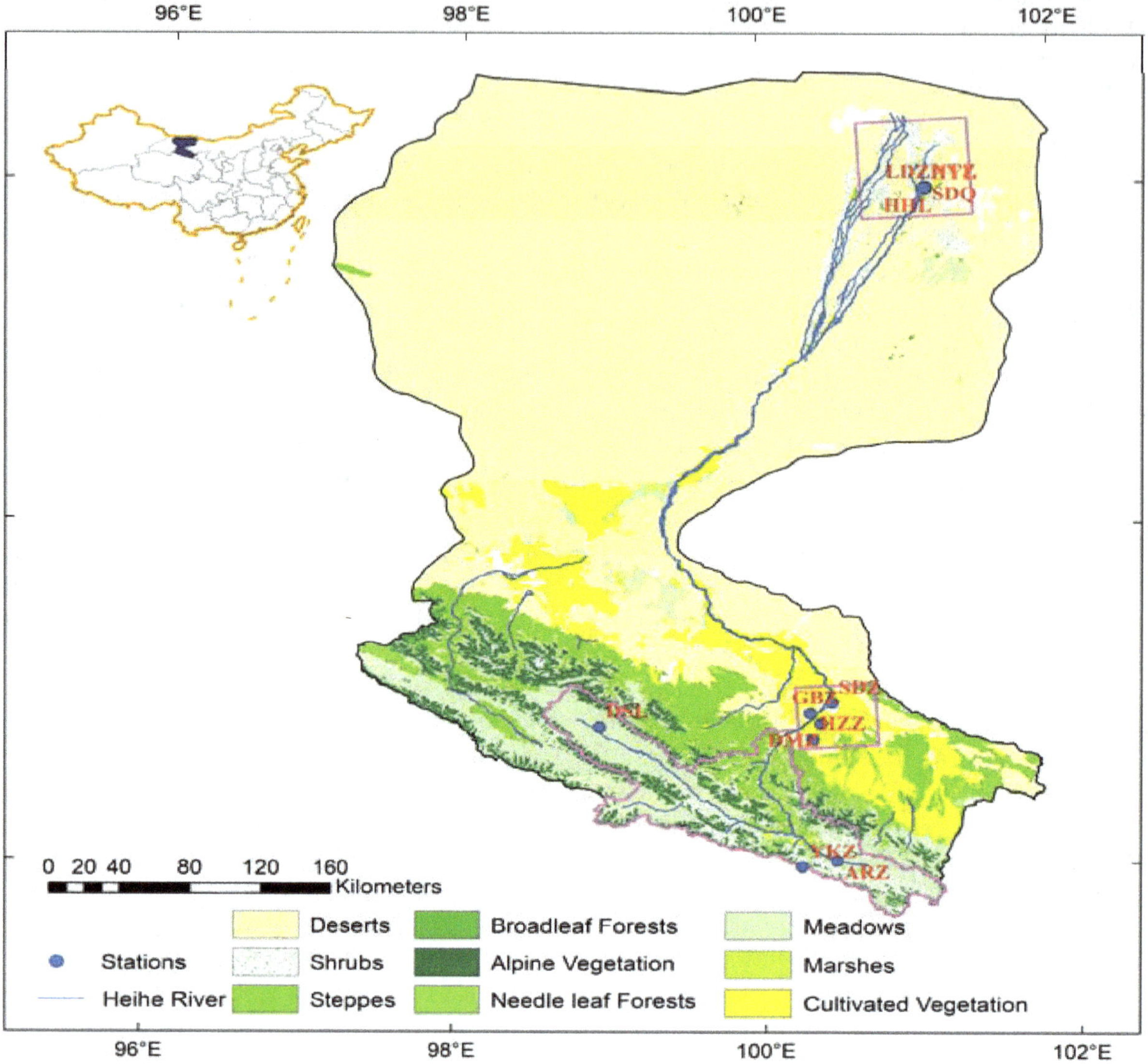

Figure 1. Locations of the flux observation sites over Heihe River Basin (HRB).

Table 1. Characteristics of the flux observation network sites used in this study. MAT (°C) represented mean annual air temperature, MAP (mm) represented mean annual accumulated precipitation, and PET (mm) represented mean annual potential evapotranspiration.

Site Code	PFTs	Year Used	Vegetation Cover	Longitude	Latitude	MAT (°C)	MAP (mm)	PET (mm)
A'rou(ARZ)	Grassland	2013–2016	alpine grassland	38.0473	100.4643	−0.29	444.70	636.18
Dashalong(DSL)	Grassland	2013–2016	alpine meadow	38.8399	98.9406	−3.91	314.43	698.07
Yakou(YKZ)	Grassland	2015–2016	alpine meadow	38.0142	100.2421	−4.68	500.79	653.16
Huazhaizi(HZZ)	Desert steppe	2012–2016	desert steppe	38.76519	100.3186	8.89	139.68	590.93
Gobi(GBZ)	Desert steppe	2012–2015	desert steppe	38.91496	100.3042	9.07	102.25	575.72
Luodi(LDZ)	Desert steppe	2013–2015	desert steppe	41.9993	101.1326	12.32	24.80	727.68
Daman(DMZ)	Cropland	2012–2016	maize	38.85551	100.3722	6.93	135.70	828.04
Nongtian(NTZ)	Cropland	2012–2016	cantaloupe	42.0048	101.1338	9.39	35.55	727.68
Shidi(SDZ)	Cropland	2012–2016	reed	38.97514	100.4464	9.19	119.9	1249.35
Huyanglin(HYL)	Forest	2013–2015	populus euphratica	41.9928	101.1236	10.33	26.00	922.91
Hunhelin(HHL)	Forest	2013–2016	mixed forest	41.9903	101.1335	10.04	35.53	1043.34
Sidaoqiao(SDQ)	Forest	2013–2016	tamarix forest	42.0012	101.1374	10.06	37.13	977.95

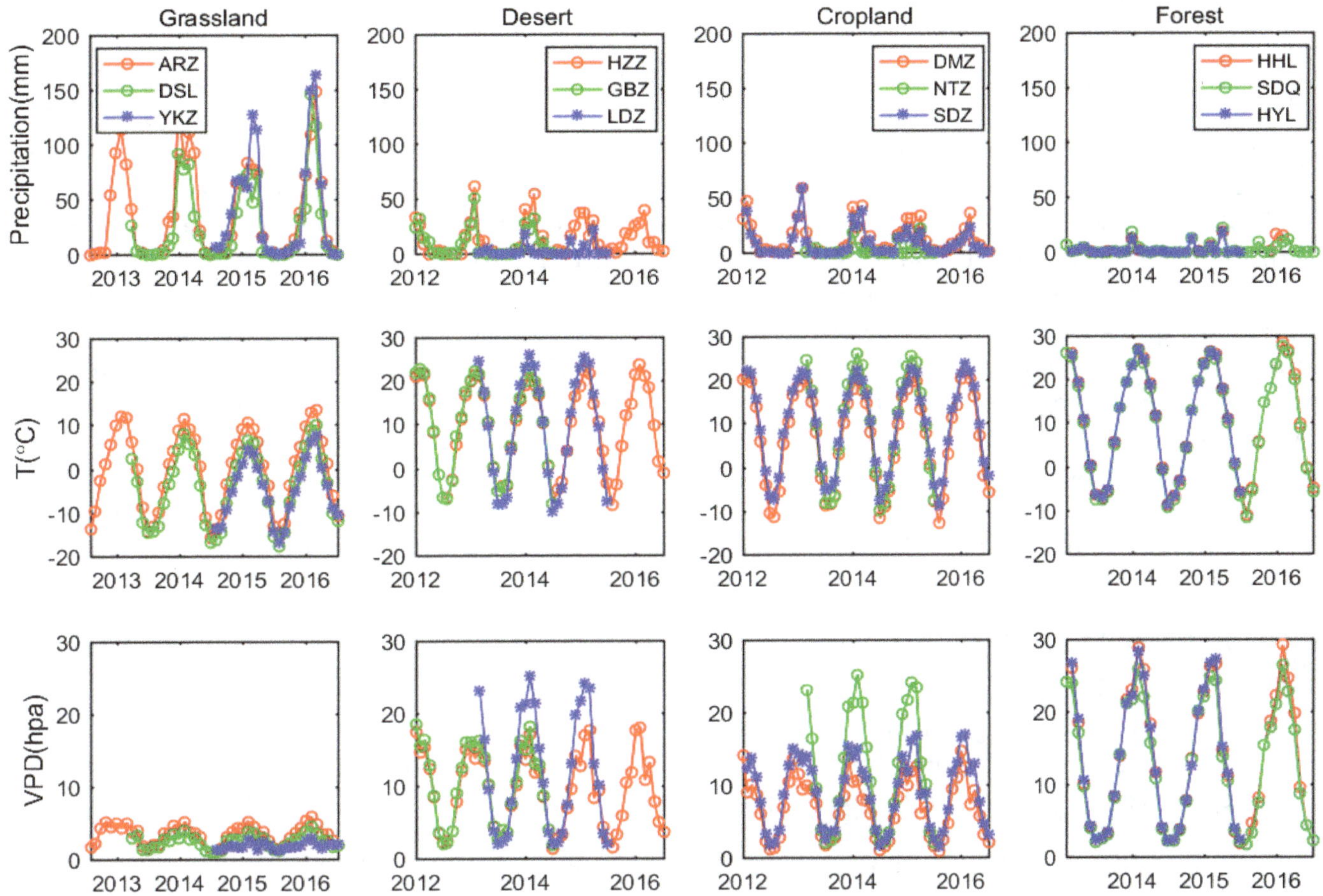

Figure 2. Plots of monthly accumulated precipitation, monthly averaged air temperature (T) and monthly averaged vapor pressure deficit (VPD) over HRB.

The open-path eddy covariance (OPEC) system was used to measure carbon and water vapor fluxes in the flux observation network. The OPEC system at each site consists of a 3D sonic anemometer (CSAT-3/Gill, Campbell Scientific Instruments Inc., USA/Gill, UK) and an open path infrared gas analyzer (Li-7500/7500A, Licor Inc., USA). The meteorological variables were measured simultaneously at each site including air temperature, rainfall, solar radiation, photosynthetically active radiation (PAR), relative humidity and soil moisture. VPD was calculated using measured relative humidity and actual vapor pressure.

The meteorological data were measured at automatic weather stations at every 10 min interval, which were carefully checked for quality and summed into 30 mins and daily timescales. The raw EC measurements of 10 Hz data were processed into half-hourly flux data using the flux processing software Eddypro (http://www.licor.com/env/products/eddy_covariance/software.html) developed by LI-COR Biosciences (Lincoln, NE, USA). The flux data processing steps included spike detection, coordinate rotation, time-lag correction, coordinate rotation, sonic virtual temperature correction, frequency-response correction, and density correction [44,45]. Then, the flux data were gap-filled using the marginal distribution sampling (MDS) method and partitioned into GPP and ecosystem respiration (Reco) following the flux partitioning algorithms from the REddyProc package [46].

2.2. MODIS Datasets

The MODIS data used in this study include MODIS GPP data (MOD17A2H products), FPAR data (MOD15A2H products), and Surface Reflectance data (MOD09A1 products) with Collection 6 at 500 m spatial resolution were downloaded directly from the Oak Ridge National Laboratory Distributed Active Center (ORNL DAAC) website. FPAR is the fraction of photosynthetically active radiation (400–700 nm) absorbed by green vegetation, which is a critical component of the MODIS GPP algorithm. To correct inferior values caused by the effects of clouds and aerosols, we reconstructed the MODIS FPAR time series data with Savizky–Golay filter algorithm [47]. Meanwhile, to validate the performance of the MODIS FPAR data in the study area, we observed the actual FPAR data of cropland and desert grassland sites in HRB using AccuPAR (METER Group, Inc., Pullman, USA) during the growing seasons of vegetation in 2012 [43], and then compared the observations with the MOD15A2H FPAR data at corresponding sites. In addition, we also used the MODIS surface reflectance data to derive vegetation indices, such as the normalized difference vegetation index (NDVI). For the site of NTZ, due to the growing season of cropland (i.e., cantaloupe) is short, the desert or low vegetation land cover was identified in the MOD15A2H product, and thus we calculated the FPAR from NDVI data following the empirical formula of: $\mathrm{FPAR} = 1.24 \times \mathrm{NDVI} - 0.168$ [48].

2.3. Description of MOD17A2H Algorithm

The MOD17A2H algorithm is based on light-use efficiency (LUE) approach [49,50], which provides global GPP estimates of 8 day temporal and 500 m spatial resolution [15]. The MODIS GPP product is calculated from the following equation:

$$\mathrm{GPP} = \varepsilon_{\mathrm{max}} \times 0.45 \times \mathrm{SW}_{\mathrm{rad}} \times \mathrm{FPAR} \times f(T_{\mathrm{min}}) \times f(\mathrm{VPD}) \tag{1}$$

where $\varepsilon_{\mathrm{max}}$ is the maximum LUE obtained from the Biome-specified Parameters Look Up Table (BPLUT) on the basis of vegetation type. The BPLUT contains values specifying minimum temperature and VPD limits, specific leaf area and respiration coefficients for the standard land cover classes [48]. $\mathrm{SW}_{\mathrm{rad}}$ is shortwave solar radiation of which 45% is photosynthetically active radiation (PAR), FPAR is the fraction of PAR absorbed by vegetation and the scale factors f(Tmin) and f(VPD) reduce $\varepsilon_{\mathrm{max}}$ under unfavorable conditions of low temperature and high VPD. The forcing data such as $\mathrm{SW}_{\mathrm{rad}}$, Tmin and VPD in the MOD17A2H GPP product were implemented by the Global Modeling and Assimilation Office (GMAO) Reanalysis data. The MODIS GPP algorithm is described in detail in previous literature [12,15,18].

2.4. Parameter Optimization and Uncertainty Analysis

The current MOD17 BPLUT is too general for local regional application [20]. The same set of parameters was applied indiscriminately to diverse types of the same ecosystems, introducing large uncertainties for the simulation of GPP in the arid region. To improve the accuracy of the GPP estimation in desert–oasis–alpine ecosystems in the arid region, we calibrated the parameters of the MOD17 model based on in situ flux tower observations using Bayesian model-data fusion approach. The model parameters were calibrated against GPP time series from the flux tower measurement network through a Bayesian data model synthesis [33,38]. According to Bayesian theory, posterior probability density functions (PDFs) of model parameters (θ) given the existing data (D), denoted $P(\theta \mid D)$, can be obtained from prior knowledge of the parameters and information generated by comparison of simulated and observed variables, and can be described as:

$$P(\theta|D) = \frac{P(\theta)P(D|\theta)}{P(D)} \tag{2}$$

where P(D) is the probability of observed GPP and P(D | θ) is the conditional probability density of observed GPP with prior knowledge, also called the likelihood function for parameter θ.

Given a collection of N measurements, the likelihood function (L) can be expressed as:

$$L = \prod_{i=1}^{N} \frac{1}{\sqrt{2\pi\sigma}} e^{\frac{(X_i - \mu_i)^2}{2\sigma}} \tag{3}$$

where σ represents the standard deviation of the data-model error, X_i represents the ith of N measurements, and μ_i is the model-derived estimates of a measurement.

In our study, we assumed the parameter priors are uniform, and the posterior PDFs for the model parameters were generated from prior PDFs P(θ) with observation data by a Markov chain Monte Carlo (MCMC) sampling technique [33]. Herein, the Metropolis–Hasting algorithm [51,52] was adopted to generate a representative sample of parameter vectors from the posterior distribution. We ran the MCMC chains with 50,000 iterations each, and regarded the first 15,000 iterations as the burn-in period for each MCMC run. All accepted samples from the runs after burn-in periods were used to compute the posterior parameter statistics of the models.

In this study, the MOD17A2H GPP algorithm contains 5 parameters: Maximum light-use efficiency (ε_{max} or LUE_{max}), temperature-constrained factors (T_{min_min}, T_{min_max},) and the water-constrained factors (VPD_{min}, VPD_{max}). The lower and upper bounds of ε_{max} (0.3–3.0 gC/m^2/day/MJ APAR) were determined from the range of ε_{max} used in PEMs [9,15,53]. Following the related References [8,15,50,54], we specified the initial bounds of these parameters: T_{min_min} (°C), T_{min_max} (°C), VPD_{min} (Pa) and VPD_{max} (Pa) as [−35,−2], [6,30], [60,1000], and [1500,6500], respectively.

2.5. Experiment Configuration and Validation

The original MOD17A2H GPP products used the GMAO Reanalysis data as the driving metrological database, and calculated the GPP with the biome based parameters look up table on a global scale. To validate and improve the performances of the MODIS GPP estimations and quantify the uncertainty of the MODIS GPP simulation algorithm (MOD17 model), we replaced the satellite-derived and meteorological inputs in the MOD17 model and compared the modeled GPP estimates with flux tower observations with the following experiment configurations: (1) We firstly assessed the performance of original MOD 17A2H GPP product at spatial resolution of 500m with the tower based GPP. The results of the model validation, in this study, is called GPP_MODIS; (2) we used in situ meteorological data to run the MOD17 algorithm to understand the influence of meteorological inputs (i.e., incoming solar radiation, minimum temperature and vapor pressure deficit) on GPP modelling rather than the GMAO Reanalysis dataset, we called this GPP_Insitu; and (3) we compared the performances between the calibration of one parameter only (ε_{max}) and calibration of all parameters of the MOD17 model to examine the sensitivity of the water and temperature-limited parameters on GPP estimation. The results are called GPP_LUEopt and GPP_Fiveopt, respectively. To understand the effects of parameter uncertainty on GPP simulation, we compared the calibrated MOD17 model algorithm with in situ meteorological inputs from the flux tower network. Similar to GPP_Insitu, GPP_LUEopt, and GPP_Fiveopt were also calculated using the in situ meteorology data. However, for GPP_LUEopt, we only optimized the parameter of ε_{max} by Bayesian approach and other parameters used the default BPLUT parameters in the MOD17 algorithm. Whereas, for GPP_Fiveopt we optimized all the five parameters using the Bayesian approach.

2.6. Statistical Analyses and Model Evaluation

We firstly calculated the daily values of EC-based GPP and then aggregated to the eight-day and annually values for seasonal and yearly GPP validation. To evaluate the performance of the MOD17A2H GPP model, we compared the modeled GPP with the flux tower-estimated GPP both in 8-day and annual time steps. We extracted the eight-day composite MODIS C6 GPP product (MOD17A2H) and the other MODIS products (e.g., MOD15A2H and MOD09A1 products) from the pixels centered on the flux towers, and compared the MODIS GPP product with the EC-based GPP observations.

The model performance (i.e., differences between simulated and tower-based GPP) were quantified by using the coefficient of determination (R^2), root mean squared error (RMSE), and relative RMSE (rRMSE):

$$\text{RMSE} = \sqrt{\frac{1}{n}\sum_{i=1}^{n}(Y_{sim,t} - Y_{obs,t})^2} \tag{4}$$

$$\text{rRMSE} = \frac{1}{\overline{Y_{obs,t}}} \times \sqrt{\frac{1}{n}\sum_{i=1}^{n}(Y_{sim,t} - Y_{obs,t})^2} \times 100 \tag{5}$$

where, Y_{sim} and Y_{obs} represent the simulated and observed GPP data, respectively, and n is the total number of samples. All the statistical analyses and results presentation are performed in Matlab R2016b software (Mathworks, Natick, MA, USA).

3. Results

3.1. Evaluation of MODIS GPP Products and MOD17 Algorithm in the Arid Region

3.1.1. Site-Specific Evaluation of MODIS GPP Products and MOD17 Algorithm

The eight-day EC flux tower GPP (GPP_obs) was compared with the results of MOD17A2H GPP (GPP_MODIS), GPP simulated with the in situ meteorology forcing data (GPP_Insitu), and GPP simulated with optimized maximum LUE parameter (GPP_LUEopt) and with optimized all five parameters (GPP_Fiveopt). As illustrated in Figure 3a, the overall eight-day MOD17A2H GPP products were significantly underestimated when compared with the EC-observed GPP. The RMSE between MOD17 products and in situ flux observations of all sites was 1.80 $gC/m^2/day$, while R^2 was 0.71 and the slope of the model was 0.49, which means the model could only contribute 71% of the tower-observed GPP. As shown in Figure 3b, when we used the in situ meteorology data to simulate the MOD17 model, a better correlation between simulation and observation was found. The model could explain 79% of the observation (the slope was 0.43, R^2 was 0.79), with a large biases close to that of MOD17A2H products. However, the modeled GPP still underestimate as compared to the observed GPP, which means that the meteorology forcing data were not the main reasons for the underestimation of GPP. By contrast, when we optimized the maximum LUE parameter (Figure 3c), a significant improvement of model performance for all sites was seen, with R^2 = 0.86, RMSE = 1.01 $gC/m^2/day$, rRMSE = 6.99%, and the slope of the regression lines was closer to the 1:1 line, which signifies the importance of the LUE parameter in GPP modelling. Furthermore, as we optimized all parameters, a better performance of the model occurred. Almost all the points were close to the 1:1 line (Figure 3d), with R^2 = 0.91, RMSE = 0.81 $gC/m^2/day$, and rRMSE = 5.59%, which indicates the best performance in these simulations.

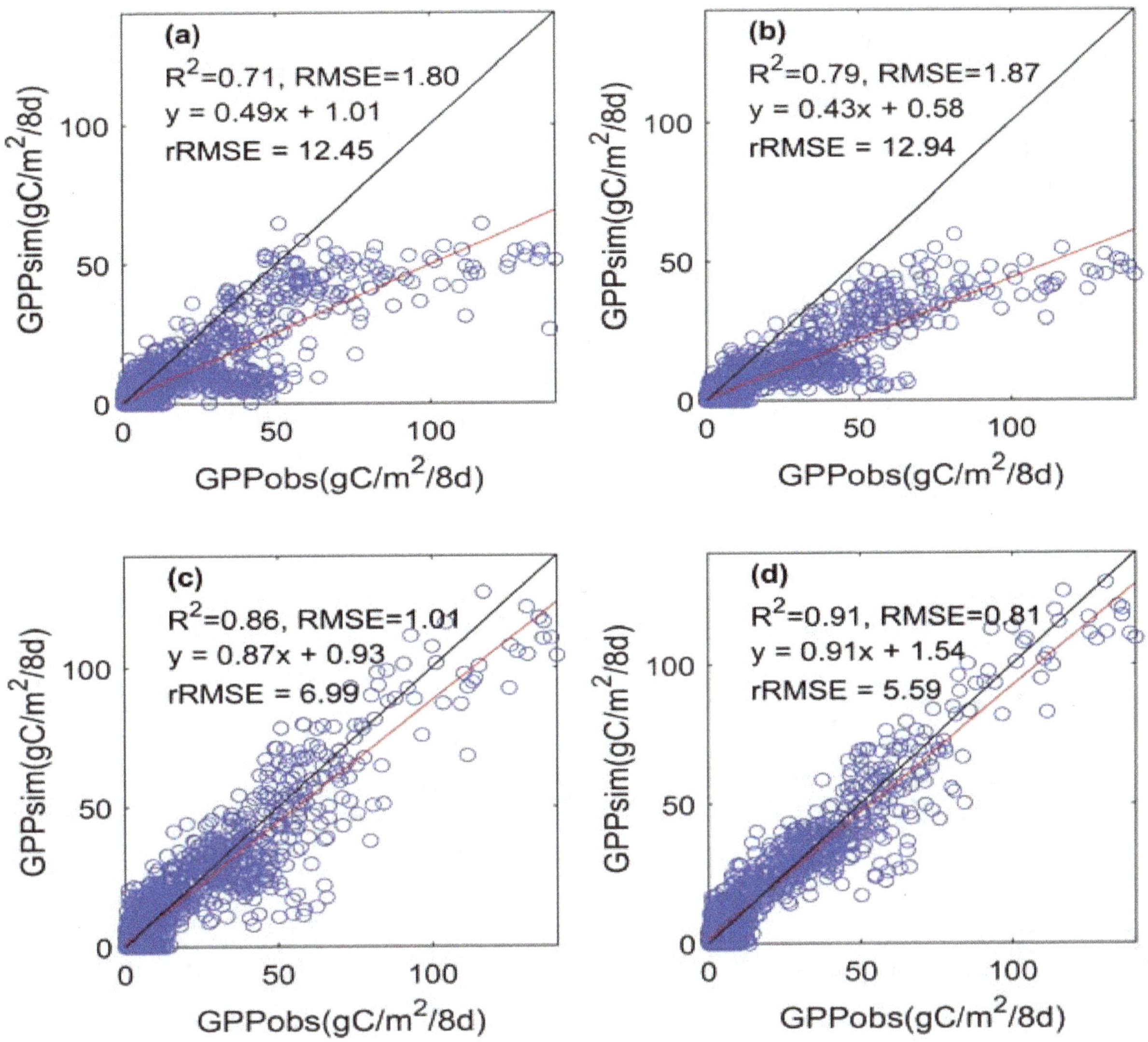

Figure 3. Comparisons of eight-day gross primary productivity (GPP) of MOD17A2H products and GPP simulations by MOD17 model with the flux tower GPP for all sites. Eight-day GPP scatter plots of the EC-observed GPP and (**a**) the original MOD17A2H products; (**b**) in situ meteorology forcing data; (**c**) only LUE optimized results; and (**d**) all parameters optimized results. The units of RMSE and rRMSE are $gC/m^2/day$ and %, respectively.

As we accumulated the observed and simulated GPP at a yearly timescale for every site, a significant underestimation of MOD17A2H GPP products also existed (Figure 4), which were similar to the results of the eight-day time scale. On an annual time scale, the simulated GPP showed a generally good agreement with the tower-observed GPP across all sites with $R^2 = 0.69$, RMSE = 347.31 $gC/m^2/y$ and rRMSE = 60.48% (Figure 4a) A better relationship was found between the modeled GPP and tower-observed GPP ($R^2 = 0.73$). The model was improved by using in situ climate forcing data. However, the modeled GPP was still underestimated as compared with observation. Moreover, the modeled GPP was significantly improved as we optimized the model parameters (Figure 4c,d). The modeled GPP was closer to the observed GPP (almost all points close to the 1:1 line), and all five parameters optimization results were better than for LUE_{max} parameter optimized only with R^2 of 0.87 and 0.92, respectively. The rRMSE was 23.93% and 19.55%, respectively, which signifies the importance of optimizing the temperature and water-constrained factors in arid regions.

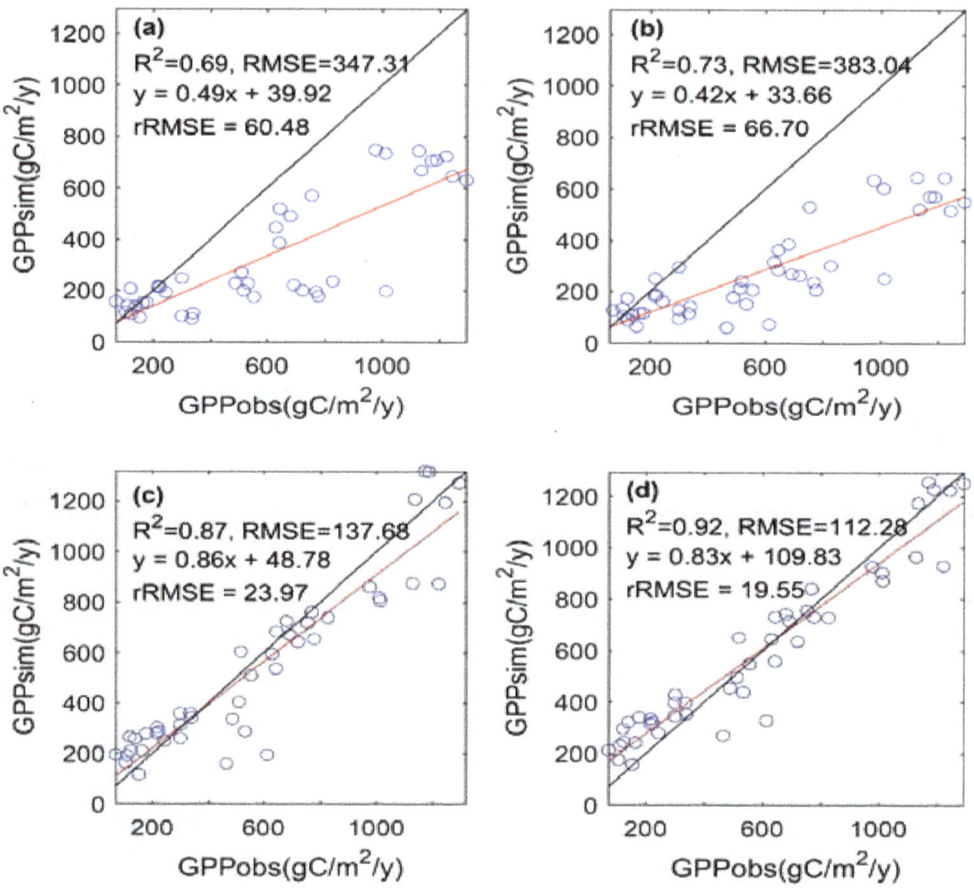

Figure 4. Comparisons of annual GPP of MOD17A2H products and GPP simulations by MOD17 model with the flux tower GPP for all sites: (**a**) Original MOD17A2H products; (**b**) in situ meteorology forcing data; (**c**) only LUE optimized results; and (**d**) all parameters optimized results. The units of RMSE and rRMSE are $gC/m^2/y$ and %, respectively.

3.1.2. Biome-Specific Evaluation of MODIS GPP Product and MOD17 Algorithm

The MOD17A2H GPP products and the other three model estimated GPP based on MOD17 algorithm were compared with the flux-derived eight-day time scale of GPP values for various biome types (Figure 5). We divided the original grassland into two types, grasslands, and desert grasslands, because of the large diversities in species and climate conditions in these sites. As shown in Figure 5, the original MOD17A2H GPP products were significantly underestimated in grassland, cropland and forest ecosystems, but not in the desert ecosystems. A good correlation between MOD17A2H GPP products and EC-observed GPP is illustrated in grassland ecosystems (R^2 = 0.82), followed by the cropland ecosystems (R^2 = 0.80) and forest ecosystems (R^2 = 0.53), while the weakest was in desert ecosystems (R^2 = 0.42). In addition, the slope of the linear regression for the scatter plot can also revealed the biases between MOD17A2H GPP and tower-observed GPP. We can see the slope of linear regression at the forest ecosystems is far from the 1:1 line, which demonstrates the largest biases between MOD17A2H GPP and the tower based GPP, followed by those of the cropland ecosystems, then the grassland and desert ecosystems. Therefore, it indicates that larger biases existed in most ecosystems in the arid regions of China, especially for the forest and cropland ecosystems. As we used the in situ forcing data, we did not find significant improvement for all biome types, and the simulations of GPP forced with in situ data in most ecosystems were still underestimated. However, as we optimized the parameters of the MOD17 model, the GPP simulation results were improved significantly in most ecosystems. The scatter points of modeled GPP and EC-measured GPP were distributed closely around the 1:1 line, indicating that the GPP simulation results can be improved after the parameter optimization of LUE_{max} and other parameters in most ecosystems in the arid region. However, a larger bias still existed even after parameter optimization. The impacts of parameter optimization on GPP simulation of desert ecosystems were less, indicating that it is difficult to effectively simulate the GPP in desert ecosystems in the current MOD17 model.

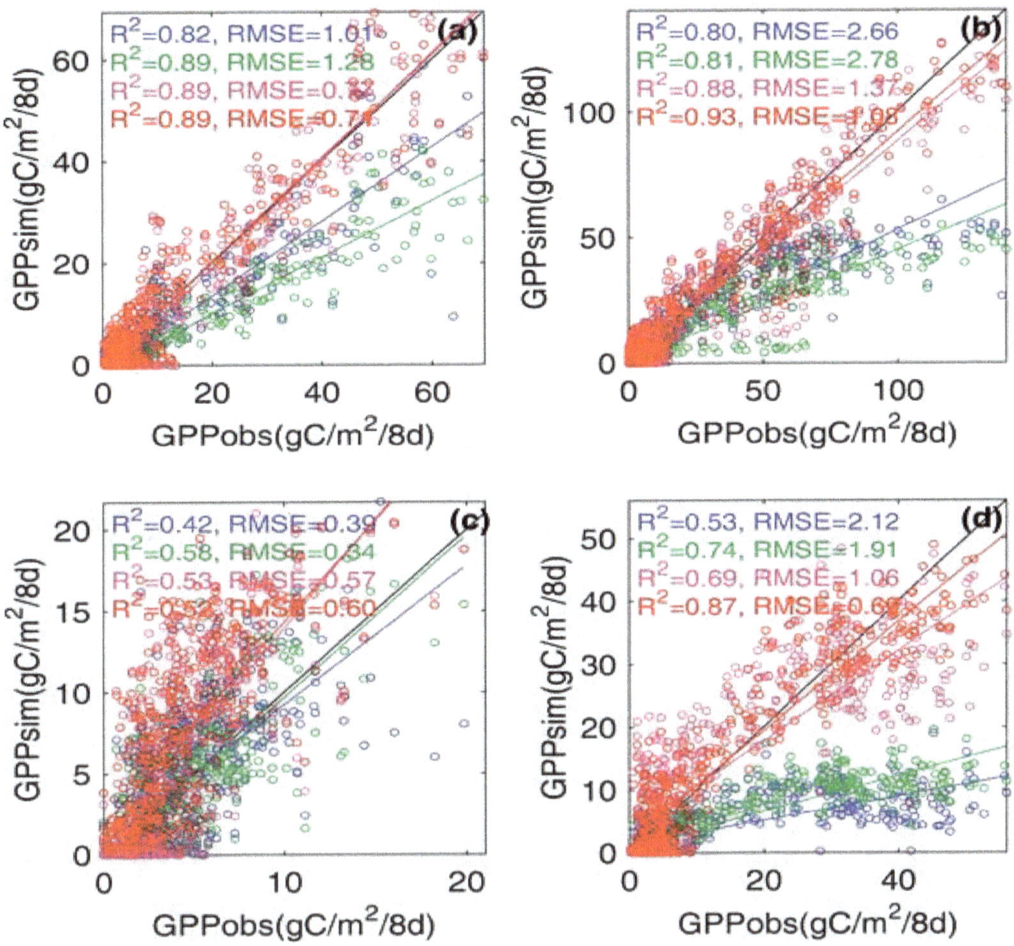

Figure 5. Comparison between eight-day GPP of MOD17A2H products and GPP simulations by MOD17 model with the flux tower GPP for the major ecosystems including: (**a**) Grassland; (**b**) cropland; (**c**) desert steppe; and (**d**) forest. The unit of RMSE is $gC/m^2/day$. The blue points represent original MOD17A2H products; green points represent in situ meteorology forcing data; pink points represent only LUE optimized results; and red points represent all parameters optimized results.

3.1.3. Site-Specific Evaluation of MODIS GPP Products and MOD17 Algorithms

The flux tower-observed GPP were compared with the original MOD17A2H GPP and GPP estimated from the MOD17 model with in situ meteorology forcing data (GPP_Insitu), LUE optimized (GPP_LUEopt) and five optimized parameters (GPP_Fiveopt) (Figure 7 and Table 2). Figure 6 illustrates the scatter plots between EC GPP and simulated GPP at the eight-day time scale at all sites. From the slope of linear regression for the scatter plot in Figure 6, most of the slope values were less than 1.0, which revealed the MOD17A2H GPP in most of the sites were obviously underestimated, as compared with the flux tower-observed GPP (Figure 6), except for the three desert grassland sites, where MODIS GPP was close to the observed GPP in most cases. However, relatively large biases existed in the desert sites. While all sites of MODIS GPP were underestimated except the desert sites, a good correlation between MOD17A2H GPP and tower-observed GPP was shown in grassland and cropland sites (coefficients of determination were greater than 0.7), followed by forest and desert ecosystems. After modelling GPP using in situ climate data, a better correlation between modeled GPP and observed GPP occurred in most sites. However, there were still apparently underestimations in most sites, which means the forcing data were not the main reason of the underestimation of GPP. Instead of the forcing data, the inappropriate BPLUT parameters were the main source of the uncertainty of GPP simulation. After the optimization of LUE and other parameters in the MOD17 model, GPP in most sites was improved significantly. Meanwhile, the performance of optimization of five parameters was better than that of only optimization of the LUE parameter. As shown in Table 2 and Figure 6, good performance of GPP simulation was observed in DMZ, ARZ, and SDZ (R^2 were greater than 0.9). However, the MODIS GPP showed a moderate performance in capturing the corresponding GPP simulation of desert ecosystems. Overall, the current MODIS GPP model correctly simulated the dynamics of GPP at most sites in the arid region. After the parameter optimization, the coefficients of determination were improved apparently, and the RMSE of most sites was less than 1 $gC/m^2/day$.

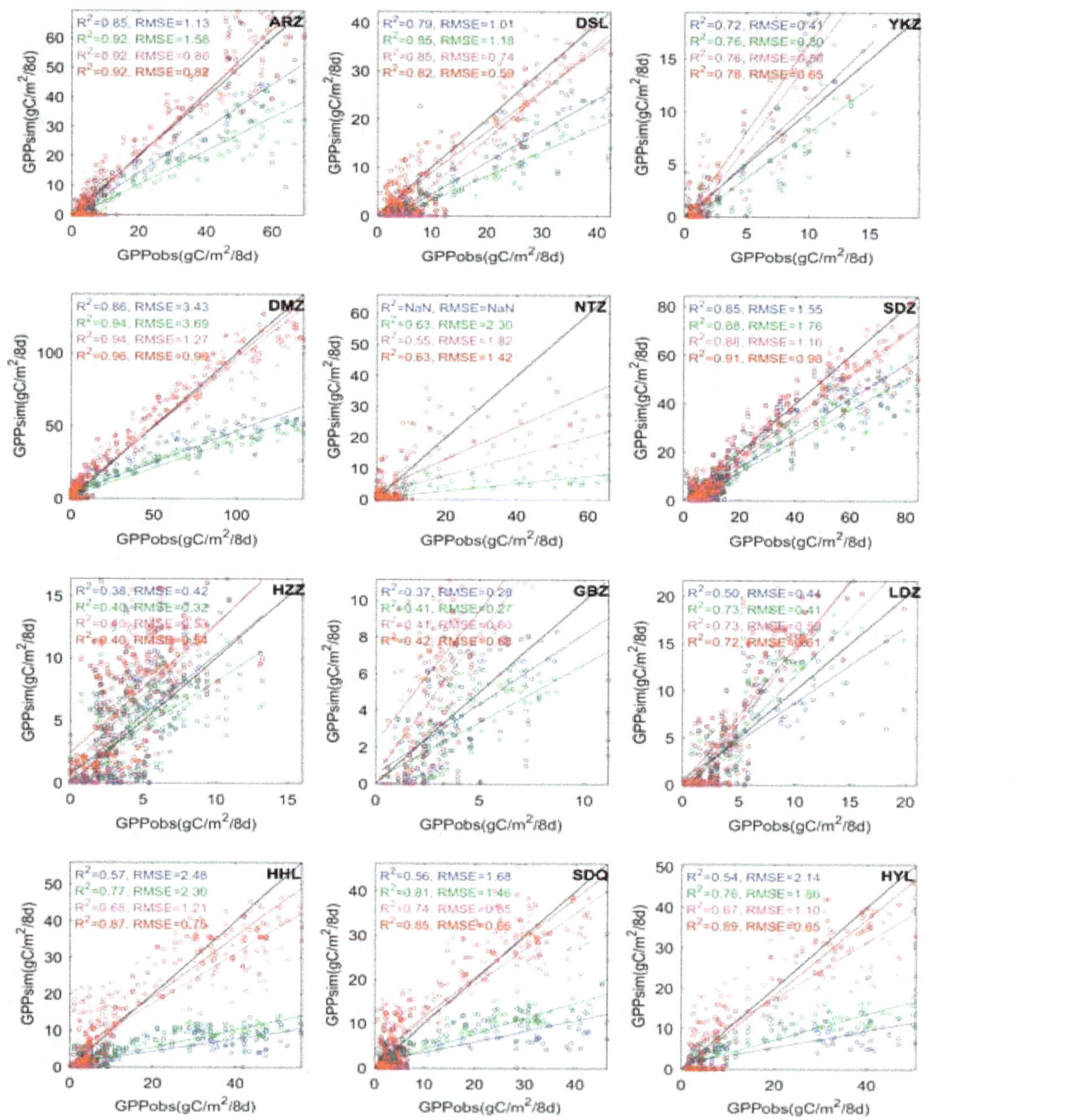

Figure 6. Time series of eight-day MODIS GPP and GPP simulations derived from the MOD17 model with the tower-estimated GPP. The full name for each site is listed in Table 1. The blue points represent original MOD17A2H products; green points represent in situ meteorology forcing data; pink points represent only maximum LUE optimized results; and red points represent all parameters optimized results.

Table 2. A summary of the performances of the MOD17 algorithm (GPP_MODIS) and the in situ metrological data forced GPP, LUEmax parameter optimized GPP (GPP_LUE), and five parameters optimized GPP (GPP_Fiveopt). GPP_LUE and GPP_Fiveopt were estimated from the in situ climate data. In the GPP_Insitu and GPP_LUE algorithms, the default values for model parameters were used in MOD17 for the original land cover types and optimal parameter values for the optimization approach.

	GPP_MODIS		GPP_Insitu		GPP_LUE		GPP_Fiveopt	
	R^2	RMSE	R^2	RMSE	R^2	RMSE	R^2	RMSE
ARZ	0.85	1.13	0.92	1.58	0.92	1.58	0.92	0.82
DSL	0.79	1.01	0.86	1.18	0.85	0.74	0.82	0.59
YKZ	0.72	0.41	0.76	0.30	0.76	0.58	0.78	0.65
HZZ	0.38	0.42	0.40	0.32	0.40	0.53	0.40	0.54
GBZ	0.37	0.28	0.41	0.27	0.41	0.60	0.42	0.68
LDZ	0.50	0.44	0.73	0.41	0.73	0.59	0.72	0.61
DMZ	0.86	3.43	0.94	3.69	0.94	1.27	0.96	0.96
NTZ	-	-	0.63	2.30	0.55	1.82	0.63	1.42
SDZ	0.85	1.55	0.88	1.76	0.88	1.16	0.91	0.98
HHL	0.57	2.48	0.77	2.30	0.68	1.21	0.87	0.75
SDQ	0.56	1.68	0.81	1.46	0.74	0.85	0.85	0.66
HYL	0.54	2.14	0.76	1.86	0.67	1.10	0.89	0.65

3.2. Uncertainty of Satellite Data in MODIS GPP Simulation over Ecosystems in the Arid Region

3.2.1. Impacts of the Accuracy of the Land Cover Classification on MODIS GPP Simulation

One of the first MODIS products used in the MOD17 algorithm is the Land Cover Product, MOD12Q1. The importance of this product cannot be overstated as the MOD17 algorithm relies heavily on land cover type through use of the BPLUT [15]. Based on the locations of the flux tower sites, we obtained the land cover type of each site from the MCD12Q1 results, and compared them with the actual land cover types. We found that MCD12Q1 misclassified most of the sites downstream of the HRB (Figure 7), which is an artificial oasis ecotone with sparse vegetation in the extremely arid region of China. For example, in our study, the land cover type of the NTZ site is a cropland ecosystem; however, it was classified as grassland in the MCD12Q1 land cover data (Figure 7). In addition, MCD12Q1 also misclassified the forest types at the sites of HHL, SDQ, and HYL as grassland.

(a) (b) (c) (d)

Figure 7. Misclassification of land cover in the MOD12Q1 products at the downstream HRB, which classified the land cover of forest (i.e., SDQ, HHL, HYL) and cropland (NTZ) as the grassland type in MOD12Q1 data. (**a**) SDQ, (**b**) HHL, (**c**) HYL, and (**d**) NTZ.

3.2.2. Impacts of Uncertainty of FPAR Data on MODIS GPP Simulation

Figure 8 showed the comparisons of AccuPAR observed FPAR data with the MOD15A2H FPAR data in the corresponding sites in HRB. We found the MODIS FPAR data was overestimated compared to the observations in the growing season of desert grassland sites, as well as the low values of FPAR in the cropland sites, and underestimated in some stages of the high values of FPAR in the cropland sites. The overestimated FPAR impacted the APAR, thus leading to an overestimated GPP. In contrast, the underestimated FPAR would underestimate the GPP. Meanwhile, a good correlation between MODIS FPAR and observed FPAR occurred in cropland sites, while in the desert grassland sites, no significant relationship was found. This revealed that the accuracy of the current MOD15A2H FPAR data in the arid region needs to be further improved, and could also be an important source of uncertainty for the estimation of GPP in desert ecosystems.

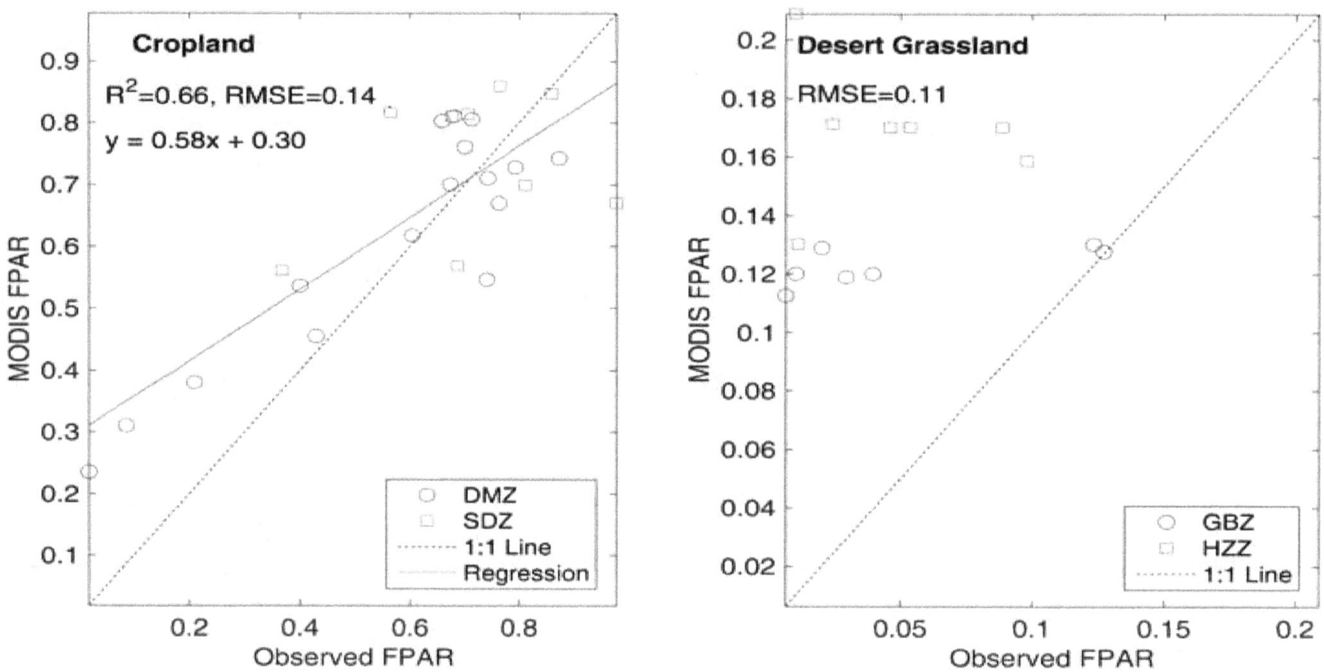

Figure 8. Validation of MODIS FPAR compared with FPAR measured by AccuPAR. The left plot is the validation at DMZ and SDZ, the right plot is the validation at GBZ and HZZ.

3.3. Uncertainty and Variability of Biophysical Parameters for Diversity Ecosystems in Arid Regions

Since the performance after calibration of all five parameters of the MOD17 model was better than after the calibration of only the parameter ε_{max}, indicating the important role of temperature and water-constrained factors in the estimation of GPP in the arid region. We thus calibrated all the parameters of MOD17 algorithms (Table 3). Our study illustrated that variability of biophysical parameters not only exist across different ecosystems, but also within the same ecosystems, such as the diverse biophysical parameters of grassland ecosystems and the desert grassland ecosystems. The current version of MOD17 BPLUT does not consider the differentials of these two types—they shared the same BPLUT parameters of grassland. However, there are different climate conditions and species in these two ecosystems in the study region. Meanwhile, there are different photosynthesis paths between C3 cropland and C4 cropland, which have many differences in their biophysical properties. However, these two types are also shared in the current version of MOD17 BPLUT.

The value of ε_{max} is biome specific, representing the maximum LUE of the corresponding vegetation in the process of photosynthesis. For a given biome type, the value of ε_{max} is constant and assigned by the MOD17 BPLUT. While the newly released version of BPLUT has corrected and updated the ε_{max} values, the ε_{max} value were still significantly underestimated in the main ecosystems in arid regions (Table 3). The mis-estimation of their values inherently further reduced the accuracy of GPP estimations.

Table 3. Prior distribution (initial value and range) and posterior distribution (mean value and 95% confidence interval) of the parameters of the MOD17 model for all sites. For the parameters ε_{max} (gC/MJ APAR), T_{min_min} (°C), T_{min_max} (°C), VPD_{min} (Pa), and VPD_{max} (Pa), we set the original values of MOD17 BPLUT as the initial values (with bold font).

Sites	ε_{max}	T_{min_min}	T_{min_max}	VPD_{min}	VPD_{max}
Prior range	(0.3,3)	(−35,−2)	(6,30)	(60,1000)	(1500,6500)
ARZ	**1.044**, 2.10(1.62,2.94)	**−8**, −10.29(−11.33,−9.29)	**12.02**, 15.52(10.16, 26.63)	**650**, 151.37(64.48,296.64)	**4300**, 3252.21(2801.68, 4009.33)
DSL	**1.044**, 1.24(0.90,1.75)	**−8**, −30.28(−34.79,−22.79)	**12.02**, 18.84(6.78, 29.44)	**650**, 430.98(81.46,961.71)	**4300**, 3960.64(1732.84, 6361.79)
YKZ	**1.044**, 1.80(0.97, 2.53)	**−8**, −15.30(−32.13,−3.62)	**12.02**, 20.49(6.94, 29.58)	**650**, 567.33(96.08,978.23)	**4300**, 4136.91(1760.66, 6397.26)
HZZ	**1.044**, 1.68(1.21, 2.42)	**−8**, −20.76(−34.33,−3.14)	**12.02**, 22.62(7.30, 29.53)	**650**, 608.86(81.31,977.20)	**4300**, 4936.72(3135.86, 6400.44)
GBZ	**1.044**, 2.46(1.87, 2.95)	**−8**, −18.93(−34.14, −2.95)	**12.02**, 13.71(6.47, 27.14)	**650**, 592.37(92.14,982.30)	**4300**, 4654.96(3150.60, 6424.90)
LDZ	**1.044**, 1.16(0.89, 2.04)	**−8**, −17.23(−34.07, −2.79)	**12.02**, 13.83(6.39, 28.65)	**650**, 568.79(98.05,977.54)	**4300**, 5006.36(3050.36, 6436.59)
DMZ	**0.860**, 2.89(2.73, 2.99)	**−8**, −3.91(−4.82,−3.08)	**12.02**, 16.37(15.66, 17.58)	**650**, 176.02(66.63,377.42)	**5300**, 6089.55(5664.46, 6457.04)
NTZ	**0.860**, 3.0(2.98, 3.0)	**−8**, −28.59 (−34.78, −11.21)	**12.02**, 8.27(6.14, 11.13)	**650**, 991.88(961.92,999.73)	**5300**, 6486.19(6432.50, 6499.42)
SDZ	**0.860**, 1.59(1.41, 1.74)	**−8**, −32.82(−34.91,−27.88)	**12.02**, 25.76(19.61, 29.68)	**650**, 237.30(66.65,519.04)	**5300**, 6329.98(5928.29, 6492.67)
HHL	**1.051**, 2.89(2.63, 3.0)	**−8**, −18.54(−26.41, −10.64)	**8.61**, 22.26(19.30, 26.62)	**650**, 771.23(229.59,992.57)	**4800**, 6435.17(6233.17, 6497.78)
SDQ	**1.268**, 2.33(1.80, 2.94)	**−8**, −18.07(−33.83, −2.82)	**9.09**, 20.25(11.59, 29.25)	**800**, 523.27(82.79,972.83)	**3100**, 5947.81(5274.70, 6461.10)
HYL	**1.165**, 2.71(2.28, 2.98)	**−6**, −6.20(−13.10,−2.44)	**9.94**, 25.11(21.11, 29.50)	**650**, 619.61(108.53,980.37)	**1650**, 6410.51(6143.55, 6496.60)

Meanwhile, the large variations in the temperature and water-constrained stress factors also existed due to the diversity of climate conditions in different parts of HRB (Table 3). For example, the climate is cold and humid in the upstream HRB, therefore, the temperature stress factor has a great impact on GPP estimation in the grassland ecosystems in the upstream HRB. However, as Table 3 reveals, the original MOD17 BPLUT overestimated the parameters of the minimum temperature stress factors and the VPD_{max} values. In comparison, the climate is extremely arid in the downstream HRB, however, the original MOD17 BPLUT underestimated the parameters of the maximum temperature stress factors and the VPD_{max} values.

In addition, the Bayesian approach can estimate the posterior distribution of model parameters, which is a useful tool to reduce model uncertainty. Using the Bayesian approach, the uncertainty of model parameters was reduced significantly for some sites (e.g., the NTZ site) located in the extremely arid region (Figure 9).

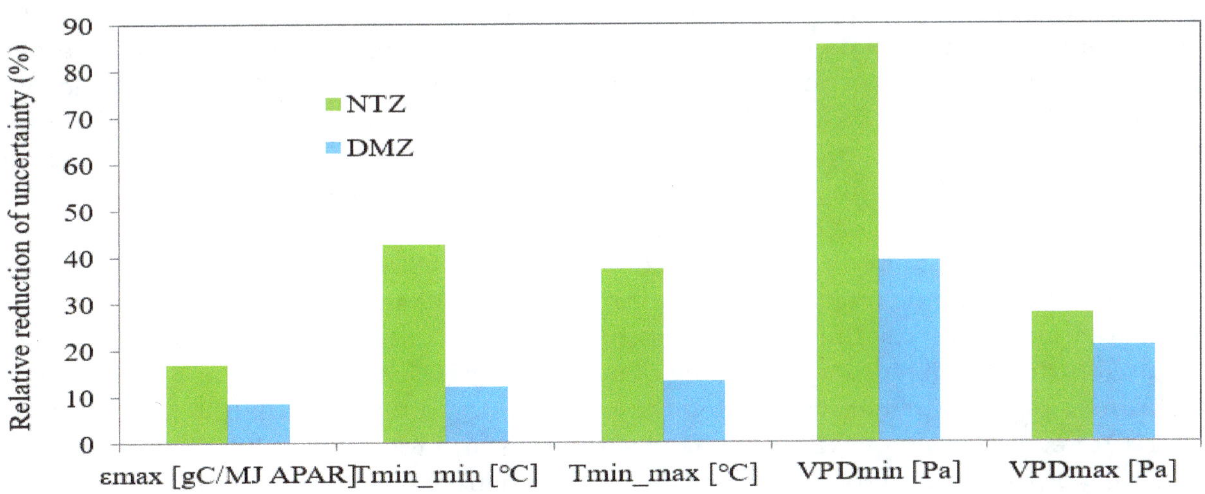

Figure 9. Relative reduction of parameter uncertainty (95% confidence interval) from prior to posterior distribution. The green bar and blue bar represent the reduction of uncertainty in model parameters for the MOD17A2H model at the NTZ and DMZ sites.

4. Discussion

4.1. Evaluations of the MOD17A2H Products over Diversity Ecosystems in the Arid Region

The MODIS Collection 6 GPP products improved the spatial resolution of GPP estimation, which means the estimated GPP is more comparable with the footprint in the areas with heterogeneous landscapes in the desert–oasis–alpine ecosystems in the arid region. Meanwhile, the MOD17A2H products updated the meteorological forcing data, FPAR data and land cover data, which highlight the better spatial resolution of 500 m. However, compared to the flux tower-based GPP data, the MOD17A2H GPP products still illustrate some limitations in the simulations of magnitude and spatial-temporal variation of GPP in the desert–oasis–alpine ecosystems in the arid region. From the slope of linear regression for the scatter plot in Figures 3a and 4a, the slope values were only 0.49, which revealed significant underestimation of the GPP in the study region. When compared to the site level of flux tower-based GPP (Figure 6), the slope values in most sites of the study regions were also less than 1.0, except for some desert ecosystem sites. This showed that the MOD17 product underestimated GPP in most high productivity sites of cropland, grassland, and forest ecosystems in the arid region, but overestimated GPP at some low productivity sites of desert ecosystems compared with tower-based GPP, consistent with the results of Reference [12] and Reference [27].

4.2. Uncertainty of Input Data in MODIS GPP Estimation in Diversity Ecosystems in the Arid Region

The accuracy of GPP estimation highly depends on the precision of all input data of the MOD17A2H GPP algorithm. Therefore, uncertainties of GPP products arise mainly from the climate drivers, parameter variability, and land cover classification [20]. There are three meteorological data types (PAR, T_{min} and VPD), as well as FPAR and land cover classification data involved in the MOD17A2H GPP algorithm, which could be the main source of error in the GPP estimates. The MOD17A2H products used GMAO Reanalysis data for direct meteorological inputs, which is an hourly time-step data set with about a half-degree spatial resolution (0.5 latitude degree by 0.67 longitude degree) generated by the Goddard Earth Observing System Model, Version 5 (GEOS-5) data assimilation system [15]. In this study, we replaced the GMAO dataset with in situ meteorological data and recalculated the MOD17 algorithm with default parameters in comparison (GPP_Insitu). Our study revealed that using the in situ forcing data can improve the relationship between modeled GPP and tower-observed GPP compared to the original MOD17A2H products both at eight-day and annual timescales (Figures 3 and 4); the determination coefficients (R^2) of these sites were slightly higher than that of the original MOD17 products (R^2 ranging from 0.71 to 0.79 for eight-day step and 0.69 to 0.73 for annual step). However, larger biases still exist between GPP_Insitu and GPP_tower. Using in situ meteorological data did not result in obvious improvements of the GPP estimation performances; on the contrary, some sites were not even as accurate as those calculated with the GMAO datasets (ARZ, DSL, DMZ, and SDZ in Table 2), which is similar to the other results from validation of the MOD17 GPP products [21,27,32]. This is caused by some missing values in the original MOD17A2H GPP products making a shorter length of model evaluations, thus reducing the model errors of the GPP_MODIS. The other implication of the results is that an improvement in meteorological data did not have a significant effect on the MODIS GPP estimation, which means the meteorological data is not the main source of uncertainty in GPP simulation in the arid region.

An accurate land cover classification map is vital to MOD17 GPP simulation [18]. Misclassification of the land cover directly determines the value of maximum light use efficiency and the other MOD17 BPLUT parameters, thus further influencing the inaccuracies of GPP calculation [20]. We validated the MOD12Q1 vegetation maps with our site observations and found the MODIS data misclassified almost all sites of forest and cropland types in the downstream HRB (Figure 7). Study suggested that the accuracies of MOD12Q1 vegetation maps are within 65–80%, and most inaccuracies are in between similar classes [55]. Since large desert–natural oasis ecosystems are distributed in the downstream HRB and most of the vegetation cover was less than 30%, the 500 m unit of MODIS land cover classification could pose a risk at such a coarse resolution. Mixed pixels, composed of varied ecosystem types, may occur in the sparsely vegetated region, thus making it difficult to describe the biophysical parameters properly. This incorrect classification of land cover types will therefore lead to an inaccurate GPP calculation.

In addition, FPAR is also an important input physiology variable in the MOD17 model, which directly modulates the essential energy input to photosynthetic processes [8,9]. In our study, we compared the MOD15A2H FPAR products with the observations in the study area, and found it significantly overestimated the ecosystems with low productivity (such as the desert ecosystems) and underestimated the ecosystems with high productivity (such as the crop ecosystems) in MODIS FPAR products in the HRB (Figure 8). This will greatly impact the energy redistribution in photosynthetic systems, and thus the GPP estimations in arid regions. Research revealed that FPAR often produces misleading signals in GPP estimations due to contamination by atmospheric characteristics [19]. The overestimation of FPAR data is caused by sparse vegetation cover and the effects of large desert cover that impacts the signals of vegetation detection in arid regions. To improve the FPAR estimation in the arid region, we can use the improved FPAR retrieval products with the multi-angle vegetation index information in the future [56].

4.3. Uncertainty and Variability of Biophysical Parameters in Modelling GPP over Diversity Ecosystems in Arid Regions

To estimate GPP across varied worldwide ecosystems, MOD17 algorithms use the biophysical variability of parameters generated from look up tables, which include five biome-specific physiological parameters in the model, i.e., ε_{max}, T_{min_min}, T_{min_max}, VPD_{min}, and VPD_{max} [56]. In this study, we optimized the maximum LUE, one of the most important parameters for GPP estimation, with the tower-based GPP. An obvious improved performance of GPP simulation can be found in Figures 3c and 4c, which show good agreements between the observed GPP and the simulated GPP with the parameter optimized (GPP_LUE). The R^2 increased from 0.71 to 0.86, and rRMSE decreased from 12.45% to 6.99%, at an eight-day timescale. Those improved performances were also seen at an annual timescale, with R^2 increasing from 0.69 to 0.87, and rRMSE decreasing from 60.48% to 23.97%. Table 2 reveals the performance of the GPP_LUE model and flux GPP observation were better than the results of only using the ground forcing data; the determination coefficients increased significantly when using optimized ε_{max} parameter, with smaller RMSE, which revealed greater improvements after performing LUE optimization.

Calibration of the maximum LUE parameter improved the performance of MODIS GPP estimation was in accordance with other studies [24,27,32]. In addition, we also investigated the potential impacts of uncertainty of the other model parameters (e.g., VPD-limited factors) on GPP monitoring, which has been overlooked by other researchers. In fact, water stress is one of the most important limiting factors controlling terrestrial primary production, especially in arid regions. Previous studies showed that the MOD17 products underestimate water stress, and thus overestimate GPP in some extremely arid regions lacking in water [57]. In our study, we found that optimizing the VPD-limited factors can further improve the performance of the GPP estimations. From the eight-day time step of the overall performances, the R^2 increased from 0.86 for the results of only optimizing the LUE parameter to 0.91 for the results of optimizing all parameters, and rRMSE decreased from 6.99% to 5.59% (Figure 3). These results were mainly distributed in some sites in the extremely arid region (i.e., SDQ, HHL, HYL etc.), which revealed the important role of the parameters of water-constrained factors in GPP simulation in arid regions.

4.4. Uncertainty of GPP Modelling of the Desert Ecosystems and Its Implications for GPP Simulation in Arid Regions

The current MOD17 model can effectively simulate GPP of main ecosystems in the arid region, however, there are still some difficulties in simulating GPP more accurately in the desert ecosystems.

Model analyses indicate the importance of arid regions in the global carbon cycle, while the models suffer from a lack of data in water-limited regions [3,4]. The large errors of GPP simulation in desert ecosystems is caused by the uncertainty of remote sensing vegetation products in regions with large heterogeneity of landscape and low vegetation cover. Moreover, the uncertainty of flux tower observation in desert ecosystems makes it is difficult to estimate a relative 'true' value of GPP [58]. To improve GPP estimation in arid regions, several directions can be explored further in the future. For example, improving the estimation of MODIS FPAR and land cover classification products in arid regions using data-driven approaches [59] and improving model structures [60] could be better choices for improving GPP simulation in arid regions.

Meanwhile, since the biome-specific look up tables (BPLUT) are constant for a given biome at any time. Since the current BPLUT of the MOD17 cannot meet the needs of accurate definition of the parameters for all ecosystems [19,61], especially for the diverse and complex ecosystems in arid regions, further research needs to be done to update these BLUPT of the model. In addition to update the parameter of ε_{max}, the water and temperature-limited parameters are also of great importance in GPP estimation, especially for the ecosystems in arid regions. As the development of eddy covariance technique, there are more than 900 EC flux sites in the world currently [62]. With the availability of these large number of flux datasets, it provided us the opportunity to retrieve the biome specific parameters for each vegetation type more reasonable, which may improve the accuracy of the current GPP simulation in the arid region.

5. Conclusions

This study validated and optimized the performance of MODIS-derived GPP compared with EC-observed GPP at seasonal and annual time scales for the main arid ecosystems relying on flux networks constructed in arid and semi-arid ecosystems in China. Our study revealed that the current MODIS GPP products were significantly underestimated, as compared with the tower-observed GPP for most types of ecosystems in the arid region of China, especially the irrigated cropland and forest ecosystems, due to uncertainty of meteorological data and model parameters. Using ground-based meteorological data and updated land use data can improve GPP estimation. In addition to the light-use efficiency parameters, the temperature-limited stress factors and the VPD-limited factors also need to be recalibrated for ecosystems in arid regions. After using the proper model parameters, great improvements to the GPP model can be performed through a Bayesian approach. However, it is difficult to estimate GPP accurately in desert ecosystems because of the uncertainty of remote sensing vegetation products in arid regions. Hence, improvements in modelling GPP in desert ecosystems are needed in future studies. Moreover, this study implies that the current MODIS-derived GPP product requires further improvements to provide accurate monitoring of terrestrial ecosystem productivity in arid regions worldwide.

Author Contributions: H.W. and X.L. designed the research, analyzed data and prepared the manuscript. M.M. and L.G. provided suggestions to data analyses. All of the authors thoroughly reviewed and edited this paper.

Acknowledgments: We thank the two reviewers' insightful and helpful comments and suggestions to our manuscript. We are grateful to Kashif Jamal for the help of English editing of our revised manuscript and appreciate the availability of the MODIS data from the Oak Ridge National Laboratory Distributed Active Center (ORNL DAAC) website.

References

1. Asner, G.P.; Archer, S.; Hughes, R.F. Net changes in regional woody vegetation cover and carbon storage in Texas dry-lands, 1937–1999. *Glob. Chang. Biol.* **2003**, *9*, 316–335. [CrossRef]
2. Bastin, J.F.; Berrahmouni, N.; Grainger, A.; Maniatis, D.; Mollicone, D.; Moore, R.; Patriarca, C.; Picard, N.; Sparrow, B.; Abraham, E.M.; et al. The extent of forest in dryland biomes. *Science* **2017**, *356*, 635–638. [CrossRef] [PubMed]

3. Ahlström, A.; Raupach, M.R.; Schurgers, G.; Smith, B.; Arneth, A.; Jung, M.; Reichstein, M.; Canadell, J.G.; Friedlingstein, P.; Jain, A.K.; et al. The dominant role of semi-arid ecosystems in the trend and variability of the land CO_2 sink. *Science* **2015**, *348*, 895–899. [CrossRef] [PubMed]
4. Poulter, B.; Frank, D.; Ciais, P.; Myneni, R.B.; Andela, N.; Bi, J.; Broquet, G.; Canadell, J.G.; Chevallier, F.; Liu, Y.Y.; et al. Contribution of semi-arid ecosystems to interannual variability of the global carbon cycle. *Nature* **2014**, *509*, 600–603. [CrossRef] [PubMed]
5. Lal, R. Carbon sequestration in dryland ecosystems. *Environ. Manag.* **2004**, *33*, 528–544. [CrossRef] [PubMed]
6. Reynolds, J.F.; Smith, D.M.; Lambin, E.F.; Turner, B.L.; Mortimore, M.; Batterbury, S.P.; Downing, T.E.; Dowlatabadi, H.; Fernández, R.J.; Herrick, J.E.; et al. Global desertification: Building a science for dryland development. *Science* **2007**, *316*, 847–851. [CrossRef] [PubMed]
7. Beer, C.; Reichstein, M.; Tomelleri, E.; Ciais, P.; Jung, M.; Carvalhais, N.; Rödenbeck, C.; Arain, M.A.; Baldocchi, D.; Bonan, G.B.; et al. Terrestrial gross carbon dioxide uptake: Global distribution and covariation with climate. *Science* **2010**, *329*, 834–838. [CrossRef] [PubMed]
8. Running, S.W.; Nemani, R.R.; Heinsch, F.A.; Zhao, M.; Reeves, M.; Hashimoto, H. A continuous satellite-derived measure of global terrestrial primary production. *Bioscience* **2004**, *54*, 547–560. [CrossRef]
9. Xiao, X.M.; Zhang, Q.Y.; Braswell, B.; Urbanski, S.; Boles, S.; Wofsy, S.; Berrien, M.; Ojima, D. Modeling gross primary production of temperate deciduous broadleaf forest using satellite images and climate data. *Remote Sens. Environ.* **2004**, *91*, 256–270. [CrossRef]
10. Yuan, W.P.; Liu, S.G.; Zhou, G.S.; Zhou, G.Y.; Tieszen, L.L.; Baldocchi, D.; Bernhofer, C.; Gholz, H.; Goldstein, A.H.; Goulden, M.L.; et al. Deriving a light use efficiency model from eddy covariance flux data for predicting daily gross primary production across biomes. *Agric. Forest Meteorol.* **2007**, *143*, 189207. [CrossRef]
11. Ma, X.; Huete, A.; Moran, S.; Ponce-Campos, G.; Eamus, D. Abrupt shifts in phenology and vegetation productivity under climate extremes. *J. Geophys. Res. Biogeosci.* **2015**, *120*, 2036–2052. [CrossRef]
12. Turner, D.P.; Ritts, W.D.; Cohen, W.B.; Gower, S.T.; Running, S.W.; Zhao, M.S.; Costa, M.H.; Kirschbaum, A.A.; Ham, J.M.; Saleska, S.R.; et al. Evaluation of MODIS NPP and GPP products across multiple biomes. *Remote Sens. Environ.* **2006**, *102*, 282–292. [CrossRef]
13. Biederman, J.A.; Scott, R.L.; Bell, T.W.; Bowling, D.R.; Dore, S.; Garatuza-Payan, J.; Kolb, T.E.; Krishnan, P.; Krofcheck, D.J.; Litvak, M.E.; et al. CO_2 exchange and evapotranspiration across dryland ecosystems of southwestern North America. *Glob. Chang. Biol.* **2017**, *23*, 4204–4221. [CrossRef] [PubMed]
14. Zhao, M.; Running, S.W.; Nemani, R.R. Sensitivity of Moderate Resolution Imaging Spectroradiometer (MODIS) terrestrial primary production to the accuracy of meteorological reanalyses. *J. Geophys. Res.* **2006**, *111*, G01002. [CrossRef]
15. Running, S.W.; Zhao, M. *Daily GPP and Annual NPP (MOD17A2/A3) Products NASA Earth Observing System MODIS Land Algorithm*; MOD17 User's Guide; University of Montana: Missoula, MT, USA, 2015; pp. 1–28.
16. Chasmer, L.; Barr, A.; Hopkinson, C.; McCaughey, H.; Treitz, P.; Black, A.; Shashkov, A. Scaling and assessment of GPP from MODIS using a combination of airborne lidar and eddy covariance measurements over jack pine forests. *Remote Sens. Environ.* **2009**, *113*, 82–93. [CrossRef]
17. Running, S.W.; Zhao, M. *MOD17A2H MODIS/Terra Gross Primary Productivity 8-Day L4 Global 500 m SIN Grid V006*; USGS: Reston, VA, USA, 2015.
18. Heinsch, F.A.; Zhao, M.; Running, S.W.; Kimball, J.S.; Nemani, R.R.; Davis, K.J.; Bolstad, P.V.; Cook, B.D.; Desai, A.R.; Ricciuto, D.M.; et al. Evaluation of remote sensing based terrestrial productivity from MODIS using regional tower eddy flux network observations. *IEEE Trans. Geosci. Remote Sens.* **2006**, *7*, 1908–1925. [CrossRef]
19. Tang, X.; Li, H.; Huang, N.; Li, X.; Xu, X.; Ding, Z.; Xie, J. A comprehensive assessment of MODIS-derived GPP for forest ecosystems using the site-level FLUXNET database. *Environ. Earth Sci.* **2015**, *74*, 5907–5918. [CrossRef]
20. Zhao, M.; Heinsch, F.A.; Nemani, R.R.; Running, S.W. Improvements of the MODIS terrestrial gross and net primary production global data set. *Remote Sens. Environ.* **2005**, *95*, 164–176. [CrossRef]
21. Niu, B.; He, Y.T.; Zhang, X.Z.; Fu, G.; Shi, P.L.; Du, M.Y.; Zhang, Y.J.; Zong, N. Tower-Based Validation and Improvement of MODIS Gross Primary Production in an Alpine Swamp Meadow on the Tibetan Plateau. *Remote Sens.* **2016**, *8*, 592. [CrossRef]
22. Fu, G.; Zhang, J.; Shen, Z.X.; Shi, P.L.; He, Y.T.; Zhang, X.Z. Validation of collection of 6 MODIS/Terra

and MODIS/Aqua gross primary production in an alpine meadow of the Northern Tibetan Plateau. *Int. J. Remote Sens.* **2017**, *38*, 4517–4534. [CrossRef]

23. Kanniah, K.D.; Beringer, J.; Hutley, L.B.; Tapper, N.J.; Zhu, X. Evaluation of Collections 4 and 5 of the MODIS Gross Primary Productivity Product and Algorithm Improvement at a Tropical Savanna Site in Northern Australia. *Remote Sens. Environ.* **2009**, *113*, 1808–1822. [CrossRef]
24. Zhang, Y.Q.; Yu, Q.; Jiang, J.; Tang, Y.H. Calibration of Terra/MODIS gross primary production over an irrigated cropland on the North China Plain and an alpine meadow on the Tibetan Plateau. *Glob. Chang. Biol.* **2008**, *14*, 757–767. [CrossRef]
25. Liu, Z.; Shao, Q.; Liu, J. The performances of MODIS-GPP and-ET products in China and their sensitivity to input data (FPAR/LAI). *Remote Sens.* **2014**, *7*, 135–152. [CrossRef]
26. Zhu, H.; Lin, A.; Wang, L.; Xia, Y.; Zou, L. Evaluation of MODIS gross primary production across multiple biomes in China using eddy covariance flux data. *Remote Sens.* **2016**, *8*, 395. [CrossRef]
27. Wang, L.; Zhu, H.; Lin, A.; Zou, L.; Qin, W.; Du, Q. Evaluation of the Latest MODIS GPP Products across Multiple Biomes Using Global Eddy Covariance Flux Data. *Remote Sens.* **2017**, *9*, 418. [CrossRef]
28. Gilabert, M.A.; Moreno, A.; Maselli, F.; Martinez, B.; Chiesi, M.; Sanchez-Ruiz, S.; Garcia-Haro, F.J.; Pérez-Hoyos, A.; Campos-Taberner, M.; Pérez-Priego, O.; et al. Daily GPP Estimates in Mediterranean Ecosystems by Combining Remote Sensing and Meteorological Data. *ISPRS J. Photogramm.* **2015**, *102*, 184–197. [CrossRef]
29. Verbeeck, H.; Samson, R.; Verdonck, F.; Lemeur, R. Parameter sensitivity and uncertainty of the forest carbon flux model FORUG: A Monte Carlo analysis. *Tree Physiol.* **2006**, *26*, 807–817. [CrossRef]
30. Kang, X.; Liang, Y.; Zhang, X.; Li, Y.; Tian, D.; Peng, C.; Wu, H.; Wang, J.; Zhong, L. Modeling gross primary production of a typical coastal wetland in china using MODIS time series and CO_2 eddy flux tower data. *Remote Sens.* **2018**, *10*, 708. [CrossRef]
31. Jin, C.; Xiao, X.; Wagle, P.; Griffis, T.; Dong, J.; Wu, C.; Qin, Y.; Cook, D.R. Effects of in-situ and reanalysis climate data on estimation of cropland gross primary production using the Vegetation Photosynthesis Model. *Agric. Forest Meteorol.* **2015**, *213*, 240–250. [CrossRef]
32. Wang, X.F.; Ma, M.G.; Li, X.; Song, Y.; Tan, J.L.; Huang, G.H.; Zhang, Z.H.; Zhao, T.B.; Feng, J.M.; Ma, Z.G.; et al. Validation of MODIS-GPP product at 10 flux sites in northern China. *Int. J. Remote Sens.* **2013**, *34*, 587–599. [CrossRef]
33. Braswell, B.H.; Sacks, W.J.; Linder, E.; Schimel, D.S. Estimating diurnal to annual ecosystem parameters by synthesis of a carbon flux model with eddy covariance net ecosystem exchange observations. *Glob. Chang. Biol.* **2005**, *11*, 335–355. [CrossRef]
34. Ricciuto, D.M.; Davis, K.J.; Keller, K. A Bayesian calibration of a simple carbon cycle model: The role of observations in estimating and reducing uncertainty. *Glob. Biogeochem. Cycles* **2008**, *22*, GB2030. [CrossRef]
35. Li, X. Characterization, controlling, and reduction of uncertainties in the modeling and observation of land-surface systems. *Sci. China Earth Sci.* **2014**, *57*, 80–87. [CrossRef]
36. Wang, H.B.; Ma, M.G.; Xie, Y.M.; Wang, X.F.; Wang, J. Parameter inversion estimation in photosynthetic models: Impact of different simulation methods. *Photosynthetica* **2014**, *52*, 233–246. [CrossRef]
37. van Oijen, M.; Rougier, J.; Smith, R. Bayesian calibration of process-based forest models: Bridging the gap between models and data. *Tree Physiol.* **2005**, *25*, 915–927. [CrossRef]
38. Wang, H.B.; Ma, M.G. Comparing the seasonal variation of parameter estimation of ecosystem carbon exchange between alpine meadow and cropland in Heihe River Basin, northwestern China. *Sci. Cold Arid Reg.* **2015**, *7*, 216–228.
39. Knorr, W.; Kattge, J. Inversion of terrestrial biosphere model parameter values against eddy covariance measurements using Monte Carlo sampling. *Glob. Chang. Biol.* **2005**, *11*, 1333–1351. [CrossRef]
40. Groenendijk, M.; Dolman, A.J.; Van der Molen, M.K.; Leuning, R.; Arneth, A.; Delpierre, N.; Gash, J.H.C.; Richardson, A.D.; Verbeeck, H.; Wohlfahrt, G. Assessing parameter variability in a photosynthesis model within and between plant functional types using global Fluxnet eddy covariance data. *Agric. Forest Meteorol.* **2011**, *151*, 22–38. [CrossRef]
41. Cheng, G.; Li, X.; Zhao, W.; Xu, Z.; Feng, Q.; Xiao, S.; Xiao, H. Integrated study of the water–ecosystem–economy in the Heihe River Basin. *Natl. Sci. Rev.* **2014**, *1*, 413–428. [CrossRef]
42. Li, X.; Cheng, G.D.; Liu, S.M.; Xiao, Q.; Ma, M.G.; Jin, R.; Che, T.; Liu, Q.H.; Wang, W.Z.; Qi, Y.; et al.

Heihe Watershed Allied Telemetry Experimental Research (HiWATER): Scientific objectives and experimental design. *Bull. Am. Meteorol. Soc.* **2013**, *94*, 1145–1160. [CrossRef]

43. Li, X.; Liu, S.M.; Xiao, Q.; Ma, M.G.; Jin, R.; Che, T.; Wang, W.Z.; Hu, X.L.; Xu, Z.W.; Wen, J.G.; et al. A multiscale dataset for understanding complex eco-hydrological processes in a heterogeneous oasis system. *Sci. Data* **2017**, *4*, 170083. [CrossRef] [PubMed]
44. Xu, Z.W.; Liu, S.M.; Li, X.; Shi, S.J.; Wang, J.M.; Zhu, Z.L.; Xu, T.R.; Wang, W.Z.; Ma, M.G. Intercomparison of surface energy flux measurement systems used during the HiWATER-MUSOEXE. *J. Geophys. Res. Atmos.* **2013**, *118*, 13140–13157. [CrossRef]
45. Wang, X.F.; Wang, H.B.; Li, X.; Ran, Y.H. Photosynthesis (NPP, NEP, Respiration). In *Observation and Measurement of Ecohydrological Processes*; Li, X., Vereecken, H., Eds.; Springer: Berlin/Heidelberg, Germany, 2018; pp. 1–30.
46. Reichstein, M.; Falge, E.; Baldocchi, D.; Papale, D.; Aubinet, M.; Berbigier, P.; Bernhofer, C.; Buchmann, N.; Gilmanov, T.; Granier, A.; et al. On the separation of net ecosystem exchange into assimilation and ecosystem respiration: Review and improved algorithm. *Glob. Chang. Biol.* **2005**, *11*, 1424–1439. [CrossRef]
47. Ma, M.G.; Veroustraete, F. Reconstructing Pathfinder AVHRR Land NDVI Time-Series Data for the Northwest of China. *Adv. Space Res.* **2006**, *37*, 835–840. [CrossRef]
48. Sims, D.A.; Rahman, A.F.; Cordova, V.D.; El-Masri, B.Z.; Baldocchi, D.D.; Flanagan, L.B.; Goldstein, A.H.; Hollinger, D.Y.; Misson, L.; Monson, R.K.; et al. On the use of MODIS EVI to assess gross primary productivity of North American ecosystems. *J. Geophys. Res* **2006**, *111*. [CrossRef]
49. Monteith, J. Solar radiation and productivity in tropical ecosystems. *J. Appl. Ecol.* **1972**, *9*, 747–766. [CrossRef]
50. Heinsch, F.A.; Reeves, M.; Votava, P.; Kang, S.; Milesi, C.; Zhao, M.; Glassy, J.; Jolly, W.M.; Loehman, R.; Bowker, C.F.; et al. *User's Guide: GPP and NPP (MOD17A2/A3) Products NASA MODIS Land Algorithm*; University of Montana: Missoula, MT, USA, 2003; pp. 1–57.
51. Metropolis, N.; Rosenbluth, A.W.; Rosenbluth, M.N.; Teller, A.H.; Teller, E. Equations of state calculations by fast computing machines. *J. Chem. Phys.* **1953**, *21*, 1087–1092. [CrossRef]
52. Hastings, W.K. Monte Carlo sampling methods using Markov chain and their applications. *Biometrika* **1970**, *57*, 97–109. [CrossRef]
53. Yuan, W.; Cai, W.; Liu, S.; Dong, W.; Chen, J.; Arain, M.A.; Blanken, P.D.; Cescatti, A.; Wohlfahrt, G.; Georgiadis, T.; et al. Vegetation-specific model parameters are not required for estimating gross primary production. *Ecol. Model.* **2014**, *292*, 1–10. [CrossRef]
54. Turner, D.P.; Ritts, W.D.; Cohen, W.B.; Gower, S.T.; Zhao, M.; Running, S.W.; Wofsy, S.C.; Urbanski, S.; Dunn, A.L.; Munger, J.W. Scaling Gross Primary Production (GPP) over boreal and deciduous forest landscapes in support of MODIS GPP product validation. *Remote Sens. Environ.* **2003**, *88*, 256–270. [CrossRef]
55. Friedl, M.A.; Sulla-Menashe, D.; Tan, B.; Schneider, A.; Ramankutty, N.; Sibley, A.; Huang, X. MODIS Collection 5 global land cover: Algorithm refinements and characterization of new datasets. *Remote Sens. Environ.* **2010**, *114*, 168–182. [CrossRef]
56. Mu, X.; Song, W.; Gao, Z.; McVicar, T.R.; Donohue, R.J.; Yan, G. Fractional vegetation cover estimation by using multi-angle vegetation index. *Remote Sens. Environ.* **2018**, *216*, 44–56. [CrossRef]
57. Mu, Q.; Zhao, M.; Heinsch, F.A.; Liu, M.; Tian, H.; Running, S.W. Evaluating water stress controls on primary production in biogeochemical and remote sensing based models. *J. Geophys. Res.* **2007**, *112*, G01012. [CrossRef]
58. Li, Y.; Wang, Y.G.; Houghton, R.A.; Tang, L.S. Hidden carbon sink beneath desert. *Geophys. Res. Lett.* **2015**, *42*, 5880–5887. [CrossRef]
59. Jung, M.; Reichstein, M.; Margolis, H.A.; Cescatti, A.; Richardson, A.D.; Arain, M.A.; Arneth, A.; Bernhofer, C.; Bonal, D.; Chen, J.; et al. Global patterns of land-atmosphere fluxes of carbon dioxide, latent heat, and sensible heat derived from eddy covariance, satellite, and meteorological observations. *J. Geophys. Res.* **2011**, *116*. [CrossRef]
60. He, M.; Ju, W.; Zhou, Y.; Chen, J.; He, H.; Wang, S.; Wang, H.; Guan, D.; Yan, J.; Li, Y.; et al. Development of a two-leaf light use efficiency model for improving the calculation of terrestrial gross primary productivity. *Agric. Forest Meteorol.* **2013** *173*, 28–39. [CrossRef]

61. Propastin, P.; Ibrom, A.; Knohl, A.; Erasmi, S. Effects of canopy photosynthesis saturation on the estimation of gross primary productivity from MODIS data in a tropical forest. *Remote Sens. Environ.* **2012**, *121*, 252–260. [CrossRef]
62. Chu, H.; Baldocchi, D.D.; John, R.; Wolf, S.; Reichstein, M. Fluxes all of the time? A primer on the temporal representativeness of FLUXNET. *J. Geophys. Res.* **2017**, *122*, 289–307. [CrossRef]

10

Self-Learning based Land-Cover Classification using Sequential Class Patterns from Past Land-Cover Maps

Yeseul Kim [1], No-Wook Park [1,*] and Kyung-Do Lee [2]

1 Department of Geoinformatic Engineering, Inha University, Incheon 22212, Korea; kim-6674@inha.edu
2 National Institute of Agricultural Sciences, Rural Development Administration, Wanju 55365, Korea; kdlee11@korea.kr
* Correspondence: nwpark@inha.ac.kr

Abstract: To improve the accuracy of classification with a small amount of training data, this paper presents a self-learning approach that defines class labels from sequential patterns using a series of past land-cover maps. By stacking past land-cover maps, unique sequence rule information from sequential change patterns of land-covers is first generated, and a rule-based class label image is then prepared for a given time. After the most informative pixels with high uncertainty are selected from the initial classification, rule-based class labels are assigned to the selected pixels. These newly labeled pixels are added to training data, which then undergo an iterative classification process until a stopping criterion is reached. Time-series MODIS NDVI data sets and cropland data layers (CDLs) from the past five years are used for the classification of various crop types in Kansas. From the experiment results, it is found that once the rule-based labels are derived from past CDLs, the labeled informative pixels could be properly defined without analyst intervention. Regardless of different combinations of past CDLs, adding these labeled informative pixels to training data increased classification accuracy and the maximum improvement of 8.34 percentage points in overall accuracy was achieved when using three CDLs, compared to the initial classification result using a small amount of training data. Using more than three consecutive CDLs showed slightly better classification accuracy than when using two CDLs (minimum and maximum increases were 1.56 and 2.82 percentage points, respectively). From a practical viewpoint, using three or four CDLs was the best choice for this study area. Based on these experiment results, the presented approach could be applied effectively to areas with insufficient training data but access to past land-cover maps. However, further consideration should be given to select the optimal number of past land-cover maps and reduce the impact of errors of rule-based labels.

Keywords: classification; self-learning; training data; crop

---

1. Introduction

Production of thematic maps such as land use/land cover and crop type maps using remote sensing data has been regarded as one of most important applications of remote sensing, as it can provide useful information with periodicity and at a variety of scales [1–4]. Since thematic maps are usually used in land surface monitoring and environmental modeling, it is critical that they be reliable [5]. For example, crop type maps are usually fed into physical models for crop yield estimation or forecasting.

Many studies have been carried out to generate a reliable thematic map from remote sensing data. From the data availability aspect, multi-sensor/source data including optical, SAR, and GIS data have been used as inputs for classification [6–9]. To properly treat input data for classification, advanced classification methodologies such as machine learning approaches and object-based classification have also been applied to either single-sensor data or multiple data sets [10–12]. Even though a proper classification methodology and appropriate data sets are applied to classification, supervised classification usually requires a large amount of high-quality training data. However, this is not always possible to obtain particularly when supervised classification is to be conducted for large or inaccessible areas. It is thus necessary to develop a new classification framework that can alleviate the difficulty of collecting a lot of training data.

To resolve this issue, several approaches have been proposed in the remote sensing community, such as semi-supervised learning (SSL) and active learning (AL) [13–22]. The idea central to these approaches is the use of unlabeled data to complement the training data [13,14]. AL and SSL are very similar in that they begin with an initial classification using a small amount of training data, followed by further classifications using the new training data derived from informative unlabeled pixels in the initial classification result [15–19]. The informative pixels are ones that provide useful information for properly modifying the decision boundary already determined from a small amount of training data, which ultimately lead to an improvement of classification accuracy. However, SSL and AL adopt different ways of extracting the new informative training data from unlabeled data. The SSL approach selects the most confident pixels from the initial classification result as new informative training data, where the most confident pixels mean ones that are likely to be classified unambiguously by a classifier and have the higher confidence [20–22]. Various SSL approaches, such as transductive support vector machine and graph-based methods, have been developed to extract new training data from the initial classification result [22]. New training data can be added directly to the training data without class assignment by an analyst because the classification algorithm itself already assigns the class labels to the most confident pixels. If the initial classification result includes many wrongly classified pixels, however, the new training data extracted from the SSL approach would be wrongly labelled, resulting in the poor classification performance [23]. In addition, the new training data with higher confidence tend to provide redundant information that is not useful for modifying the decision boundary. Thus, there might be no significant improvement to the classification accuracy, compared with AL [23]. Conversely, the most informative pixels in the AL approach are defined as ones that a classifier fails to properly classify, which correspond to pixels with higher uncertainty or lower confidence in the initial classification result. After these pixels are extracted, an analyst then manually assigns their class labels. Since the analyst designates class labels for uncertain or ambiguous pixels, these new training data can positively contribute to modifying the decision boundary. However, it is difficult to apply AL to areas where prior information on land-cover classes is not readily available to facilitate the analyst's interpretation. If the class label assigned by the analyst is incorrect, the accuracy of the classification may deteriorate.

Recently, several studies have proposed combining AL and SSL to take full advantage of both approaches [24,25]. Muñoz-Marí et al. [24] proposed a semiautomatic approach that integrated a hierarchical clustering tree with active queries to generate land-cover maps. Based on hierarchical clustering with a small amount of training data, the most coherent pixels were exploited and an active learning query was applied to extract the most informative pixels. Dópido et al. [25] also developed a SSL approach that adapted AL methods to integrate self-learning. Pixels adjacent to initial training data were selected as candidates for new training data. AL first extracted the most informative pixels from the adjacent pixels, and then these pixels were used as the new training data. In both approaches, the large number of training data could be selected from unlabeled data and a significant improvement in classification accuracy was obtained for hyperspectral image classification. However, despite utilizing spectral and spatial similarities to assign class labels to the most informative pixels within the self-learning framework, there was still uncertainty or difficulty with the class assignment.

Regarding the issue of class assignment, supplementary information from past land-cover maps [19] and predefined rules [26] in an area of interest could be incorporated into self-learning frameworks. For example, information on crop cultivation systems, such as crop rotations, could be effectively used as a kind of temporal contextual information. Although this information could facilitate the collection of additional high-quality training data, the new training data extracted from self-learning tends to be over-sampled for specific class labels that occupy the largest proportion of the study area. As a result, the biased training data might degrade the classification accuracy [19]. To the best of our knowledge, little emphasis has been placed on both class assignment and extraction of unbiased training sets in self-learning approaches for remote sensing data classification.

In this paper, a new self-learning approach is presented for crop classification that can collect a large number of labeled training data without analyst intervention. Rule information on class changes is first generated from past land-cover maps, and then the class labels for the new training data are assigned based on the rules. The impact of the rule information on classification accuracy is also investigated by changing the number of past land-cover maps used. The methodological developments and applicability of this self-learning approach are demonstrated by a crop classification experiment using time-series Moderate Resolution Imaging Spectroradiometer (MODIS) Normalized Difference Vegetation Index (NDVI) data sets and cropland data layers (CDLs) as classification inputs and supplementary data, respectively.

2. Materials and Methods

2.1. Study Area

A classification experiment was conducted in the crop cultivation areas of Kansas State, USA, in 2015 (Figure 1). The reason for the choice of the study area was two-fold: Kansas is known as one of the main production areas of winter wheat, in addition to various crops such as corn, sorghum, and soybean [27]. Thus, it was possible to examine how well the self-learning approach of this study could discriminate between complex land-cover types. The second reason was the availability of past time-series land-cover maps. The CDLs, provided by the National Agricultural Statistics Service (NASS) of the United States Department of Agriculture (USDA) [28], were used to both extract the cultivation rules of cropping systems in the study area, and validate the classification results.

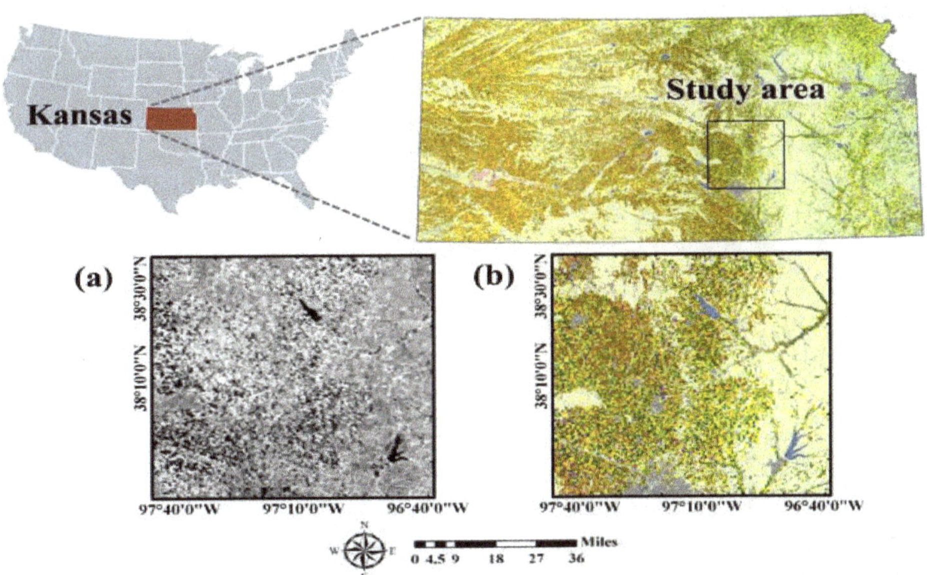

Figure 1. Location of the study area and data sets: (**a**) Moderate Resolution Imaging Spectroradiometer (MODIS) Normalized Difference Vegetation Index (NDVI) on 9 May 2015, at a 250 m spatial resolution and (**b**) cropland data layer (CDL) 2015 at a 30 m spatial resolution.

2.2. Data Sets

2.2.1. MODIS NDVI Data

Time-series MODIS NDVI data sets were used as classification inputs in this study. The MODIS-based vegetation index has been widely used for large-scale crop classification because it provides time-series information at a 250 m spatial resolution [1,29,30]. Similar to Kim and Park [31], we aimed for early crop map production, prior to the release of the CDL 2015 data, as part of crop acreage estimation. Thus, a total of 13 MODIS 16-day composite NDVI data sets from January to July 2015 were experimentally used to account for time-series variations of various crops in the study area. To minimize the effects of clouds in the 16-day composite data sets, the Savitzky-Golay filter [32] was applied prior to classification (Figure 1a). The study area consisted of a 400 by 400-pixel array with a spatial resolution of 250 m.

2.2.2. Landsat Data

In this study, the classification output was at a 250 m spatial resolution, which was the same as the MODIS NDVI data. It is often difficult to collect training data for supervised classification from mid-resolution remote sensing data. Thus, Landsat data sets were used to supplement the MODIS NDVI data sets for initial training data collection. The training data were collected through visual analysis of a total of 33 Landsat-7 ETM+ and Landsat-8 OLI images obtained from April to August.

The class types and the number of training data per class are shown in Table 1. To mimic a situation where many training data were not available, only a small amount of training samples were collected, which occupied approximately 0.26% of the study area. Supervised classification was conducted using the initial training data sets of 10 class types. The main purpose of classification in this study was to accurately classify the major crops; to facilitate this, some minor crops such as alfalfa and hay, in addition to class types such as water, city, forest, and grass were merged as general grain/hay and non-crop classes, respectively, for evaluation of the classification results (Table 1). Besides the collection of the initial training data, the Landsat data sets were also used for visual comparison and confirmation of classification results.

Table 1. The list of class types for supervised classification and merged classes, and the number of initial training data per each class.

No	Class	Merged Class	Number of Training Data
1	Corn	Corn	45
2	Sorghum	Sorghum	25
3	Soybean	Soybean	45
4	Winter wheat	Winter wheat	100
5	Alfalfa	Grain/hay	20
6	Other hay	Grain/hay	15
7	Water	Non-crop	37
8	City	Non-crop	28
9	Forest	Non-crop	20
10	Grass	Non-crop	85
	Total		420

2.2.3. Past Land-Cover Maps

CDLs, which have been produced annually using time-series Landsat images and field surveys by the USDA NASS [27,28], provide the state-level crop types at a spatial resolution of 30 m. In this study, as for past land-cover maps, CDLs prior to 2015 were used to define rule information for assigning the class labels for the self-learning process. From a preliminary test, meaningful rule information for minority classes such as sorghum and other hay could not be extracted from CDLs prior to 2010. Thus, five years of CDLs, from 2010 to 2014, were considered to extract the rule information for classification of the 2015 data. In addition, the CDL in 2015, which was not used for classification, was used to extract reference data sets for computing classification accuracy statistics.

As classification was conducted at a 250 m spatial resolution, there was a mismatch of spatial resolution between CDLs and MODIS data sets. The CDLs were upscaled to a 250 m spatial resolution by assigning the most frequently occurring CDL class label within each 250 m pixel to that corresponding pixel. Due to mixed pixel effects, some pixels in the upscaled CDL had higher uncertainty and significantly affected the classification accuracy. To prevent this, only reliable pixels with higher confidence were considered as reference data sets. More specifically, pixels were selected if the fraction of the most prevalent class label within the pixel was greater than 0.8. In the end, a total of 64,000 pixels for six merged classes were used to compute accuracy statistics (Table 2).

Table 2. Number of reference data.

Class	Number
Corn	6263
Sorghum	1864
Soybean	6016
Winter wheat	14,582
Grain/hay	651
Non-crop	34,624
Total	64,000

2.3. *Classification Methodology*

The proposed classification methodology theoretically adopts the AL concept, which tries to improve classification performance by adding the most informative pixels with higher uncertainty selected from unlabeled data to new training data. However, unlike the conventional AL approach which requires analysts to manually assign class labels to the most informative pixels, this new classification methodology is based on a self-learning concept by using rule information derived from past land-cover maps (e.g., past CDLs of the study area). This approach can assign class labels to the most informative pixels in an automated manner. The whole procedure of the self-learning approach employed in this study is presented in Figure 2.

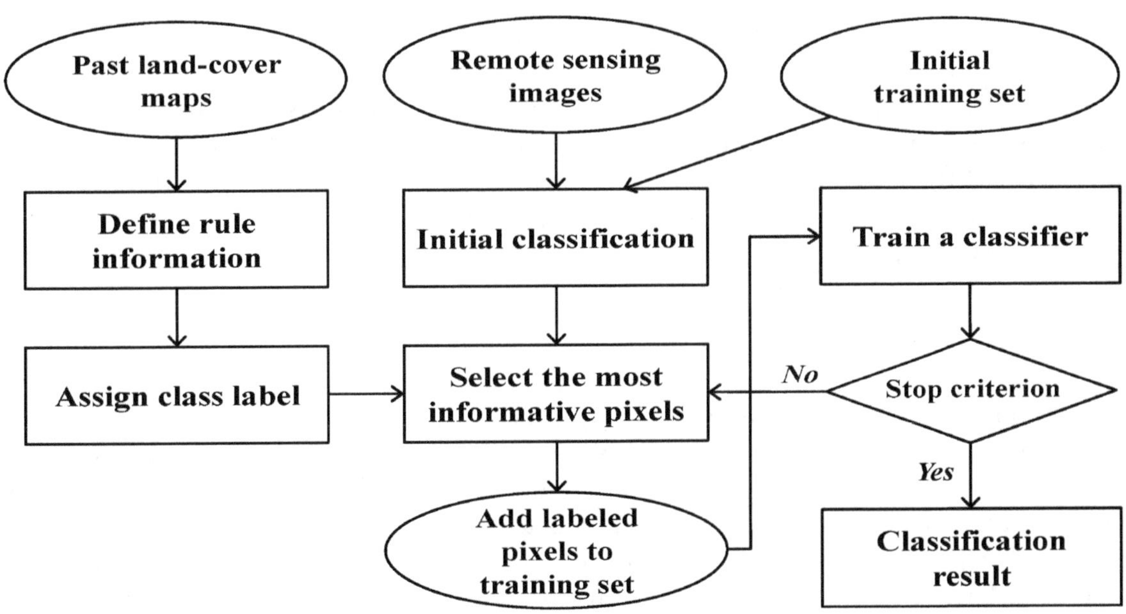

Figure 2. Flow chart of the classification procedures presented in this study.

2.3.1. Initial Classification

In the first processing step, an initial classification with a small amount of training data is carried out. For this process, a support vector machine (SVM), which has been widely applied in supervised classification of remote sensing data [33,34], is applied as the main classifier. SVM tries to find an optimal hyperplane (i.e., decision boundary) that provides the maximum margin [35]. It has been reported that the SVM is superior to other conventional classifiers when a small amount of training sites and many features are used for classification [33,34].

The class-wise *posteriori* probability from the SVM classification is used for the next step, as a kind of index that defines the uncertainty of the initial classification result. Since the SVM does not directly provide probability estimates, the *posteriori* probabilities were computed using pairwise coupling [36]. Several binary classifiers for each possible pair of classes (i.e., one-versus-one) are first created. The probability for each class is then estimated using pairwise coupling [36]. Suppose that D and r_{ij} are the feature vector and the estimates of $P(\omega_i \mid \omega_i \text{ or } \omega_j, D)$ for a certain class (ω_i) by a binary classifier, respectively. Then, the class-wise probability ($P(\omega_i \mid D)$) for multi-class classification is estimated by solving the system [36] as follows:

$$P(\omega_i|D) = \sum_{j:j\neq i}\left(\frac{P(\omega_i|D) + P(\omega_j|D)}{M - 1}\right)r_{ij},\ \forall_i,\ \text{subject to} \sum_{i=1}^{M} P(\omega_i|D) = 1,\ P(\omega_i|D) \geq 0,\ \forall_i. \quad (1)$$

where M is the total number of classes in the study area.

2.3.2. Selection of Informative Pixels

The next step is to select candidate pixels to be used as the new training data from the initial SVM classification results. Pixels with higher uncertainty tend to be located near a hyperplane determined by the initial classification, and therefore are more likely to be mixed pixels [24,26]. If the class label of these pixels is correctly defined, then further classification with these pixels as new training data could properly modify the decision boundary, resulting in improved classification performance.

Among various approaches for the selection of pixels with high uncertainty [37–39], the breaking ties (BT) algorithm [40], which is simple to implement, was adopted in this study. The BT algorithm first computes the difference between the largest and the second largest *posteriori* probabilities (ΔP) as,

$$\Delta P = P(\omega_1|D) - P(\omega_2|D) \quad (2)$$

where $P(\omega_i \mid D)$ is the *posteriori* probability of the ith class (ω_i) computed using Equation (1). ω_1 and ω_2 are the classes with the largest and the second largest *posteriori* probabilities, respectively.

The larger the difference between these two *posteriori* probabilities meant the decision to select the pixel is less ambiguous. The pixels with smaller differences are extracted as the ones with the higher uncertainties. The pixels with high uncertainty selected through the BT algorithm are used as candidates in the initial training data.

2.3.3. Generation of Rule Information and Prediction of Class Labels

Once informative pixels have been selected, the next critical step, which is the core part of the self-learning approach, is to define the class labels for the candidates. Rule information on sequential land-cover changes are first defined by comparing past land-cover maps (i.e., CDLs in this study), then the rule information is used to assign the class labels to the selected candidates. The rule information extracted in this study can be regarded as a form of temporal contextual information. Some land-cover classes, such as urban and water, tend to remain unchanged. Conversely, some crops are likely to change to others or become fallow by certain cropping systems in the study area. For example, crop rotations such as a corn-soybean rotation are very common in the USA. If such temporal change information is properly characterized, it could be used to predict the class labels of the candidates for new training data.

The main concept of defining rule information on sequential land-cover patterns is illustrated in Figure 3. Suppose that pixels #1 and 3 undergo bi-annual crop rotations in Figure 3. If these pixels are selected as candidates for new training data, their class labels can be predicted for T_n by considering the sequential rotation patterns. Conversely, pixels #2 and 4 remain unchanged during the same time period, and are predicted to be unchanged for T_n.

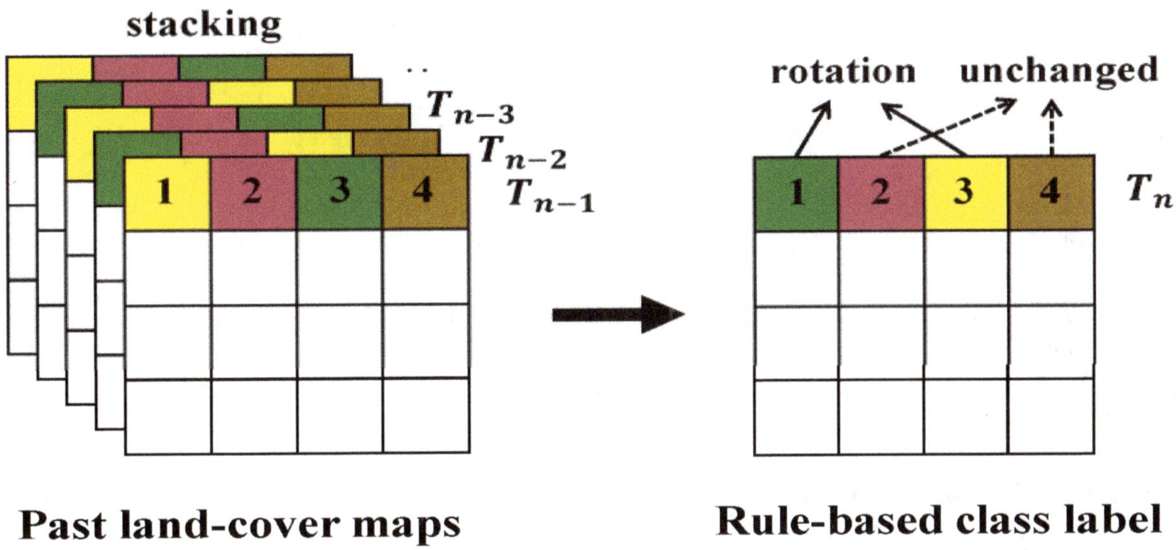

Figure 3. Illustrations for the process of generating rule-based class labels. Past land-cover maps are stacked and compared, where T_n is the reference year under consideration.

Our goal is to predict the class label for the considered time (T_n) using past land-cover maps, which will be used to define the class labels for the candidates of new training data. When past k years (T_{n-i}, $i = 1, \cdots, k$) are considered, the rule information on sequential land-cover patterns for T_n can be formulated as,

$$\left\{\omega^{T_{n-k}}(\mathbf{x}),\ \omega^{T_{n-k-1}}(\mathbf{x}),\ \cdots,\ \omega^{T_{n-2}}(\mathbf{x}),\ \omega^{T_{n-1}}(\mathbf{x})\right\} \rightarrow \omega^{T_n}(\mathbf{x}) \tag{3}$$

where $\omega^{T_j}(\mathbf{x})$ denotes the class at a location $(\mathbf{x})$ for the year T_j.

In this study, a simple but efficient heuristic approach is applied to predict the class labels from sequential land-cover patterns using past land-cover maps. Under the assumption that the typical sequential patterns of land-cover changes in the study area could be maintained, the class label can be predicted for the considered time. This assumption has been routinely adopted for classification processes using temporal contextual information. More specifically, by overlaying past time-series land-cover maps, sequential land-cover patterns are first identified at each pixel and unique sequence rules from the specific sequences of land-covers in successive years are then defined. Each unique sequence rule has its own non-overlapping combination of sequential land-cover patterns. Pixels with different colors in Figure 3 have their corresponding unique sequence rules. Based on unique sequence rules that provide useful information on the prediction of class labels, the class labels for the considered time can be predicted by adopting above assumption. For example, suppose that a certain pixel has a unique sequence such as corn-soybean-corn-soybean-corn, which reflects a corn-soybean rotation, like pixel #1 in Figure 3. For any pixel with this unique sequence rule, the class label for T_n is predicted as soybean to account for a corn-soybean rotation.

2.3.4. Iterative Classification

Once the new training data are added to the initial training set, they are used as inputs for a SVM classifier and this procedure is repeated until a predefined stopping criterion is satisfied. The iterative classification stopped when the percentage of changed pixels between current and previous classification results was less than 5%.

2.4. Experimental Setting

When applying the heuristic approach to self-learning, several practical issues arise. The first issue is regarding the optimal number of land-cover maps to generate unique sequence rules. If too many land-cover maps are used, stable rules are identified, but superfluous complex rules are also generated. By contrast, using too few land-cover maps may result in simplistic but unstable sequential patterns. The effects of the number of past land-cover maps was investigated in this study.

The second issue is a biased sampling problem. The pixels selected as new training data are likely to be biased to a specific class type. This is because the new training data selected by the BT algorithm come from specific boundaries containing a large number of training samples [25,39]. The inclusion of biased pixels that favor a specific major class type into the new training data may result in the over-estimation of that class type, and the overall degradation of classification accuracy. In this study, random under-sampling (i.e., restriction of the number of newly labeled pixels) of the training data assigned to specific class types was applied to obtain unbiased training data.

The last issue is regarding the quality or reliability of class labels predicted from unique sequence rules. The self-learning approach requires no analyst intervention, so the reliability of the class labels predicted from the unique sequence rules is critical for classification performance. In this study, the unique sequence rules are built from upscaled 250 m CDLs, not from the original 30 m CDLs. Thus, it is necessary to use pixels with high confidence in the upscaled CDLs. Since the most frequent class within each 250 m pixel was assigned to that corresponding pixel during upscaling, the confidence in the class assignment at 250 m could be derived from fractions of the assigned class. To obtain more reliable rules, the most confident pixels in all 250 m CDLs, which have higher fraction values, were used to build the sequence rules.

3. Results

3.1. Generation of Rule-Based Class Labels

To use the most confident pixels in 250 m CDLs for the rule generation, we used only pixels whose fractions of classes assigned to the 250 m CDLs from 2010 to 2014 exceeded a specific thresholding value to define rule information. When a thresholding value of greater than 70% was applied, few pixels were extracted for most classes except for winter wheat and non-crop. Thus, the rule information was finally generated using only pixels whose fractions were greater than 60%.

Overlaying many past CDLs generates too many unique sequence rules that have similar but not identical class sequences. It is very difficult to predict the single class label from complex rules because there are some possible class labels in 2015. To reduce the uncertainty attached to a class label assignment, all possible rules were not considered for the generation of rule-based class labels.

After analyzing typical cropping characteristics in the study area, we selected some unique sequence rules that could provide predictable information on a class label assignment. Winter wheat–fallow rotation has been known as the common cropping system in Kansas [41]. The winter wheat–fallow rotation system allows the accumulation of soil moisture in the cultivation area during the fallow periods. Due to soil erosion potential, however, winter wheat–summer crops such as corn, sorghum, and soybean rotations are being widely planted [41–43]. Of these crops, corn-soybean rotations dominate in Kansas.

A total of 21 rules were finally defined to predict class labels in 2015 (Table 3). Not all 21 rules represent the frequent patterns. Some frequent patterns (e.g., rules #4 and 21 in Table 3) were selected, but other patterns that were less frequent but facilitated the prediction of class labels in 2015 (e.g., rules #6 and 9 in Table 3) were also selected. Although a simple heuristic approach was applied to generate rule information, the sequential patterns of land-covers between 2010 and 2014 in Table 3 well reflect the above predominant crop rotation sequences in Kansas. Typical sequence rules in the study area include winter wheat–fallow rotation, winter wheat-summer crop rotations, and summer crop rotations, as well as continuously growing crops. In addition, grain/hay and non-crop classes including water and urban remain unchanged.

Table 3. Sequential patterns of land-covers between 2010 and 2014 (C: corn, S: sorghum, SB: soybean, WW: winter wheat, A: alfalfa, OH: other hay, FA: fallow, W: water, U: urban, FO: forest, G: grass). The class labels in 2015 predicted from CDLs spanning the past five years are shown in bold.

No	2010	2011	2012	2013	2014	Predicted Label for 2015
1	C	C	C	C	C	**C**
2	S	S	S	S	S	**S**
3	SB	SB	SB	SB	SB	**SB**
4	WW	WW	WW	WW	WW	**WW**
5	C	S	C	S	C	**S**
6	S	C	S	C	S	**C**
7	C	SB	C	SB	C	**SB**
8	SB	C	SB	C	SB	**C**
9	S	SB	S	SB	S	**SB**
10	WW	C	WW	C	WW	**C**
11	C	WW	C	WW	C	**WW**
12	WW	SB	WW	SB	WW	**SB**
13	SB	WW	SB	WW	SB	**WW**
14	WW	FA	WW	FA	WW	**FA**
15	FA	WW	FA	WW	FA	**WW**
16	A	A	A	A	A	**A**
17	OH	OH	OH	OH	OH	**OH**
18	W	W	W	W	W	**W**
19	FO	FO	FO	FO	FO	**FO**
20	U	U	U	U	U	**U**
21	G	G	G	G	G	**G**

As mentioned in Section 2.3, the effectiveness of the sequential patterns of land-covers depends on the number of CDLs used. To investigate this, the following different cases were considered to generate the rules: (1) using CDLs from 2010 to 2014 (5 years), (2) using CDLs from 2011 to 2014 (4 years), (3) using CDLs from 2012 to 2014 (3 years), and (4) using CDLs from 2013 to 2014 (2 years).

The class labels in 2015 of pixels in which sequential patterns of land-cover changes between 2010 and 2014 matched to the 21 rules were predicted as the corresponding labels of the rightmost column in Table 3. The rule-based class label images predicted from these four different cases are given in Figure 4. By superimposing the new training data candidates on the predicted label image, the class labels of the candidates were assigned automatically. Note that the rule-based class labels were not assigned to all pixels in the study area because some sequence rules were not considered and only the most confident pixels in CDLs were used to define rule information.

As the number of CDLs used to define the sequence rule decreased, the proportion of pixels in which the class labels in 2015 could be predicted increased accordingly (e.g., 23.38% (37,415 pixels) and 35.31% (56,496 pixels) for using past five-year and two-year CDLs, respectively). The fewer the CDLs, the more areas that were assigned to certain crop types such as corn, soybean, and winter wheat. By contrast, if the number of CDLs increased, relatively few areas had the rule-based class label and many areas remained unlabeled. Note that the number of pixels with rule-based class labels is much larger than that of initial training pixels (i.e., 37,415 versus 420). These rule images were separately used for further classification procedures and their classification performance were compared.

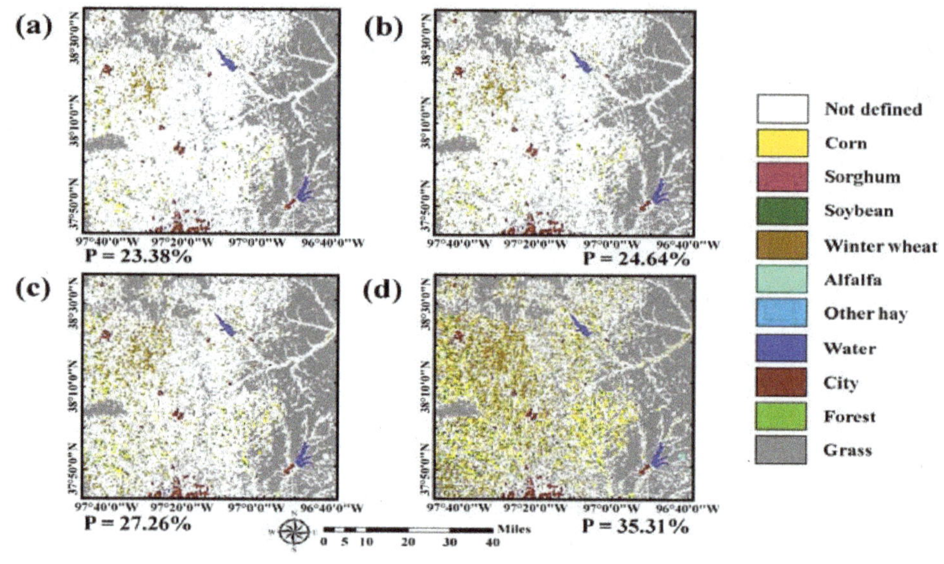

Figure 4. Rule-based class label images predicted from past CDLs: (**a**) 5-year (2010 to 2014), (**b**) 4-year (2011 to 2014), (**c**) 3-year (2012 to 2014), and (**d**) 2-year (2013 to 2014). P (the percentage value below each predicted class label image) denotes the proportion of pixels in which the class label for 2015 could be predicted.

3.2. Initial Classification Result

Before a self-learning procedure was employed, an initial classification was first performed using the initial training data. The qualitative and visual assessment of the initial classification results was conducted using time-series Landsat images (Figure 5). Two subareas were identified as an over-estimation of sorghum and a clustered pattern of winter wheat. The clustered pattern of winter wheat was attributed to the inclusion of more training data than the other class types in the western part of the study area. Confusion between winter wheat and alfalfa, which showed similar temporal NDVI variations in winter, could have also contributed to the clustered pattern of winter wheat in the western part. In addition, sorghum and soybean, which are the typical summer crops in Kansas, showed similar temporal NDVI variations, which led to an over-estimation of sorghum. Confusion between grain/hay and grass was also observed in the initial classification result

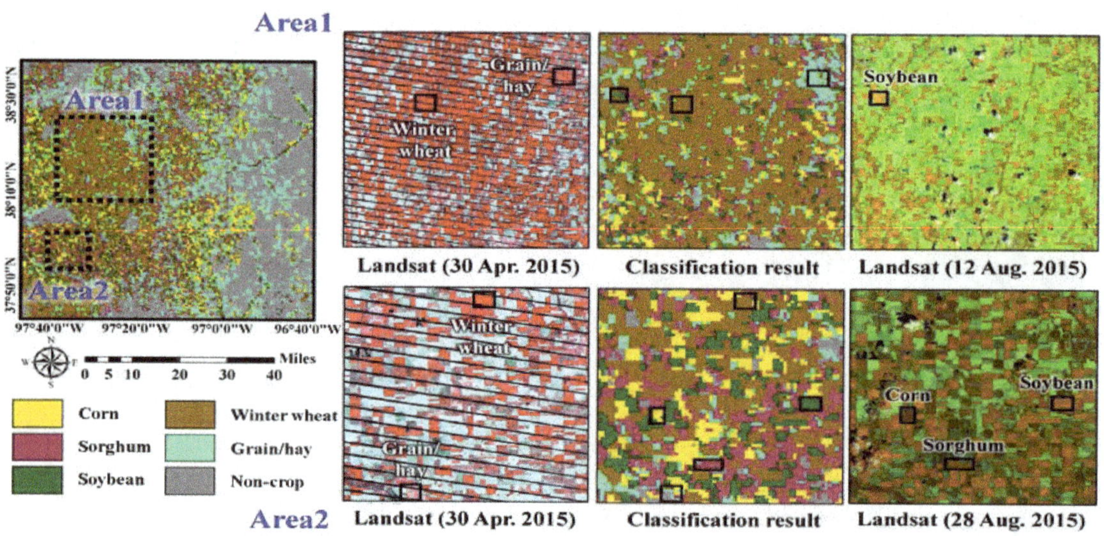

Figure 5. Initial classification result, and its visual comparison with Landsat images in two subareas.

3.3. Self-Learning Classification Result

To select new training data candidates from the initial classification result, the BT algorithm was applied to the *posteriori* probabilities from a SVM classifier. The pixels that had a difference between the largest *posteriori* probability and the second largest *posteriori* probability of less than 0.05 were selected as the most informative pixels with higher uncertainty. Then, the class labels of the selected candidates were assigned to the rule-based class labels predicted from past CDLs.

If no restriction on the number of added training data was given, a large number of pixels were selected for winter wheat that is the major crop in the study area. As mentioned in Section 2.4, adding too many training data for the majority class (e.g., winter wheat) might result in the over-estimation of that class. To prevent this, another criterion was applied to restrict the number of added training data. Based on a trial and error approach, the number of training data assigned to the majority class was randomly under-sampled, and the total number of newly added training data was set to maximum 300 pixels per iteration. The variations of the number of updated training data for iterative classification are listed in Table 4. Since the number of new training pixels to be added into the previous training set was restricted, the difference of the total number of new training data was not great. However, the locations of the newly added training data were different, which led to different classification results for four CDL combination cases. Self-learning procedures for all combination cases were terminated after three or four iterations, which implied that most of pixels were mainly labeled during the first three or four iterations, and there was no significant change in the subsequent iterations.

Table 4. Number of new training data at each iteration for four different past CDL combination cases.

Class	Past 5-Year CDLs			Past 4-Year CDLs			Past 3-Year CDLs				Past 2-Year CDLs			
Iteration	1	2	3	1	2	3	1	2	3	4	1	2	3	4
Corn	90	150	194	91	129	156	90	123	156	216	88	128	162	187
Sorghum	25	50	79	38	63	86	39	60	90	111	42	62	76	86
Soybean	90	150	195	90	132	167	91	128	165	217	86	126	162	182
Winter wheat	160	200	220	155	155	210	163	201	216	241	160	185	211	246
Alfala	26	36	72	36	56	74	35	45	55	73	35	43	43	63
Other hay	15	15	15	15	15	15	15	15	15	15	15	15	15	15
City	57	92	112	57	83	101	57	77	87	107	52	77	91	101
Water	47	72	107	53	93	118	50	95	105	130	62	96	116	126
Forest	40	63	80	28	62	85	30	70	77	77	35	60	70	76
Grass	170	192	227	157	192	222	150	180	228	267	145	205	255	269
Total	720	1020	1310	720	980	1234	720	994	1194	1454	720	997	1201	1351

The classification results based on a self-learning approach are presented in Figure 6. When compared with the initial classification result in Figure 5, over-estimation of sorghum and grain/hay was greatly reduced in the four classification results. The four classification results showed similar patterns overall: crop areas mainly in the west, and grain/hay and non-crop areas in the east. However, distributions of crop areas were locally different. In particular, over-estimation of soybean and under-estimation of grain/hay were observed in the two-year CDLs classification result, compared with the others. This could be attributed to the fact that the number of new training pixels assigned to alfalfa and other hay was relatively smaller than that of other CDL combination cases, as shown in Table 4. Conversely, sorghum was under-estimated in the five-year CDLs classification result. Therefore, it is expected that these different classification patterns from four CDL combination cases would result in the different classification accuracy assessment results.

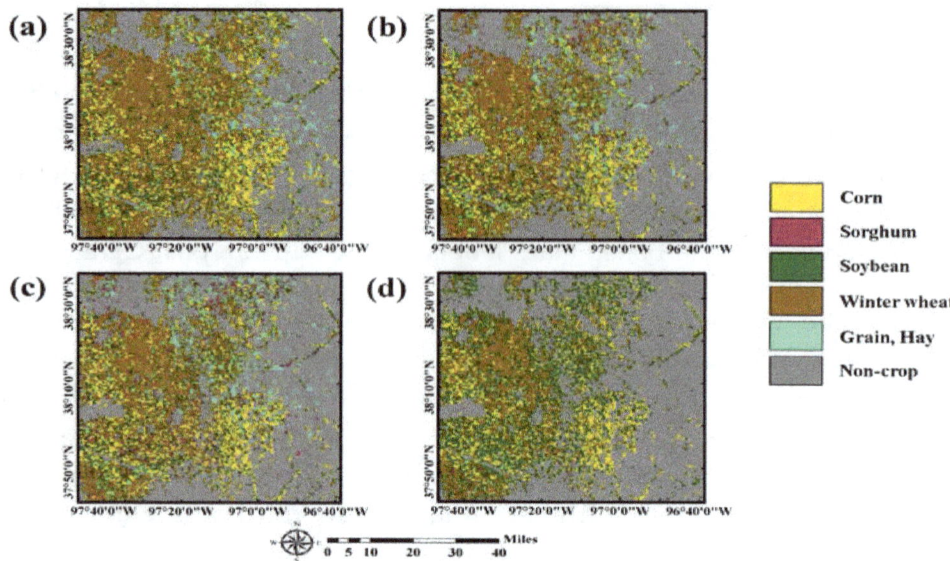

Figure 6. Final classification results of a self-learning approach with different past CDLs: (**a**) 5-year (2010 to 2014); (**b**) 4-year (2011 to 2014); (**c**) 3-year (2012 to 2014); and (**d**) 2-year (2013 to 2014).

3.4. *Accuracy Assessment*

For the classification accuracy assessment, accuracy statistics such as overall accuracy, Kappa coefficient, and class-wise accuracy were computed by comparing the classification result and the reference data set in Table 2. Figure 7 shows the variations of overall accuracy for each iteration of different CDL combination cases. As shown in Figure 7, the overall accuracy increased as the number iterations increased. As a result, the self-learning approach presented in this study gave a better overall accuracy than the initial SVM classification for all different CDL combination cases. An increase of about 5.52 to 8.34 percentage points in overall accuracy was obtained by adding new training data with rule-based class labels. Based on a McNemar test [44], the improvement of overall accuracy was statistically significant at the 5% significance level. When comparing the overall accuracy values of different CDL combination cases, the best and worst (84.42% versus 81.60%) were obtained from the three-year CDLs and two-year CDLs, respectively. The case of the four-year CDL completed with fewer iterations, yet appeared to be on a trajectory to compete with the case of the 3-year CDLs which showed the best classification accuracy.

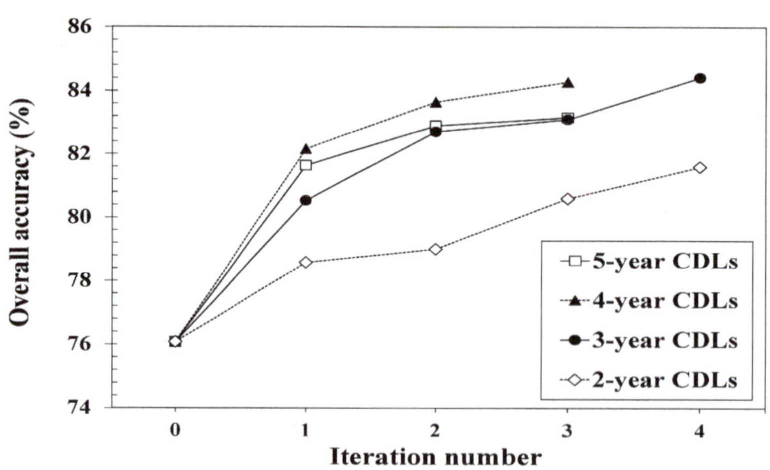

Figure 7. Overall accuracy versus the iteration number for the four different CDL combination cases. Iteration 0 indicates the initial classification.

The confusion matrices for the initial classification and self-learning classification with past CDLs are listed in Table 5. Table 6 also summarizes the accuracy statistics, including overall accuracy, Kappa coefficient, and class-wise accuracy, with respect to the initial classification and the four different CDL combination cases.

Table 5. Confusion matrices for initial classification and self-learning classification for four different CDL combination cases.

Initial Classification

Classification \ Reference	Corn	Sorghum	Soybean	Winter Wheat	Grain/Hay	Non-Crop	Sum
Corn	3698	49	144	435	28	567	4921
Sorghum	630	1186	1329	725	32	1223	5125
Soybean	387	161	3420	222	7	152	4349
Winter wheat	394	309	856	11,500	77	766	13,902
Grain/hay	436	33	66	1333	469	3500	5837
Non-crop	718	126	201	367	38	28,416	29,866
Sum	6263	1864	6016	14,582	651	34,624	64,000

Self-learning classification with past 5-year CDLs

Classification \ Reference	Corn	Sorghum	Soybean	Winter Wheat	Grain/Hay	Non-Crop	Sum
Corn	4596	124	524	308	35	1334	6921
Sorghum	0	1020	92	124	0	40	1276
Soybean	443	466	3775	369	9	708	5770
Winter wheat	640	0	1176	12,938	166	1600	16,520
Grain/hay	33	3	7	128	330	378	879
Non-crop	551	251	442	715	111	30,564	32,634
Sum	6263	1864	6016	14,582	651	34,624	64,000

Self-learning classification with past 4-year CDLs

Classification \ Reference	Corn	Sorghum	Soybean	Winter Wheat	Grain/Hay	Non-Crop	Sum
Corn	4308	112	467	359	27	461	5734
Sorghum	53	916	100	449	6	134	1658
Soybean	374	200	4147	459	13	76	5269
Winter wheat	337	303	545	11,372	66	781	13,404
Grain/hay	62	5	15	435	326	304	1147
Non-crop	1129	328	742	1508	213	32,868	36,788
Sum	6263	1864	6016	14,582	651	34,624	64,000

Self-learning classification with past 3-year CDLs

Classification \ Reference	Corn	Sorghum	Soybean	Winter Wheat	Grain/Hay	Non-Crop	Sum
Corn	5244	137	136	546	24	1384	7471
Sorghum	124	1043	290	683	5	329	2474
Soybean	338	153	5110	760	20	295	6676
Winter wheat	211	236	355	10,431	41	465	11,739
Grain/hay	42	8	23	619	333	285	1310
Non-crop	304	287	102	1543	228	31,866	34,330
Sum	6263	1864	6016	14,582	651	34,624	64,000

Self-learning classification with past 2-year CDLs

Classification \ Reference	Corn	Sorghum	Soybean	Winter Wheat	Grain/Hay	Non-Crop	Sum
Corn	5099	76	496	321	19	270	6281
Sorghum	68	635	130	582	0	9	1424
Soybean	464	655	4651	3245	51	690	9756
Winter wheat	201	168	336	8530	29	359	9623
Grain/hay	1	1	0	22	13	3	40
Non-crop	430	329	403	1882	539	33,293	36,876
Sum	6263	1864	6016	14,582	651	34,624	64,000

Table 6. Comparison of the accuracy statistics for the different classification results. UA and PA denote user's accuracy and producer's accuracy, respectively. The best case is shown in bold.

		Initial Classification	Self-Learning			
			Past 5-Year CDLs	Past 4-Year CDLs	Past 3-Year CDLs	Past 2-Year CDLs
Overall accuracy		76.08	83.16	84.28	**84.42**	81.60
Kappa coefficient		0.65	0.74	0.75	**0.76**	0.71
Corn	PA	59.05	73.38	68.78	**83.73**	81.41
	UA	75.15	66.41	75.13	70.19	**81.18**
Sorghum	PA	**63.63**	54.72	49.14	55.95	34.07
	UA	23.14	**79.94**	55.25	42.16	44.59
Soybean	PA	56.85	62.75	65.61	**84.94**	77.31
	UA	78.64	65.42	**78.71**	76.54	47.67
Winter wheat	PA	78.86	**88.73**	77.99	71.53	58.50
	UA	82.72	78.32	84.84	**88.86**	88.64
Grain/hay	PA	**72.04**	50.69	50.08	51.15	2.00
	UA	8.03	**37.54**	28.42	25.42	32.50
Non-crop	PA	82.07	88.27	94.35	92.03	**96.16**
	UA	**95.14**	93.66	89.34	92.82	90.28

As indicated in Figure 7 and Table 6, overall, adding new training data via self-learning showed the best overall accuracy and Kappa coefficient. Except for producer's accuracy for sorghum and grain/hay and user's accuracy for non-crop, the class-wise accuracy for the self-learning approach is superior to that for the initial classification.

Despite the poorest overall accuracy, the initial classification result gave relatively higher producer's accuracy for sorghum and grain/hay, but the accuracy was relatively lower than other classes. As sorghum and grain/hay are minority classes in the study area, their highest producer's accuracy could not lead to the significant improvement in overall accuracy. As shown in Figure 5 (e.g., northern and eastern parts in the study area), over-estimation of those classes decreased omission errors and resulted in this high producer's accuracy. However, user's accuracy (the probability that the probability that a pixel classified into a given class represents the actual class [45]) was very low for sorghum and grain/hay, which indicates very poor reliability of these two classes in the initial classification map. Most pixels of these two classes were misclassified into soybean or grass, as shown in Table 5. The accuracy for these two classes was improved by adding new training data. For sorghum, the case of the five-year CDLs showed a significant increase of approximately 56.80 percentage points in user's accuracy. The most significant improvement of about 29.51 percentage points in user's accuracy for grain/hay was also achieved when using past five-year CDLs. Producer's accuracy of non-crop was the highest in the initial classification result. Despite the best accuracy of non-crop in the initial classification, this accuracy was mainly due to under-estimation of non-crop areas in the classification (see the confusion matrix in Table 5). Meanwhile, improved accuracy of major crops such as winter wheat, corn, and soybean were obtained from self-learning with past CDLs and led to the significant improvement in overall accuracy, compared with the initial classification. In summary, the improved overall accuracy of the self-learning approach was attributed to both an increase of the number of majority classes that were correctly classified and the decrease of misclassification of sorghum and grain/hay.

When the accuracy of self-learning classification with different CDL combination cases was compared, the self-learning with the five-year CDLs did not show the best classification accuracy. The case of the three-year CDLs showed the best overall accuracy and Kappa coefficient, and the case of the four-year CDLs was the second best. The poorest overall accuracy was obtained from the case of the two-year CDLs. In addition, there was no one CDL combination case where class-wise accuracy was always superior to the initial classification across all classes. Improved classification of each case resulted from the contribution of different land-cover types. In the case of the three-year CDLs, an increase of correctly classified pixels of corn and soybean led to the best overall accuracy. The second best overall accuracy in the case of the four-year CDLS was mainly due to correct classification of soybean and non-crop. An improvement in classification accuracy of cases of the five-year and two-year CDLs, compared to the initial classification, was attributed to an increase of correct classification of winter wheat and non-crop, respectively.

The core component of the self-learning approach is to derive rule-based class labels from sequential land-cover patterns in order to assign predefined class labels to the candidates for new training data. Thus, the accuracy of the predefined class label greatly affects the classification performance. To investigate this effect, further analysis was conducted by analyzing the accuracy of rule-based class labels derived from past CDLs in Figure 4. Since the true land-cover map (i.e., the CDL in 2015) was available, the rule-based class labels were directly compared with it.

The accuracy assessment results of rule-based class labels are listed in Table 7. Except for the case of the two-year CDLs, the overall accuracy of all cases was very high. As the number of CDLs for deriving sequential land-cover patterns increased, the corresponding accuracy of the rule-based class labels also increased. However, this high overall accuracy was obtained by the contribution of very high accuracy of non-crop which is one of majority classes in the study area. Regardless of different CLD combination cases, non-crop and sorghum showed the best and worst accuracy values, respectively. Unlike the rules on crop rotations, non-crop was unambiguously predicted to remain unchanged from the unique sequence rule, which led to the most accuracy of the rule-based label of non-crop. The decrease in the class-wise accuracy for crops in different CDL combination cases was due to the fact that sequential patterns of land-cover changes derived from past land-cover maps during too short a period (e.g., the two-year CDLs) were not sufficient to generate accurate rule-based class labels.

Table 7. Accuracy of rule-based class labels for four different past CDL combination cases.

	Past 5-Year CDLs	Past 4-Year CDLs	Past 3-Year CDLs	Past 2-Year CDLs
Overall accuracy	98.42	97.50	94.17	81.49
Kappa coefficient	0.94	0.92	0.86	0.69
Corn	91.25	86.73	76.27	52.99
Sorghum	32.35	51.02	38.08	21.65
Soybean	80.84	78.51	65.75	39.83
Winter wheat	92.42	89.47	82.02	65.72
Grain/hay	85.45	82.69	84.88	77.10
Non-crop	99.91	99.91	99.91	99.90

Despite the best accuracy of rule-based class labels of the five-year CDLs, however, the best classification accuracy was not obtained. This result can be attributed to the number of pixels that were assigned to rule-based class labels. The more land-cover maps that were used resulted in fewer pixels having rule-based class labels (see Figure 4). This was because more strict and stable rules were only extracted in cases that used more past-land cover maps. Although some candidate pixels with higher uncertainty were selected, their class labels cannot be assigned because no rule-based class labels were available at those pixels. As the most uncertain candidates were ignored, less uncertain candidate pixels might be selected as new training data. As a result, the selected training data might not be informative pixels. To verify these explanations, the interquartile range (IQR) of ΔP in Equation (2) at new training pixels was computed to measure the spread of uncertainty (Table 8). The smaller IQR implies the selection of more uncertain pixels with lower ΔP. As expected, the case of the five-year CDLs did not show the smallest IQR values for all classes. The smallest IQRs for corn and soybean in the case of the three-year CDLs indicate that the most informative pixels with higher uncertainty were selected as new training data, resulting in an improvement of accuracy for corn and soybean, and the best overall accuracy. From these interpretation results, the three-year CDLs were efficient for the study area because the accuracy was similar or better than the other cases. To derive a guideline on the selection of the optimal number of past land-cover maps, it is necessary to conduct more experiments on other sites using the different temporal length.

Table 8. Interquartile range of uncertainty values at new training pixels for four different past CDL combination cases.

Class	Past 5-Year CDLs	Past 4-Year CDLs	Past 3-Year CDLs	Past 2-Year CDLs
Corn	0.256	0.257	0.153	0.267
Sorghum	0.184	0.256	0.208	0.161
Soybean	0.273	0.261	0.208	0.231
Winter wheat	0.321	0.237	0.349	0.277
Grain/hay	0.123	0.189	0.101	0.091
Non-crop	0.512	0.654	0.437	0.303

Based on all accuracy evaluation results, it can be concluded that by adding the most informative pixels with rule-based class labels, the decision boundary could be positively revised, consequently leading to an accuracy improvement. It was also found that the selection of the most informative pixels was more important for classification performance than the accuracy of rule-based class labels.

4. Discussion

4.1. Active Learning Versus Self-Learning

AL requires analyst intervention for labeling of the most informative pixels to be used for further classification. Similarly, the self-learning approach also selects pixels with high uncertainty as the most informative ones, but the class labels of the most informative pixels are defined from unique sequence rules of time-series past land-cover maps, without any analyst intervention. Thus, manual labeling load can be reduced, which is the main advantage of the self-learning approach. When classification is conducted for large areas (e.g., state or country units) or inaccessible areas, this advantage can be greatly highlighted. However, the self-learning approach does not always aim at obtaining better classification accuracy than AL because in some cases, manual labeling by analyst might be more accurate than automatic labeling in the self-learning approach.

To investigate how classification performance of self-learning is compatible with AL, an additional comparative experiment was conducted. To mimic analyst intervention, manual labeling by analyst was replaced by defining the class label of the most informative pixels to that of corresponding pixels in the 2015 CDL. The same rule for the selection of informative pixels in the self-learning approach was also applied to AL for a fair comparison. The overall accuracy and Kappa coefficient of the AL classification result were 84.99% and 0.757, respectively. When we compared these accuracy statistics of AL with self-learning with the three-year CDLs that showed the best accuracy, the difference in overall accuracy was only 0.57 percentage points (84.99% versus 84.42%). The Kappa coefficient of AL was also very similar to that of self-learning (0.757 versus 0.759). Therefore, the classification accuracy of self-learning, which is compatible to that of AL, confirms the effectiveness of the presented approach in this study.

4.2. Generation of Sequence Rules

The generation of reliable sequence rules from past land-cover maps is essential in the self-learning approach. This study applied a heuristic approach to predict the class labels of candidates for new training data. Recently, a machine learning approach was presented to build rules on crop rotations for early crop type mapping before the crop season [46]. A Markov logic network (MLN), which can combine learning from data with expert knowledge, was applied to model crop rotations. The assessment results based on different temporal length and spatial coverages revealed that the MLN showed an accuracy of up to 60%, particularly the good prediction accuracy even for a large region with heterogeneous climatic conditions and soils. In this study, a heuristic approach for the rule generation has been tested on a relatively small area. From our previous study in Illinois State [19], the rule-based class labels, which were combined with the AL-based classification results, could contribute to an improvement in classification accuracy in a large area, which indicates the applicability of rule information. Similar to the approach in Osman et al. [46], we will test whether the rules built from data in a small area can be transferred to other landscapes or large areas with diverse crop rotations, as well as the applicability of the MLN. Since a wrong label assignment greatly affects classification performances [47], the effects of class label noise and the accuracy of existing land-cover maps should also be investigated.

4.3. Practical Issues

For crop classification, annual change patterns should be used in order to properly account for various cropping systems. However, from a practical viewpoint, the collection of many consecutive annual land-cover maps is a demanding task, and not always possible. If limited land-cover maps are available, or if the time interval between sequential land-cover maps is more than two years, another approach within the self-learning framework should be developed. Instead of using deterministic hard class labels, a probabilistic approach based on transition probability can be applied. In terms of temporal contextual information, the transition probabilities between considered land-cover classes can be defined using expert knowledge. These probabilities can then be combined with the conditional probabilities based on spectral or scattering features within a probabilistic framework. If this probabilistic approach is combined with the self-learning approach, errors from rule-based class labels could be reduced. Future studies will investigate these aspects.

Other practical issue is regarding the selection of new training pixels from candidates. To alleviate the bias towards majority classes, new training pixels were selected from random under-sampling. For the comparison purposes, under-sampling of pixels with highest uncertainty was tested additionally. The candidate pixels were first sorted in a descending order by their uncertainty values and the pixels with highest uncertainty were then selected as new training pixels. When past three CDLs were used for the rule generation, the overall accuracy of this different under-sampling approach was 81.44%, which is slightly lower than that of random under-sampling (84.42%). This different classification accuracy is due to the number and class types of newly added training pixels. Relatively many pixels for sorghum which has the lower accuracy of rule-based class labels were selected as new training pixels, whereas fewer pixels for winter wheat and non-crops with high accuracy were selected. This different selection of new training pixels resulted in the lower classification accuracy. Despite the relatively higher accuracy of random under-sampling, majority classes still affected the classification accuracy, as mentioned in Section 3.4. To avoid selecting redundant pixels, the representativeness and diversity of the most uncertain pixels should be considered as criteria for the identification of the most informative pixels. To this end, entropy and spatial density measures, and clustering can be applied [48,49]. The application of these criteria will be tested within a self-learning framework.

5. Conclusions

A self-learning classification approach, which can select the most informative labeled pixels as new training data, was presented in this study. The proposed approach differs from the AL approach in that no analyst intervention was required. The class labels for new candidate pixels were predicted from representative sequence rules selected from sequential change patterns of past land-cover maps. A classification experiment in crop cultivation areas demonstrated that this method could be used to properly define the class labels of unlabeled informative pixels. By progressively adding these informative labeled pixels into the training data, misclassification based purely on spectral information from a small number of training data could be greatly reduced, and higher classification accuracy was achieved. To strengthen the advantage of the self-learning approach, more extensive classification experiments in other regions with a wide variety of land-cover types and climatic conditions and different availability of past land-cover maps will be included in future work.

Acknowledgments: This work was carried out with the support of "Cooperative Research Program for Agriculture Science & Technology Development (Project No. PJ009978)," Rural Development Administration, Republic of Korea. The authors thank three anonymous reviewers for providing invaluable comments on the original manuscript.

Author Contributions: Yeseul Kim and No-Wook Park designed this study. Yeseul Kim performed data processing and prepared the draft of this manuscript. No-Wook Park contributed to methodological developments and edited the manuscript. Kyung-Do Lee made important contributions to interpretations of crop rotations and classification results.

References

1. Wardlow, B.D.; Egbert, S.L. Large-area crop mapping using time-series MODIS 250 m NDVI data: An assessment for the U.S. Central Great Plains. *Remote Sens. Environ.* **2008**, *112*, 1096–1116. [CrossRef]
2. Corcoran, J.M.; Knight, J.F.; Gallant, A.L. Influence of multi-temporal remotely sensed and ancillary data on the accuracy of random forest classification of wetlands in Northern Minnesota. *Remote Sens.* **2013**, *5*, 3212–3238. [CrossRef]
3. Jia, K.; Liang, S.; Wei, X.; Yao, Y.; Su, Y.; Jiang, B.; Wang, X. Land cover classification of Landsat data with phenological features extracted from time series MODIS NDVI data. *Remote Sens.* **2014**, *6*, 11518–11532. [CrossRef]
4. Kong, F.; Li, X.; Wang, H.; Xie, D.; Li, X.; Bai, Y. Land cover classification based on fused data from GF-1 and MODIS NDVI time series. *Remote Sens.* **2016**, *8*, 741. [CrossRef]
5. Park, N.-W.; Kyriakidis, P.C.; Hong, S. Spatial estimation of classification accuracy using indicator kriging with an image-derived ambiguity index. *Remote Sens.* **2016**, *8*, 320. [CrossRef]
6. Walter, V.; Fritsch, D. Automatic verification of GIS data using high resolution multispectral data. *Int. Arch. Photogramm. Remote Sens.* **1998**, *32*, 485–490.
7. Waske, B.; Benediktsson, J.A. Fusion of support vector machines for classification of multisensor data. *IEEE Trans. Geosci. Remote Sens.* **2007**, *45*, 3858–3866. [CrossRef]
8. Attarchi, S.; Gloaguen, R. Classifying complex mountainous forests with L-band SAR and Landsat data integration: A comparison among different machine learning methods in the Hyrcanian forest. *Remote Sens.* **2014**, *6*, 3624–3647. [CrossRef]
9. Gessner, U.; Machwitz, M.; Esch, T.; Tillack, A.; Naeimi, V.; Kuenzer, C.; Dech, S. Multi-sensor mapping of West African land cover using MODIS, ASAR and TanDEM-X/TerraSAR-X data. *Remote Sens. Environ.* **2015**, *164*, 282–297. [CrossRef]
10. Heumann, B.W. An object-based classification of Mangroves using a hybrid decision tree-support vector machine approach. *Remote Sens.* **2011**, *3*, 2440–2460. [CrossRef]
11. Sonobe, R.; Tani, H.; Wang, X.; Kobayashi, N.; Shimamura, H. Random forest classification of crop type using multi-temporal TerraSAR-X dual-polarimetric data. *Remote Sens. Lett.* **2014**, *5*, 157–164. [CrossRef]
12. Wieland, M.; Pittore, M. Performance evaluation of machine learning algorithms for urban pattern recognition from multi-spectral satellite images. *Remote Sens.* **2014**, *6*, 2912–2939. [CrossRef]
13. Zhu, X. *Semi-Supervised Learning Literature Survey*; Technical Report 1530; Department of Computer Sciences, University of Wisconsin-Madison: Madison, WI, USA, 2005.
14. Settles, B. *Active Learning Literature Survey*; Technical Report 1648; Department of Computer Sciences, University of Wisconsin-Madison: Madison, WI, USA, 2010.
15. Tuia, D.; Pasolli, E.; Emery, W.J. Using active learning to adapt remote sensing image classifiers. *Remote Sens. Environ.* **2011**, *115*, 2232–2242. [CrossRef]
16. Ruiz, P.; Mateos, J.; Camps-Valls, G.; Molina, R.; Katsaggelos, A.K. Bayesian active remote sensing image classification. *IEEE Trans. Geosci. Remote Sens.* **2014**, *52*, 2186–2196. [CrossRef]
17. Huang, X.; Weng, C.; Lu, Q.; Feng, T.; Zhang, L. Automatic labeling and selection of training samples for high-resolution remote sensing image classification over urban areas. *Remote Sens.* **2015**, *7*, 16024–16044. [CrossRef]

18. Wan, L.; Tang, K.; Li, M.; Zhong, Y.; Qin, A.K. Collaborative active and semisupervised learning for hyperspectral remote sensing image classification. *IEEE Trans. Geosci. Remote Sens.* **2015**, *53*, 2384–2396. [CrossRef]
19. Kim, Y.-S.; Yoo, H.-Y.; Park, N.-W.; Lee, K.-D. Classification of crop cultivation areas using active learning and temporal contextual information. *J. Korean Assoc. Geogr. Inf. Stud.* **2015**, *18*, 76–88. (In Korean) [CrossRef]
20. Bruzzone, L.; Persello, C. A novel context-sensitive semisupervised SVM classifier robust to mislabeled training samples. *IEEE Trans. Geosci. Remote Sens.* **2009**, *47*, 2142–2154. [CrossRef]
21. Uhlmann, S.; Kiranyaz, S.; Gabbouj, M. Semi-supervised learning for ill-posed polarimetric SAR classification. *Remote Sens.* **2014**, *6*, 4801–4830. [CrossRef]
22. Chapelle, O.; Schölkopf, B.; Zien, A. *Semi-Supervised Learning*; The MIT Press: Cambridge, UK, 2006.
23. Leng, Y.; Xu, X.; Qi, G. Combining active learning and semi-supervised learning to construct SVM classifier. *Knowl.-Based Syst.* **2013**, *44*, 121–131. [CrossRef]
24. Muñoz-Marí, J.; Tuia, D.; Camps-Valls, G. Semisupervised classification of remote sensing images with active queries. *IEEE Trans. Geosci. Remote Sens.* **2012**, *50*, 3751–3763. [CrossRef]
25. Dópido, I.; Li, J.; Marpu, P.R.; Plaza, A.; Dias, J.M.B.; Benediktsson, J.A. Semisupervised self-learning for hyperspectral image classification. *IEEE Trans. Geosci. Remote Sens.* **2013**, *51*, 4032–4044. [CrossRef]
26. Blum, A.; Mitchell, T. Combining Labeled Data and Unlabeled Data with Co-training. In Proceedings of the Eleventh Annual Conference on Computational Learning Theory, Madison, WI, USA, 24–26 July 1998; pp. 92–100.
27. Boryan, C.; Yang, Z.; Mueller, R.; Craig, M. Monitoring US agriculture: The US department of agriculture, national statistics service, cropland data layer program. *Geocarto Int.* **2011**, *26*, 341–358. [CrossRef]
28. CropScape. Available online: https://nassgeodata.gmu.edu/CropScape (accessed on 1 March 2017).
29. Wardlow, B.D.; Egbert, S.L. A comparison of MODIS 250-m EVI and NDVI data for crop mapping: A case study for southwest Kansas. *Int. J. Remote Sens.* **2010**, *31*, 805–830. [CrossRef]
30. Conrad, C.; Colditz, R.R.; Dech, S.; Klein, D.; Vlek, P.L.G. Temporal segmentation of MODIS time series for improving crop classification in Central Asian irrigation systems. *Int. J. Remote Sens.* **2011**, *32*, 8763–8778. [CrossRef]
31. Kim, Y.; Park, N.-W.; Hong, S.; Lee, K.; Yoo, H.Y. Early production of large-area crop classification map using time-series vegetation index and past crop cultivation patterns. *Korean J. Remote Sens.* **2014**, *30*, 493–503. (In Korean) [CrossRef]
32. Chen, J.; Jönsson, P.; Tamura, M.; Gu, Z.; Matsushita, B.; Eklundh, L. A simple method for reconstructing a high-quality NDVI time-series data set based on the Savitzky-Golay filter. *Remote Sens. Environ.* **2004**, *91*, 332–344. [CrossRef]
33. Melgani, F.; Bruzzone, L. Classification of hyperspectral remote sensing images with support vector machines. *IEEE Trans. Geosci. Remote Sens.* **2004**, *42*, 1778–1790. [CrossRef]
34. Mathur, A.; Foody, G.M. Crop classification by support vector machine with intelligently selected training data for an operational application. *Int. J. Remote Sens.* **2008**, *29*, 2227–2240. [CrossRef]
35. Vapnik, V. *Statistical Learning Theory*; Wiley: New York, NY, USA, 1998.
36. Wu, T.-F.; Lin, C.-J.; Weng, R.C. Probability estimates for multi-class classification by pairwise coupling. *J. Mach. Learn. Res.* **2004**, *5*, 975–1005.
37. Tuia, D.; Ratle, F.; Pacifici, F.; Kanevski, M.F.; Emery, W.J. Active learning methods for remote sensing image classification. *IEEE Trans. Geosci. Remote Sens.* **2009**, *47*, 2218–2232. [CrossRef]
38. Tuia, D.; Volpi, M.; Copa, L.; Kanevski, M.; Muñoz-Marí, J. A survey of active learning algorithms for supervised remote sensing image classification. *IEEE J. Sel. Top. Signal Process.* **2011**, *5*, 606–617. [CrossRef]
39. Li, J.; Bioucas-Dias, J.M.; Plaza, A. Hyperspectral image segmentation using new Bayesian approach with active learning. *IEEE Trans. Geosci. Remote.* **2011**, *49*, 3947–3960. [CrossRef]
40. Luo, T.; Kramer, K.; Goldgof, D.B.; Hall, L.O.; Samson, S.; Remsen, A.; Hopkins, T. Active learning to recognize multiple types of plankton. *J. Mach. Learn. Res.* **2005**, *6*, 589–613.
41. Acosta-Martínez, V.; Mikha, M.M.; Vigil, M.F. Microbial communities and enzyme activities in soils under alternative crop rotations compared to wheat-fallow for the Central Great Plains. *Appl. Soil Ecol.* **2007**, *37*, 41–52. [CrossRef]

42. Culman, S.W.; DuPont, S.T.; Glover, J.D.; Buckley, D.H.; Fick, G.W.; Ferris, H.; Crews, T.E. Long-term impacts of high-input annual cropping and unfertilized perennial grass production on soil properties and belowground food webs in Kansas, USA. *Agric. Ecosyst. Environ.* **2010**, *137*, 13–24. [CrossRef]
43. Wardlow, B.D.; Egbert, S.L.; Kastens, J.H. Analysis of time-series MODIS 250 m vegetation index data for crop classification in the U.S. Central Great Plains. *Remote Sens. Environ.* **2007**, *108*, 290–310. [CrossRef]
44. Foody, G.M. Thematic map comparison: Evaluating the statistical significance of differences in classification accuracy. *Photogramm. Eng. Remote Sens.* **2008**, *70*, 627–633. [CrossRef]
45. Lillesand, T.M.; Kiefer, R.W.; Chipman, J.W. *Remote Sensing and Image Interpretation*, 6th ed.; Wiley: Hoboken, NJ, USA, 2008.
46. Osman, J.; Inglada, J.; Dejoux, J.-F. Assessment of a Markov logic model of crop rotation for early crop mapping. *Comput. Electron. Agric.* **2015**, *113*, 234–243. [CrossRef]
47. Pelletier, C.; Valero, S.; Inglada, J.; Champion, N.; Sicre, C.M.; Dedieu, G. Effect of training class label noise on classification performances for land cover mapping with satellite image time series. *Remote Sens.* **2017**, *9*, 173. [CrossRef]
48. He, T.; Zhang, S.; Xin, J.; Zhao, P.; Wu, J.; Xian, X.; Li, C.; Cui, Z. An active learning approach with uncertainty, representativeness, and diversity. *Sci. World J.* **2014**, *2014*, 827586. [CrossRef] [PubMed]
49. Demir, B.; Minello, L.; Bruzzone, L. An effective strategy to reduce the labeling cost in the definition of training sets by active learning. *IEEE Geosci. Remote Sens. Lett.* **2015**, *11*, 79–83. [CrossRef]

11

Selection of the Key Earth Observation Sensors and Platforms Focusing on Applications for Polar Regions in the Scope of Copernicus System 2020–2030

Estefany Lancheros [1,*], Adriano Camps [1,*], Hyuk Park [1], Pedro Rodriguez [2], Stefania Tonetti [3], Judith Cote [4] and Stephane Pierotti [4]

1 Universitat Politècnica Catalunya-BarcelonaTech & IEEC, Campus Nord, 08034 Barcelona, Spain; park.hyuk@tsc.upc.edu
2 Thales Alenia Space, 28760 Madrid, (España)Madrid, Spain; pedro.r@thalesaleniaspace.com
3 Deimos Space S.L.U, 28760 Madrid, Spain; stefania.tonetti@deimos-space.com
4 Thales Alenia Space, 06150 Toulouse, France; judith.cote@thalesaleniaspace.com (J.C.); stephane.pierotti@thalesaleniaspace.com (S.P.)
* Correspondence: estefany@tsc.upc.edu (E.L.); camps@tsc.upc.edu (A.C.)

Abstract: An optimal payload selection conducted in the frame of the H2020 ONION project (id 687490) is presented based on the ability to cover the observation needs of the Copernicus system in the time period 2020–2030. Payload selection is constrained by the variables that can be measured, the power consumption, and weight of the instrument, and the required accuracy and spatial resolution (horizontal or vertical). It involved 20 measurements with observation gaps according to the user requirements that were detected in the top 10 use cases in the scope of Copernicus space infrastructure, 9 potential applied technologies, and 39 available commercial platforms. Additional Earth Observation (EO) infrastructures are proposed to reduce measurements gaps, based on a weighting system that assigned high relevance for measurements associated to Marine for Weather Forecast over Polar Regions. This study concludes with a rank and mapping of the potential technologies and the suitable commercial platforms to cover most of the requirements of the top ten use cases, analyzing the Marine for Weather Forecast, Sea Ice Monitoring, Fishing Pressure, and Agriculture and Forestry: Hydric stress as the priority use cases.

Keywords: Earth Observation; satellite; sensors; platform; radiometer; SAR; GNSS-R; VIS/NIR imager; polar; weather; ice; marine

---

1. Introduction

The Copernicus system, previously known as Global Monitoring for Environmental Security (GMES), is a revolutionary program of the European Union (EU) to address the end-user requirements over six thematic services: Atmosphere, Marine, Land, Climate Change, Emergency Management, and Security. Copernicus is supported by the space and in situ components. The space segment is based on a set of Earth Observation (EO) satellites known as the Sentinels and some contributing missions. Contributing missions with space infrastructure are the Earth Explorer missions [1] operated by the European Space Agency (ESA), the meteorological missions operated by the European Organization for the Exploitation of Meteorological Satellites (EUMETSAT), and EO missions operated by the European Union (EU), third countries, and commercial providers.

Currently, there are seven Sentinels satellites in orbit: Sentinel-1A and Sentinel-1B with C-band Synthetic Aperture Radar (SAR) for land and ocean observation, Sentinel-2A and Sentinel-2B with high resolution optical imager called Multi-Spectral Imager (MSI) for land and vegetation observation, Sentinel-3A and Sentinel-3B with a suite of instruments such as Synthetic Aperture Radar altimeter (SRAL), and medium resolution optical imager: Ocean and Land Colour Imager (OLCI) and Sea and Land Surface Temperature Radiometer (SLTR) for ocean and land observation, and Sentinel-5P with cross-nadir scanning sounder called Tropospheric Monitoring Instrument (TROPOMI) for atmospheric chemistry and aerosol studies. Future Sentinel missions that will be launched in the next decade are Sentinel-4 for atmospheric chemistry as hosted payload over Meteosat Third Generation-Sounding (MTG-S); Sentinel-5 will be launched as hosted payloads over MetOp-Second generation (MetOp SG) for atmospheric chemistry, aerosol and spectral irradiance studies; and Sentinel-6 will be launched in a Low Earth Orbit (LEO) inclined over the equator for ocean altimetry as an international program between ESA, the National Aeronautics and Space Administration (NASA), the National Centre for Space Studies (CNES), EUMETSAT, and the National Oceanic and Atmospheric Administration (NOAA). Additionally, the third and fourth units of Sentinel-1C/D, Sentinel-2C/D, and Sentinel-3C/D will have planned to launch for the continuity of these programs.

At present, Earth Explorer missions are: Soil Moisture and Ocean Salinity (SMOS) launched on 2 November 2009 for sea surface salinity and soil moisture monitoring; this is considered as a potential gap because this mission has no continuity; Atmospheric Dynamics Mission—Aeolus (ADM-AEOLUS) launched on 22 August 2018, with an Atmospheric Laser Doppler Instrument (ALADIN) for contribution to aerosol observation and wind profile. Future Earth Explorer missions are: EarthCARE mission with a suite of instruments such as a Atmospheric Lidar (ATLID), Broad-Band Radiometer (BBR), Cloud Profiling Radar (CPR), and Multi-Spectral Imager (MSI) for cloud, aerosol, and radiation process studies; Biomass mission with a interferometric and polarimetric P-band SAR for biomass and glacier topography study; and FLEX mission with a FLORIS instrument for photosynthetic activity monitoring. Additionally, the ESA has chosen two potential Earth Explorer candidates missions [2], the Far-infrared Outgoing Radiation Understanding and Monitoring (FORUM) with measure in the 15–100 micron range, and Sea-Surface Kinematics Multi-scale (SKIM) monitoring with a multi-beam radar altimeter with a wide swath. These two candidates considered will spend the next two years being studied thoroughly and only one will be implemented.

State of the art of the meteorological contributing missions of Copernicus are MetOp in Low Earth Orbit (LEO), and Meteosat Second Generation (MSG) in Geostationary orbit (GEO). For the incoming decade (2020 to 2030), these programs will have continuity because new missions will be launched such as Meteosat Third Generation (MTG) and MetOp Second Generation (SG).

For Sentinel expansion, the ESA has identified six possible candidates with phase A/B under preparation for the expansions to the Copernicus space component [3], such as Sentinel-7 Anthropogenic CO_2 monitoring mission, Sentinel-8 High Spatio-Temporal Resolution Land Surface Temperature (LST) Monitoring Mission (companion to Sentinel-2 C/D), Sentinel-9 with two components: Polar Ice and Snow Topographic Mission, and Polar Weather payload on a Highly Elliptical Orbit, and Sentinel-10 with a Hyperspectral Imaging Mission. Other possible candidates for the expansion of Copernicus are Passive Microwave Imaging Mission, and L-Band SAR mission. In parallel, a recent study of the Copernicus Market [4] mentioned that the agriculture, ocean monitoring, oil, and gas are a potential market in terms of Copernicus impact and user benefits. The approach followed is to identify the user's needs, identifying the gaps and potential areas for improvement in the Copernicus EO infrastructure, taking into account the future instruments and missions. This form could analyse if the plans of the extension of Copernicus support the emergent needs.

The European Commission (EC) has led a revolutionary programme aiming at securing and exploiting space infrastructure to meet future demands and societal needs. The H2020 Operational Network of Individual Observation Node (ONION) project identified the main needs of the space segment infrastructure of the Copernicus system and identified the key technology challenges to be faced in the future, taking into account the user requirements at the center of the design process. The ONION project analyzed the user needs and ranked the top 10 use cases [5]. Each use case is associated with a Copernicus service, and they are formed by a set of measurements required to meet the users' needs. The measurements are the geophysical products derived from satellite observations. In addition, the measurement gaps and user requirements were identified and defined by the ONION project (Table 1) [5,6], taking into account if, in the coming decade, the Copernicus and contributing missions satisfy the user requirements. This work focuses on the identification of the potential sensor technologies and platforms to meet those needs detected. The capability of the different technologies is evaluated according to current trends in the design of small satellites. These technologies are presented in view of the novel developments in spacecraft and sensor miniaturization, reduced power consumption, measurement requirements, and data quality, in order to cover the user requirements [6], so as to obtain competitive and cost-effectiveness services.

The 20 measurements with gaps detected [6] in the top ten use cases are: (1) Ocean surface currents, (2) dominant wave direction, (3) significant wave height, (4) horizontal wind speed over the sea surface, (5) sea ice type, (6) iceberg tracking, (7) sea ice cover, (8) sea ice extent, (9) sea ice drift, (10) sea ice thickness, (11) atmospheric pressure over the sea surface, (12) sea surface temperature, (13) ocean chlorophyll concentration, (14) ocean imagery and water leaving radiance, (15) color dissolved organic matter, (16) detection of water stress in crops, (17) estimation of crop evapotranspiration, (18) surface soil moisture, (19) crop growth and condition, and (20) monitoring system vessels. Marine for Weather Forecast, Sea Ice Monitoring, Fishing Pressure, and Agriculture and Forestry: Hydric Stress use cases involved all the measurements with observations gaps detected over Copernicus space infrastructure in the period 2020–2030. The Marine for Weather Forecast, Sea Ice Monitoring, and Fishing Pressure use cases are ranked as the emerging observation needs. These use cases required measurements that are of crucial importance for a wide range of activities from maritime traffic, fishery, environment, food and medicine supply for populations at high latitudes, as well as for oil and gas operations. Another high priority use case with observation gaps (Table 1) is the Agriculture and Forestry: Hydric Stress. The key measurements to cover for this use case are important to study the hydrological cycles, agriculture production, climatology, and meteorology. With the objective to cover these 20 measurements with gaps, we designed a methodology that focuses on the critical technologies to complement Copernicus observation gaps.

The methodology applied to select the appropriate sensors and platforms is sketched in Figure 1. First, a survey of the commercial small platform capabilities is presented in terms of mass, payload power, communications, pointing knowledge, and control. Second, the state-of-the-art sensors in terms of mass, power consumption, swath, and data rate is presented. Each sensor or technology is then studied to cover the observation gaps. Based on the survey of the instrument capabilities and data quality, a summary of the existing, and emerging in EO sensors is given, including the scientific and technological limitations in terms of spatial resolution, accuracy, and swath. Within these bounds, the potential instruments are selected according to the available commercial small platforms. The reference instruments are evaluated based on the variables with gaps that can be measured using a scoring method. This scoring method assigns a high score to the sensors that present lower power consumption, lower mass, and high data quality (better accuracy, smaller spatial resolution, and/or wider coverage). Finally, the most relevant instrument technologies compatible with small platforms are identified to complement the existing Copernicus Services for the selected use cases.

Table 1. The top ten use cases.

	Use Case [5]	Copernicus Services Related	2020–2030 Copernicus Instrument/Mission [7]	Contributing Instrument/Mission [8]	Measurements with Gaps Detected [6]
1	Marine for Weather Forecast	Marine	SAR-C/Sentinel-1 SRAL/Sentinel-3 OLCI/Sentinel-3 Poseidon-4/Sentinel-6	PALSAR-3/ALOS-4 SAR-2000 S.G/CSG SAR/HRWS SAR-X/TSX-NG SAR-X/PAZ SWIM/CFOSAT ASCAT/MetOp SCA/MetOp-SG	Wind speed over sea surface (horizontal), Ocean surface currents, Dominant wave direction, Dominant wave period, Significant wave height, Atmospheric pressure over sea surface.
2	Sea Ice Monitoring: Extent, Thickness	Marine	SAR-C/Sentinel-1 SLTR, OLCI, SRAL /Sentinel-3	PALSAR-3/ALOS-4 SAR-2000 S.G/CSG SAR/HRWS SAR-X/TSX-NG SAR-X/PAZ SWIM/CFOSAT ASCAT/MetOp SCA/MetOp-SG MSI/Earth-CARE IASI and AVHRR-3/MetOp METimage,IASI-NG/MetOp-SG	Sea surface temperature, Sea ice cover, Sea ice type, Sea ice thickness, Iceberg tracking, Sea ice drift, Sea ice extent, Wind speed over sea surface horizontal, Ocean surface currents, Dominant wave direction, Dominant wave period, Significant wave height.
3	Fishing Pressure, Stock Assessment	Marine	OLCI/Sentinel-3 SAR-C/Sentinel-1	SEVERI/MSG MSI/Earth-CARE IASI and AVHRR-3/MetOp METimage,IASI-NG/MetOp-SG FCI/MTG-I IRS/MTG-S FLORIS/FLEX	Color dissolved organic matter, Ocean imagery and water leaving radiance, Ocean chlorophyll concentration, Monitoring system- vessels.
4	Land for Infrastructure Status Assessment	Security	SAR-C/Sentinel-1 MSI/Sentinel-2 OLCI/Sentinel-3	SAR-2000 S.G/CSG SAR-X/TSX-NG HRWS-SAR/HRWS SAR-X/PAZ DESIS/ISS DESIS HYC/PRISMA P-BAND SAR/BIOMASS HSI/EnMap FCI/MTG-I HiRAIS/Deimos-2 NAOMI/SPOT-7 REIS/RapiEye	None
5	Agriculture and Forestry: Hydric Stress	Land	SAR-C/Sentinel-1 MSI/Sentinel-2 SLTR, OLCI/Sentinel-3	SAR-2000 S.G/CSG SAR-X/TSX-NG HRWS-SAR/HRWS SAR-X/PAZ DESIS/ISS DESIS HYC/PRISMA P-BAND SAR/BIOMASS ASCAT/MetOp SCA/MetOp-SG MSI/EartCARE HSI/EnMap FCI/MTG-I HiRAIS/Deimos-2 NAOMI/SPOT-7 REIS/RapiEye SEVERI/MSG MSI/Earth-CARE IASI and AVHRR-3/MetOp METimage,IASI-NG/MetOp-SG FCI/MTG-I IRS/MTG-S FLORIS/FLEX	Surface soil moisture, Crop grow and conditions, detection of water stress in crops, Estimation of crop evapotranspiration.

Table 1. *Cont.*

	Use Case [5]	Copernicus Services Related	2020–2030 Copernicus Instrument/Mission [7]	Contributing Instrument/Mission [8]	Measurements with Gaps Detected [6]
6	Land for Basic Mapping: Risk Assessment	Emergency Management	SAR-C/Sentinel-1 MSI/Sentinel-2 OLCI/Sentinel-3	SAR-2000 S.G/CSG SAR-X/TSX-NG HRWS-SAR/HRWS SAR-X/PAZ DESIS/ISS DESIS HYC/PRISMA P-BAND SAR/BIOMASS HSI/EnMap FCI/MTG-I HiRAIS/Deimos-2 NAOMI/SPOT-7 REIS/RapiEye	Surface soil moisture.
7	Sea Ice Melting Emissions Assessment	Marine	SAR-C/Sentinel-1 SLTR, OLCI, SRAL/Sentinel-3	PALSAR-3/ALOS-4 SAR-2000 S.G/CSG SAR/HRWS SAR-X/TSX-NG SAR-X/PAZ SWIM/CFOSAT ASCAT/MetOp SCA/MetOp-SG MSI/Earth-CARE IASI and AVHRR-3/MetOp METimage,IASI-NG/MetOp-SG	Sea surface temperature, Sea ice cover, Sea ice type, Sea ice thickness.
8	Atmosphere for Weather Forecast	Atmosphere	SAR-C/Sentinel-1 Sentinel-4/MTG-S Sentinel-5/MetOp-SG TROPOMI/Sentinel-5p	ASCAT/MetOp SCA/MetOP-SG SEVERI/MSG MSI, CPR/Earth-CARE IASI and AVHRR-3/MetOp METimage,IASI-NG/MetOp-SG FCI/MTG-I IRS/MTG-S	Wind speed over sea surface (horizontal), Wind vector over sea surface (horizontal), Atmospheric pressure over sea surface.
9	Climate for Ozone Layer and UV	Climate Change	SLTR, OLCI/Sentinel-3 Sentinel-4/MTG-S Sentinel-5/MetOp-SG TROPOMI/Sentinel-5p	SEVERI/MSG MSI, CPR/Earth-CARE GOME-2, IASI, AVHRR-3/MetOp METimage,IASI-NG/MetOp-SG FCI/MTG-I IRS/MTG-S HYC/PRISMA UVAS/Ingenio	None
10	Natural Habitat and Protected Species Monitoring	Land	SAR-C/Sentinel-1 MSI/Sentinel-2 OLCI, SLTR/Sentinel-3 Sentinel-4/MTG-S Sentinel-5/MetOp-SG TROPOMI/Sentinel-5p	ASCAT/MetOp SCA/MetOP-SG SEVERI/MSG MSI, CPR/Earth-CARE IASI and AVHRR-3/MetOp METimage,IASI-NG/MetOp-SG FCI/MTG-I IRS/MTG-S HYC/PRISMA FLORIS/FLEX	Surface soil moisture.

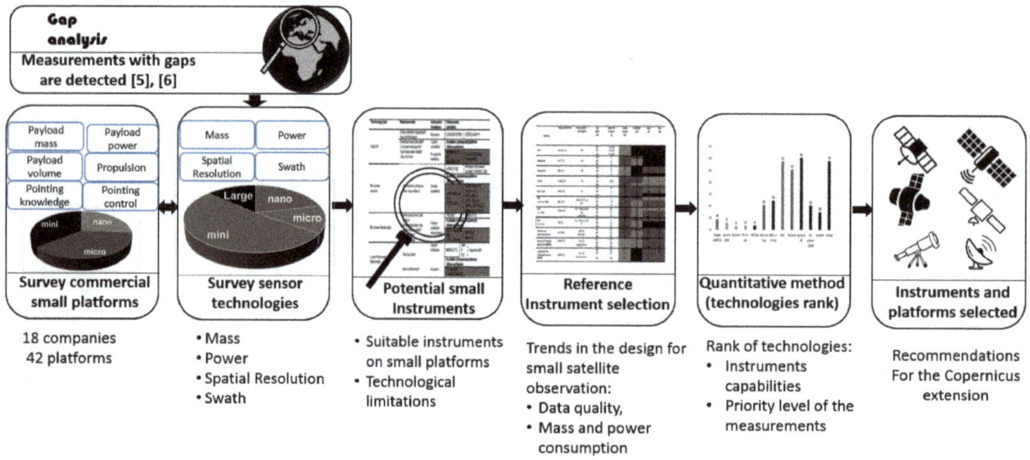

Figure 1. Design process to select payload and platform according to the requirements.

2. Survey of Commercial Small Platforms

This section presents the results of a comprehensive survey of commercial Low Earth Orbit (LEO) small platforms for EO, in order to properly select the platforms for each technology. To do this, the capabilities and limitations of the small commercial buses are taken into account. A total of forty-two commercial platforms from eighteen different companies have been identified, and their information has been compiled from company websites and conferences proceedings (Appendix A).

These small platforms cover a wide range of payload mass and power. They are categorized into three groups nano-, micro-, and mini-satellites. Table 2 summarizes their typical parameters. These platforms support payload masses from 1 kg to 600 kg [9], payload powers (orbital, average) from 1 W to 1500 W [10], downlink up 15 Mbps (S-band) [11], 100 Mbps (X-band) [12], and 1.2 Gbps (K-band) [13]. In this context, the recent evolution of the capability of micro- and mini-class platforms, and the payload miniaturization have demonstrated being a true competitor of large spacecrafts for some applications. Table 3 summarizes the capabilities of CubeSat EO platforms (3U, 6U, and 27U). Nanosatellites are now becoming popular thanks to the CubeSat standard. Typical CubeSat missions can be implemented in 1 to 3 years, with typical budgets from 200 K to 1 M $ USD, including launch.

On the other side, ESA has promoted the development of a generic Small Geostationary Platform [14] (SmallGEO or SGEO) industrialized by OHB [15]. This flexible and modular platform has a lifetime of up to 15 years, a payload mass of up to 400 kg, and a payload power of up to 4 kW [16 This platform was originally proposed to help European industries in the commercial telecom satellite market. However, the Earth Observation domain can also benefit from the capability of this platform in terms of available power and payload mass. In this way, an analysis of the EO technologies that are appropriate for use in small platforms is conducted in the next section.

Table 2. Summary of survey of commercial small platforms capabilities.

Classification	Satellite Mass [kg]	Max. Payload Mass [kg]	Max. Payload Power (average) [W]	Max. Data Rate (Downlink)
Nano	<10	≤3 [17]	≤15 [11,18]	≤15 Mbps [19]
Micro	10–100	≤54 [20]	≤150 [21]	≤160 Mbps [22]
Mini	100–1000	≤600 [9]	≤1500 [10]	≤1.2 Gbps [13]

Table 3. Summary of survey of commercial CubeSat platforms capabilities.

Classification	Approximate Size [cm]	Payload Mass [kg]	Payload Power [average/peak] [W]	Payload Data Rate [downlink]	References
3U	$10 \times 10 \times 30$	<3	≤15	≤15 Mbps	[17]
6U	$10 \times 20 \times 30$	<12	≤20	≤15 Mbps	[23]
27U	$30 \times 30 \times 30$	<50	≤90	≤100 Mbps	[20]

3. Survey of Earth Observation Sensors and Measurements Requirements to Cover the Future Gaps on Copernicus

EO satellites have revolutionized the study of the environment, and are contributing to a more rational use of the natural resources, and environmental protection. The applications of the data supplied by these systems are enormous: disaster monitoring, weather forecast, maritime safety, marine resources monitoring, forestry, vegetation state, water cycle, energy budget, pollution control, water quality, climate change, and security; using radars, microwave and optical/IR radiometers, optical imagers or scanners. Table 4 presents the generic classification of the remote sensors. Instruments are classified in the following four categories: active or passive, either microwave or optical. Optical sensors measure the signals received around the visible part of the spectrum, from the Ultra-Violet (UV) to the Thermal Infrared (TIR). Microwave sensors use the signals in the microwave and millimetre-wave parts of the spectrum, typically from 1 GHz to 1 THz. Passive systems are based on the collection of the electromagnetic waves that are emitted/scattered by external sources, such as the Sun or other bodies. On the other hand, active systems such as radars and lidars, transmit an electromagnetic wave, either radio or laser, and measure the scattered/reflected signal from the Earth's surface or atmosphere. Microwave sensors do not rely on the Sun as source of illumination. These particular characteristics are especially important in Polar Regions that have extended dark periods in winter. In addition, microwaves are mostly unaffected by the cloud cover, except in some specific bands. This feature makes microwave sensors more suitable than optical sensors in these regions.

This section presents a survey of the selected EO technologies. In order to identify the potential EO sensors to improve the Copernicus space infrastructure, EO technologies are analyzed in depth based on the measurements with identified gaps, and the technological limitations. A total of 77 instruments have been surveyed, and their parameters (mass, power consumption, spatial resolution, swath, frequency bands, aperture, and orbit altitude) have been compiled from the Observing Systems Capability Analysis and Review (OSCAR) Tool [24], the Earth Observation Portal Directory [25], and companies websites (Appendix B). The best instruments in terms of data quality and suitable for the small platform are identified for each technology.

Table 4. Instrument categorization: potential instruments to complement the Copernicus system [6].

<table>
<tr><th></th><th colspan="2">PASSIVE</th><th colspan="2">ACTIVE</th></tr>
<tr><td rowspan="3">MICROWAVE</td><td>Radiometer</td><td>• Imager
• Sounder</td><td>Real Aperture Radar</td><td>• Altimeter
• Scatterometer</td></tr>
<tr><td>Signals of Opportunity (SoOp)</td><td>• GNSS-R [a]
• Receiver of SoOP [b]</td><td rowspan="2">Synthetic Aperture Radar</td><td rowspan="2">• Altimeter
• Imager</td></tr>
<tr><td>Receiver</td><td>• Automatic Identification System (AIS)</td></tr>
<tr><td rowspan="2">OPTICAL</td><td>Radiometer</td><td rowspan="2">• Multispectral
• Hyperspectral</td><td colspan="2" rowspan="2">Lidar</td></tr>
<tr><td>Sounder</td></tr>
</table>

[a] multi-static radar using satellite navigation signals of opportunity (SoOp). [b] e.g., Direct Broadcast Satellite (DBS) television at Ku-band or X-band.

3.1. Passive Microwave

3.1.1. Microwave Imagers (MWIm)

The main applications of Microwave Imagers (MWIm) are atmospheric (X, K, Ka, and milimiter waves bands), oceanographic (C, X, K, and Ka bands), vegetation and soil moisture monitoring (P, L, S, C and X bands). High frequency microwave radiometers are particularly well suited for small platforms because of the antenna size constraints. These types of instruments can measure: wind speed [26,27], sea ice thickness [28,29], and sea ice cover [30], among other variables. Table A4 presents the features of some microwave radiometers, in terms of frequency bands, spatial resolution, antenna size, swath, mass, power consumption, and data rate. Assuming only one payload per platform, the affordable platforms (nano, micro, mini, and large) for the instruments are identified according to the power and mass requirements. This information is valuable in order to choose the potential instruments that will complement the Copernicus Space segment, trying to make them compatible with the smallest possible platforms, while fulfilling the user requirements. The measurement gaps that can be covered with this technology are: horizontal wind speed over the sea surface (MWIm with channels around 7, 10, 19, 37 GHz or 19 and 37 GHz), sea ice monitoring (cover, type, drift, MWIm with channels around 7, 10, 19, 37, and 90 GHz), sea ice thickness (MWIm with channels around 1.4 GHz), soil moisture (MWI with channels around 1.4 GHz, or 7 GHz, or 11 GHz), and sea surface temperature (MWIm with channels around 7 and/or 10 GHz).

According to Table A4, two microwave imagers capable of measuring the variables with gaps have been identified. These are selected because they are suitable for small platforms and present good data quality, to cover the user requirements.

- A Tropical Rainfall Measuring Mission Microwave Imager (TMI) like instrument is capable of measuring wind speed (at 10.65, 19.35 and 37 GHz), sea ice cover (at 19.35, 37, and 85.5 GHz), and sea surface temperature (at 10.65 GHz). Modified versions of TMI for micro- or mini-platforms achieving a 10 km spatial resolution using an aperture size (inflatable antenna) of 3.4 m @ 10.65 GHz from 600 km height will suit LEO polar Sun-Synchronous Orbit (SSO, $\sim$14 orbits/day) reducing the revisit time to 3 h in the Polar Regions. The required number of satellites was optimized in [31].
- The available L-Band microwave sensors, such as Microwave Imaging Radiometer using Aperture Synthesis (MIRAS) and Soil Moisture Active-Passive (SMAP) are suitable for mini-platforms. L-band microwave radiometers are capable of measuring the variables with the detected gaps, such as sea ice thickness and soil moisture. Sea ice thickness presents gaps in the revisit and latency times. The revisit time required is 24 h, and a latency time of 1 h. Surface soil moisture monitoring presents gaps in the accuracy 0.01 m^3/m^3 and the latency time 1 h.

3.1.2. Microwave Sounders (MWS)

In the last few years, intensive work has been conducted to develop missions to prove the feasibility of using microwave sounders on nano-platforms, such as MicroMas [32], and the Earth Observing Nanosatellite-Microwave (EON-MW) [33]. The measurement with gaps that can be analyzed with this technology is the atmospheric pressure over the sea surface.

Table A5 presents a survey of the representative current and future missions with microwave sounders capable of measuring the atmospheric pressure over the sea surface. The gaps for this variable are the revisit and the latency times. To fill these gaps, a constellation of microwave sounders based on CubeSats missions could observe fast weather phenomena requiring high revisit time (3 h or less). A good example of CubeSat mission is EON-MW. The payload is a dual-reflector radiometer with a mass of 4 kg, an antenna size of 11 cm, and spatial resolution of 30 km on altitude of 600 km at 54 GHz.

3.1.3. Signals of Opportunity (SoOp): GNSS-R, and Receiver of SoOp

The utmost sensors used for oceanography (SARs and radar altimeters) have features that make them difficult to board on nano-satellites, most notably the power requirements, and the antenna size. An attractive option to explore the sea surface topography is the use of reflected Global Navigation Satellite Systems (GNSS) signals [34,35]. GNSS reflectometry is a favourable technique to perform some ocean measurements with small satellites [36]. The advantage of this technique is the capability to operate in all-weather conditions with a spatial resolution of $\sim$25 km. In the last two decades, a big effort has been made to develop models that prove the feasibility of using GNSS signals, proving to be successful for sea surface, altimetry measurements [37,38], wind speed [39,40], soil moisture [41–45], ice thickness [46], ice cover [47], and others. A few characteristics of GNSS-R missions have been identified and summarized in Table A6.

The current and planned missions using GNSS-R technology are presented in Table A6, such as TechDemosat-1 (TDS-1) [48], the Cyclone Global Navigation Satellite System (CYGNSS) [36], and FSSCAT [49,50].

TDS-1 was launched in June 2014 and it includes a GNSS-R payload with a mass of around 1.5 kg and approximately 10 W power consumption . It demonstrated the capabilities of GNSS-R for low power, low cost, and low mass. This payload measures complete delay-Doppler Maps (DDM) providing scientific-quality data [51]. The CYGNSS mission takes advantage of a constellation of eight microsatellites (weighting 17.6 kg) that provide nearly gap-free Earth coverage over Equatorial regions, with an average revisit time of seven hours and a median revisit time of three hours. CYGNSS was launched on December 2016. FFSCAT is a tandem mission of two 6U Cubesats (^{3}Cat-5/A and ^{3}Cat-5/B) featuring a hybrid microwave radiometer/GNSS- Reflectometer and a hyperspectral imager. FSSCAT will be the first nanosatellite mission to complement the Copernicus program [49]. Its main focus is over Polar Regions, and it will be launched in 2019.

The European Space Agency (ESA) conducted the studies of a space-borne demonstrator called Passive Reflectometry and Interferometry System In-Orbit Demonstrator (PARIS IoD) [52–54]. PARIS IoD was later reincarnated into the GEROS experiment on board the International Space Station [55], but it was never implemented.

Novel techniques using signals of opportunity, such as from Direct Broadcast Satellite (DBS) television at Ku- or X-bands, can be used to measure precipitation and winds over the sea surface [56], and these signals are sensitive to detect fluctuations of the sea surface roughness.

In this regard, the SGR-ReSI [57] payload onboard TDS-1 is selected as a possible candidate to cover the measurements with gaps such as wind speed over the sea surface (horizontal), sea ice cover, sea ice thickness, and soil moisture [6].

3.1.4. Receiver: Automatic Identification System (AIS)

Although not an EO technique, Automatic identification systems (AIS) could also be a potential technology for emergency and management for the Copernicus services. AIS is an automatic tracking system used by ships and vessel traffic services. The AIS is a standardized receiver using two channels in the maritime VHF band. It has a positioning system with electronic navigation sensors such as a gyrocompass or rate of turn indicator. The main advantages of this system are the accuraccy of the position, course, and speed information. Additionally, the International Maritime Organization (IMO) has normative guidelines to put AIS on board for all passenger ships larger than 300 GT. Additionally, the latency can be reduced thanks to an update rate of $\sim$3 min. In addition, it is suitable for nano-satellites [58] (low size, low power, low weight, and these can be translated into low system cost) (Table A7).

3.2. *Passive Optical*

This type of technology has shown its feasibility for small missions [59,60]. For example, for an optical instrument in the visible part of the spectrum, with a ground resolution better than 10 m, and an aperture of 10 cm (CubeSat size), the altitude of the satellite should be less than 500 km.

The data provided by passive optical instruments, from the ultraviolet to the far-infrared wavelengths can be used for weather forecast, vegetation, atmosphere, ocean and land studies. The main limitation of optical sensors is that data cannot be acquired in night-time (visible and near infrared parts of the spectrum) or cloudy conditions, and cloudy weather is very frequent in Polar Regions.

In this manuscript, the classification of optical sensors as radiometer imager and atmospheric sounders, and its subclassification between multispectral and hyperspectral is studied. Radiometer imagers measure the intensity of electromagnetic radiation in the visible or infrared bands, and sounders measure the vertical distribution of atmospheric parameters such as pressure, temperature, and humidity. Multispectral instrument refers to a maximum number of tens of bands, and hyperspectral radiometers consist of hundreds of narrow and continuously distributed bands (10–20 nm).

3.2.1. Radiometer: Multispectral and Hyperspectral

Table A8 presents the features of the available multispectral and hyperspectral radiometers instruments, in terms of wavelength, spatial resolution, aperture size, swath, mass, power consumption, and data rate. The variables of interest that can be measured with optical sensors for the Marine for Weather Forecast, Sea Ice Monitoring, Fishing Pressure, and Agriculture and Forestry: Hydric stress use cases are the Sea Surface Temperature (SST), atmospheric pressure over the sea surface, ocean chlorophyll concentration, ocean imagery and weather leaving radiance, Color Dissolved Organic Matter (CDOM), detection of water in crops, estimation of crop evapotranspiration and the sea ice cover.

A good example of multispectral radiometer on micro-platform is AVHRR/3 [61] and also has good performance, and it could support the measurements with detected gaps, such as SST, ocean chlorophyll concentration, ocean imagery and weather leaving radiance, CDOM, detection of water in crops, estimation of crop evapotranspiration, sea ice cover, and atmospheric pressure over the sea surface (it can be inferred through measurements in the infrared band).

3.2.2. Sounder: Multispectral and Hyperspectral

A good example of hyperspectral infrared sounder capable of measuring atmospheric pressure over the sea surface on CubeSat is EON-IR [62]. This instrument is under development with spatial resolution comparable to legacy sounders such as Infrared Atmospheric Sounding Interferometer (IASI), Atmospheric Infra-Red Sounder (AIRS), and Cross-track Infrared Sounder (CrIS).

Table A9 presents the details of the available multispectral and hyperspectral sounders instruments, in terms of spatial resolution, aperture size, swath, mass, power consumption, and data rate. For each optical sensor, it classifies (nano-, micro-, mini-, and large-satellite) according to the payload power and mass that can support the available commercial platforms summarized in Table 2.

3.3. *Active Microwave*

Several missions have been launched with active microwave instruments that can be grouped into three main families: Scatterometers, Synthetic Aperture Radars (SAR), and Radar Altimeters (RA). This section describes the variables of interest that can be measured with satellite-based active microwave sensors: wind speed, and direction over the sea surface using radar scatterometers, SAR and SAR altimeters; sea level, significant wave height, wave and wind speed using RA; and dominant wave direction, significant wave height and sea ice cover by SAR. Then, each variable is presented with the available active microwave technology, and the new trends of these sensors in small satellites.

3.3.1. Real Aperture Radar Altimeter

Radar altimeters measure the distance of the Earth's surface underneath the spacecraft by measuring the time between transmitting the signal and receiving the echo. Microwave radar altimeters have been used for a wide range of applications that can be grouped as: (a) geodesy and geophysics, study the Earth's shape and size, on the ground as well as on the sea surface [63]; (b) ocean applications (ocean surface currents, wind speed, significant wave height); (c) ice sheets and sea ice (sea ice thickness, and glacier topography) [64]; (d) climate (ocean topography and the heat exchange with the atmosphere); and (e) hydrology.

Nowadays, altimeter constellations on small platforms are deemed important, since they bring improved temporal resolution, and some ocean phenomena can only be perceived if subject to an almost continuous observation. At the same time, a shorter revisit time represents an increase in the spatial coverage and a finer spatial sampling grid. Equally, SSO should be avoided because of the errors associated with solar tidal effects.

Examples of recent altimetry missions are presented in Table A10. Typical requirements are: 100 W average power consumption, 1.2 m antenna diameter, 61 kg payload mass. The implementation on nano- platforms for radar altimeters may partially degrade the quality of the measurements. Additionally, nadir looking altimeters do not provide a wide swath. In this way, constellations of small satellites embarking a compact nadir altimeter [65] could improve the temporal/spatial sampling and therefore closing the gap with current planned missions.

3.3.2. Real Aperture Radar Scatterometers

Current and planned scatterometers missions have been identified and are summarized in Table A11. Earth Observation missions based on scatterometers typically operate at C-, and Ku-bands, and present spatial resolutions from 10 to 50 km. Current and future contributing missions to the Copernicus system with radar scatterometer are: ASCAT and SCA, ASCAT/Metop-A/B/C (2007 to 2021), with global coverage every 1.5 days and 12.5 km spatial resolution for basic sampling, SCA/Metop-SG-B1/B2/B3 (from 2022 to 2030) with near global coverage every 1.5 days, from 15 to 20 km of spatial resolution with sampling at 6.25 km intervals.

The main variables derived from radar scatterometer data are wind speed and vector over sea surface [66], but scatterometers are also capable to obtain surface soil moisture indices [67], leaf area index [68], snow water equivalent, snow cover [69], and sea ice extent measurements [70] Table A11 shows the characteristics of the radar scatterometer. The power consumption of these sensors is in the range of 210–540 W, and mass is in the range from 260 to 600 kg. According to the requirements of power consumption, size and mass, this payload can be carried over mini- or large-satellites.

3.3.3. Synthetic Aperture Radar (SAR) Altimeter

SAR altimeter differs from real aperture radar altimeter (conventional) in that it exploits coherent processing of groups of transmitted pulses, while conventional altimeters is exploited to make the most efficient use of the power reflected from the surface. The SAR altimeter offers many potential improvements over conventional altimetry for measurements, since it increases the resolution and offers multilook processing.

Currently, three mini-satellites are dedicated to altimetry with SAR processing, such as SARAL, Sentinel-3A, and Sentinel-3B. The planned missions are Sentinel 6 (Jason-CS). Table A12 summarizes the main characteristics of radar altimeters with SAR processing. Typical requirements are similar to the conventional altimeters for mini-platform: 100 W average power consumption, 1.2 m antenna diameter, 63 kg payload mass.

The geophysical variables of interest to analyze with SAR altimeter are ocean surface currents, significant wave height, dominant wave direction, sea ice cover, sea ice type, sea ice thickness, and horizontal wind speed over the sea surface.

3.3.4. Synthetic Aperture Radar (SAR) Imager

Spaceborne SAR imager sensors have been widely used for ocean monitoring (e.g., sea-ice cover, oil spills monitoring, sea-ice type, wave direction, dominant wave period, sea level, etc.), and land applications (e.g., soil moisture indices, vegetation monitoring, classification, fire fractional cover, fraction of vegetation over land, landslides and motion risk assessment, permafrost, and others) to support the environment management, with resolutions comparable to those of optical systems. The manufacturing and implementation related to a small SAR satellite mission have opened a market for a new technology which has recently been developed: the constellations of small SAR satellites, being the principle of Fractionated and Federated Satellites (FSS) [71], and/or bistatic SARs as companion satellites (e.g., SAOCOM [72]).

The use of SARs imager in small satellites poses some major challenges, such as the antenna dimensions and power requirements of the system. Another challenge is how to generate the power required by this sensor, reducing the transmitted power, resulting in a narrow swath and therefore increasing the revisit time. In this line, SARs are now feasible in small platforms—for example, NovaSAR-S [73] and ICEYE's Synthetic Aperture Radar [74]. NovaSAR-S is a novel platform for small synthetic aperture Radar (S-band) development by Surrey Satellite Technology Ltd. (Guildford, United Kingdom), with a mass of 500 kg and peak power of 1.8 kW. The antenna is a microstrip patch phased array with size of 3 × 1 m. ICEYE's Synthetic Aperture Radar is a microsatellite developed by ICEYE, with a satellite mass of 100 kg, and phase array antenna at X-band. According to the frequency band of the SAR, beyond 2028, there will be no X-band SAR mission in orbit, but there will be L- and C-band SARs mission (Figure 2). On this subject, the frequency band selected for SAR instrument is X-band, in order to obtain a smaller instrument and cover the frequency gap.

The geophysical variables of interest to analyze with SAR imager are iceberg tracking, sea ice cover, sea ice type, sea ice thickness, sea ice drift, sea ice extent, wind speed, ocean surface currents, dominant wave direction, dominant wave period, wind speed, and significant wave height. Nevertheless, single, large SAR satellites are not compatible with the requirements of 3 h of revisit time. Constellations of small SAR Satellites are under development or implementation stages [74]. In contrast, large SAR Satellites have been in orbit for years. Small SAR satellites can replace large SAR, for some specific applications requiring medium resolution imagery and smaller areas covered (due to power limitations). If the frequency band is higher (X-band), the spatial resolution and swath wide can be adjusted, therefore reducing the size and mass of the system. Table A13 presents a survey of the representative SAR image missions and classifies each instrument into mini or large according to capabilities of commercial platforms surveyed in the previous chapter.

Figure 2. Frequency bands of future (2020–2030) European Union (EU) mission carrying Synthetic Aperture Radar (SAR) imager instruments.

3.4. Active Optical

Lidar

Active Optical Instruments or Lidars use pulsed laser emissions to measure atmospheric profiles and Earth surface applications such as vegetation height. Due to the short wavelengths, the laser pulse propagation through the atmosphere is scattered and attenuated by air molecules and aerosols. On the Earth's surface, the vegetation and canopy also cause scattering. A small portion of the scattered light is sent back to the instrument which collects, and detects it. Subsequently, the electric signal is digitized through a Lidar signal numerical processing. Over the ocean, the variables that can be measured with Lidars are sea ice thickness, sea level and ocean dynamic topography.

Lidars can be divided into two broad categories: (i) atmospheric profilers producing also the total column content for atmospheric composition, i.e., particles layers and key trace gases, and (ii) altimeters with decimeter to meter accuracy for topography retrieval and canopy vertical distribution. The objectives of relevant Lidars are:

- Surface topography, ice sheet [75], and canopy [76] (e.g., ICESat-1).
- Climate and Radiation Budget by profiling clouds and aerosols optical and microphysical properties (e.g., NASA CALIPSO since 2006 [77], and ESA/JAXA EarthCARE [78], to be launched in 2021).
- Atmospheric dynamics or horizontal winds, (e.g., ESA Atmospheric Dynamic Mission ADM-Aeolus [79] was launched on 22 August 2018). Lidar instruments present the following main characteristics:
- Operating wavelengths in the UV, VIS, NIR, and SWIR; possible dual-wavelength, polarimetry, and two receivers (for Mie and Rayleigh scattering).
- Spatial resolution in the range of 100 m to a few tens of centimeters for LIDAR altimeters.
- Non-scanning, either nadir-viewing or oblique.

Doppler LIDARs generally operate in the UV to track aerosol and air molecules and it are used for track aerosol and air molecules. Backscatter LIDARs are typically operated at one or two wavelengths (UV or VIS + NIR), often with amount of polarizations cross-talk into a succession of atmospheric backscatter measurements (rotatable half-wave plate) to discriminate between spherical and non spherical particles in the atmosphere, the nadir view brings the capability to measure aerosol profiles, cloud top height and atmospheric discontinuities, and the multi-beam to perform a large swath. Lidars altimeter operated at two wavelengths (VIS + NIR) can measure with very high vertical resolution and horizontal resolution (for sea-ice elevation, and ice boundaries). Differential absorption LIDARs (DIAL) operate at one wavelength centered on the absorption peak of one trace gas (e.g., O_3, H_2O and CO_2). The main limitation of this technology is the narrow swath. The variable with a gap that can be analyzed with Lidar is the sea ice thickness.

Table 5 summarizes all technologies discussed in this section: radiometer imager, radiometer sounder, GNSS-R, AIS, scatterometers, altimeters, altimeter with SAR processing, SARs imager, Passive optical and Lidars. The measurements with gaps that can be measured for each technology are identified. The studied technologies are feasible on small platforms taking into account the survey of the commercial platform addressed in the previous section. Now, the best technology option needs to be analyzed, based on the future observations required by the Copernicus space infrastructure.

Table 5. Mapping of the potential technologies to cover measurements with gaps.

Technology Type			Measurements
Microwave	Passive	Radiometer Imager (X-, K-, Ka-, W-bands)	Wind speed over sea surface (horizontal) [b] Sea ice cover [b] Sea ice type [a] Sea ice drift [a] Sea surface temperature [a]
		Radiometer Imager (L-band)	Soil moisture at the surface [c] Sea ice cover [b] Sea ice thickness [a] Crop growth & condition
		Radiometer Sounder (50–60 GHz)	Atmospheric pressure (over sea surface) [c]
		Signals Oportunity: GNSS-R	Soil moisture [b] Sea ice thickness [a] Dominant wave direction [b] Wind speed over the sea surface (horizontal) [a] Significant wave height [b] Sea ice cover [b] Ocean surface currents [b]
		Signals Opportunity: Receiver of SoOp (X, Ku-band)	Wind speed over sea surface [a]
		Receiver: Automatic Identification System (AIS)	Monitoring system: vessels [c]
	Active	Real Aperture Radar: Altimeter	Ocean surface currents [c] Significant wave height [b] Dominant wave direction [b] Sea ice thickness [a] Wind speed over sea surface (horizontal) [a]
		Real Aperture Radar: Scatterometer	Wind speed over sea surface (horizontal) [c] Sea ice extent [a] Sea ice cover [a]
		Synthetic Aperture Radar (SAR): Altimeter	Ocean surface currents [c] Significant wave height [b] Dominant wave direction [b] Sea ice type [b] Sea ice cover [b] Sea ice thickness [a] Wind speed over sea surface (horizontal) [a]
		Synthetic Aperture Radar (SAR): Imager	Ocean surface currents [c] Iceberg tracking [c] Sea ice drift [c] Sea ice extent [c] Sea ice type [c] Sea ice cover [c] Dominant wave direction [b] Dominant wave period [b] Significant wave height [b] Sea ice thickness [a] Wind speed over sea surface [a] Ocean imagery and water leaving radiance
Optical	Passive	Multispectral radiometer (VIS/NIR/TIR)	Ocean chlorophyll concentration [c] (λ: 442.5, 490, 510, 560 nm) Ocean imagery and water leaving radiance [c] (λ: 485, 560, 660, 2100 nm) Color Dissolved Organic Matter (CDOM) [c] (λ: 442.5, 490, 510, 560, 665 nm) Sea surface temperature [c] (λ: 3.7, 4.05, 8.55, 11, 12 μm) Sea ice cover [a] (λ: 640, 1610 nm) Detection of water stress in crops [c] Estimation of crop evapotranspiration [c]
		Hyperspectral radiometer (VIS/NIR)	CDOM [c] Sea ice cover [b]
		Sounder (IR)	Atmospheric pressure over sea surface [c] Sea surface temperature [c]
	Active	Lidar	Sea ice thickness [b]

The data relevance of the instrument depends on its ability and limitations to obtain the measurements: [a] Marginal relevance; [b] medium relevance; [c] high relevance.

4. Potential Instrument, Suitable Platforms, and Technological Limitations

After the survey of the suitable EO technologies in terms of the spatial resolution, swath, mass and power consumption, in this section, the suitable small commercial platforms and technological limitations of the potential sensors are identified. Tables 6 and 7 show the potential technologies studied in this work, with the suitable platforms and limitations with respect to the needs detected in the horizon 2020–2030. Platforms are selected according to their capacity to support the instrument mass and power consumption (available commercial platforms surveyed, Tables A1–A3). Additionally, it takes in to account the platforms with minor categorization (e.g., nano-, micro-, or mini-platforms), that satisfy both requirements. Special attention has been paid to the possibility to use new techniques and smaller platforms, focusing on the quality of the measurements as compared to the ones generated by full-fledged payloads onboard large spacecrafts. Indeed, since a small platform also means less volume, mass, power and data rate for the payload, the measurements are usually of reduced quality. Depending on the mission (i.e., environmental data), this may be compensated by more frequent data acquisitions (exchange between measurement quality and revisit time), yet to be evaluated on a case-by-case basis. A brief the potential instruments, suitable platforms, and technological limitations are explained below:

- GNSS-R (1.4 kg, 12 W) instruments are suitable for nanosatellites (3U or 6U). Table 6 presents sample available commercial platforms for the SGR-ReSi [57], such as the Endeavour-3U [18] and the MAI-3000 [17]. Endeavour by Tyvak Nanosatellite Technology Inc. (San Luis Obispo, CA. United States of America), is a 3U platform with 15 W of average payload power, 3 deg of pointing control. MAI-3000 by Maryland Aerospace, is a 3U platform with 12 W of payload power and 3 kg of available payload mass. The main limitation of GNSS-R altimetry data is the poorer (decimetric) resolution and accuracy ($\sim$20 cm for SSH, and 2 m/s for wind speed) are offset by the much larger number of simultaneous observations from different specular reflection points [80].

- Another good example are microwave sounders on small-platforms such as EON-MW [33], for measuring the atmospheric pressure over the sea surface. However, the antenna system must be redesigned to achieve the spatial resolution required. For a 10 km spatial resolution, at 50 GHz, the require antenna aperture is 36 cm, from an altitude of 600 km. Table 6 summarizes a list of the available commercial micro-platforms suitable for this instrument.

- Microwave imagers at X-, K-, Ka-, and W- bands are particularly well suited for implementation on small platforms (Table 6). TMI is a light instrument suitable for mini-satellites, with X-, K-, Ka-, and W- bands capable of measuring and covering the gaps for wind speed, sea ice cover, sea ice type, and sea surface temperature variables. For sea surface temperature, microwave radiometers improve the coverage in polar regions because of their all weather capabilities. In order to obtain a spatial resolution of 10 km at 18.7 GHz from 600 km height, a 2.2 m antenna is required. On the other hand, an SSM/I type of instrument with a modified antenna, could be implemented in a micro-platform in order to cover wind speed over the sea surface, sea ice cover, and sea ice measurements, with the required performance. L-band radiometers contribute to sea ice thickness monitoring, agriculture (soil moisture) and forestry measurements. Those instruments are suitable for mini-platforms (Table 6). The main limitation is their coarse resolution. Inflatable antennas must be used to reduce the footprint size, or aperture synthesis techniques could be implemented [81]. ELiTeBUS 1000 [10] by Thales Alenia Space (Cannes, France) is an available commercial small-platform suitable for this instrument. ELiTeBUS 1000 is a platform for Medium Earth Orbit (MEO) and Low Earth Orbit (LEO) orbit with 1000 to 1500 W of available payload power.

- Scatterometers contribute to the Marine for Weather Forecast and Sea Ice Monitoring use cases. The instrument taken as a reference is the SCAT on board the CFOSAT mission [25,82], the power consumption of this sensor is less than 200 W, and the mass less than 200 kg. According to the power consumption and mass requirements, this payload can be carried on board mini-platforms (Table 7). Scatterometers are valuable sensors for wind measurements. However, the main limitations are the coarse accuracy and spatial resolution of the data. However, their wide swath and the possibility of scatterometer constellations open the door to improve the accuracy and spatial resolution, combining the data from multiple passes of different satellites.

- For radar altimeters, the accuracy of the measurements depends on the Pulse Repetition Frequency (PRF), which is directly driven by the power available on-board to the payload. Since the power available on-board decreases with solar panel size, the accuracy of the measurements on a small satellite is also expected to be degraded as compared to that of large satellites. For example, if the power consumption is reduced by a factor of 4, the PRF is reduced roughly by the same factor, and the Root-Mean-Square (RMS) error increases by a factor of 2. For the Jason-2 altimeter (power consumption $\sim$70 W), a reduction of its power consumption to 1 W, would increase the sensor error level from 2 cm to $\sim$16.7 cm, which is actually comparable to GNSS-R [55,82]. It is easy to understand that the types of products that can be generated with this accuracy are different from the ones generated with an SRAL radar altimeter, but one must also consider that the number of radar altimeters with a transmitted power of 1 W that can be manufactured and launched at the same cost as for a high accuracy radar altimeter is much larger. These few examples illustrate the fact that the quality and frequency of the measurements have to be considered in the overall comparison process. In some cases, the concept of operations may partially be compensated by the degradation of the quality of the individual measurements (e.g., part-time measurement instead of systematic measurement if the power available on board is the main parameter driving the performance of the measurement).

- SAR sensors are one of the most effective instruments for ocean, land, and ice observation. A good example of miniaturization of this technology is the Severjamin-M instrument (Meteor-M N missions) [83], an X-band SAR with power consumption of 1 kW and a mass of 150 kg, including the mass of the antenna of 40 kg. The main technological limitation is the narrow swath, but this could be compensated with a constellation of SAR satellites.

- Optical payloads are characterized in terms of image quality such as the Ground Sampling Distance (GSD), the Modulation Transfer Function (MTF), and the Signal-to-Noise Ratio (SNR). To be able to interpret an image (e.g., in the maritime surveillance, the capability to estimate the type of a boat), the GSD is not sufficient, since a degraded MTF (i.e., blurred image) or a degraded SNR (noisier image) would prevent it. Ensuring a good MTF and SNR for a given GSD requires a minimum aperture for the optical instrument, and reducing it below this minimum value will limit the type of applications. Image quality is also limited by the platform's attitude control system, i.e, any jitter in the pointing will blur the image. This has also to be taken into account as smaller platforms exhibit poorer performances.

Table 6. Mapping of potential passive sensors and platforms to meet the user requirements.

Technology Type	Measurements	Instrument Limitations	Instruments Identified	
Microwave Radiometer Imager (X-, K-, Ka-, W-bands) or (K-, Ka-, W-bands)	Wind speed over sea surface Sea ice cover Sea ice type Sea ice drift Sea surface temperature (at X-band)	Coarse spatial resolution and accuracy	TMI [25]	SSM/I [84]
			Available commercial platform (Non-exclusive)	
			NAUTILUS (NEMO-150) [85] SSTL-150 ESPA [86] BCP-100 [87] TET-XL [13]	SN-50 [21] Altair [20]
Microwave Radiometer Imager (L-band)	Surface soil moisture Sea ice cover Crop growth & condition Sea ice thickness	Coarse spatial resolution	MIRAS [25,88]	SMAP [25,89] Aquarius [25]
			Available commercial platform (Non-exclusive)	
	Sea ice thickness	Accuracy	ELiTeBUS 1000 [10] LEOStart-2 BUS [90]	
Microwave Radiometer sounder (50-60 GHz)	Atmospheric pressure (over sea surface)	Coarse spatial resolution	ATMS [25]	Miniature microwave sounder EON-MW [33]
			Available commercial platform (Non-exclusive)	
			SSTL-300/-600 [92,93] SN-200 [94] Eagle [90] TET-XL [13]	NEMO / DEFIANT [85] SSTL-12/-X50/-100 [22,91,92] SMALL SAT 27U [12] SN-50 [21] Altair [20] LEOS-30 [95] BCP-50 [96]
Signals of Opportunity (SoOp): GNSS-R	Surface soil moisture Ocean surface currents Sea ice thickness Significant wave height Wind speed over sea surface	Accuracy	SGR-ReSI [57]	GEROS-ISS [80]
	Dominant wave direction Surface soil moisture	Coarse spatial resolution	**Available commercial platform (Non-exclusive)**	
	Sea ice cover	No specific limitation	Endeavour-3U [18] MAI-3000 [17]	ELiTeBUS 1000 [10] LEOStart-2 BUS [90]

Table 6. *Cont.*

Technology Type	Measurements	Instrument Limitations	Instruments Identified	
Receiver: AIS	Monitoring system vessels	No specify limitation	SD AIS Receiver [58]	NAIS [97]
			Available commercial platform (Non-exclusive)	
			GOMX 2U/3U [98] THUNDER (3U), GRYPHON (GNB) [85] Endeavour-3U [18] MAI-3000 [17] SMALL SAT 6U [12]	GOMX 3U [98] SMALL SAT 6U [12] MAI-3000 [17] Endeavour-3U [18]
Multispectral radiometer (VIS/MWIR/TIR)	Ocean chlorophyll concentration Ocean imagery and water leavin radiance CDOM Sea surface temperature Sea ice cover	Cloud sensitivity Day light only	AVHRR/3 [25]	VIRS [25]
			Available commercial platform (Non-exclusive)	
	Detection of water stress in crops Estimation of crop evapotranspiration	Coarse spatial resolution Cloud sensitivity Day light only	SSTL-12 [22] SSTL-X50 [91] SN-50 [21] Altair [20]	SN-50 [21] Altair [20]
Hyperspectral radiometer (VIS/NIR)	Sea ice cover CDOM	Cloud sensitivity Day light only	CHRIS [25]	COMIS [25]
			Available commercial platform (Non-exclusive)	
			LEOS-50/-100 [95] Small sat 12 U and 27U [12] SSTL-12/-X50/-100 [22,91,92] BCP-50 [96] Altair [20] SN-50 [21]	MAI-6000 [23] NEMO [85] LEOS-30 [95] DEFIANT [85] SMALL SAT 12U [12]
Hyperspectral sounder (IR)	Atmospheric pressure (over sea surface) Sea surface temperature	Cloud sensitivity	EON-IR[25]	CrIS [25]
			Available commercial platform (Non-exclusive)	
			MAI-6000 [23] NEMO, DEFIANT [85] LEOS-30/-50/-100 [95] SN-50 [21] Altair [20] SMALL SAT 16U [12] SSTL-X50/-100 [91,92] BCP-50 [96]	DAUNTLESS [85] SN-200 [94] Eagle-1M, LEOStart-2 BUS [90] LEOSTART-500XO [9] SSTL-600 [92] ELiTeBUS 1000 [10]

The background color in the Table indicates the platform suitable for the instrument according to the power and mass requirements: very lightgray: nano-platform; light gray: micro-platform; gray:mini-platform.

Table 7. Mapping of potential active sensors and platforms to meet the user requirements.

Technology Type	Measurements	Instrument Limitations	Instruments Identified	
Real Aperture Radar scatterometer	Wind speed over sea surface (horizontal) Sea ice extent Sea ice cover	Accuracy	RapidScat [24]	SCAT [25,82]
			Available commercial platform (Non-exclusive)	
			SSTL-600 [92] LEOSTART-500XO [9] LEOStar-2 BUS [90] EliTeBUS 1000 [10]	DAUNTLES [85] BCP-100 [87] SN-200 [94] Eagle-1M [90] SSTL-600 [92] LEOSTART-500XO [9] LEOStar-2 BUS [90] EliTeBUS 1000 [10]
Real Aperture Radar Altimeter and/or SAR Altimeter	Ocean surface currents Significant wave height Dominant wave direction Wind speed over sea surface (horizontal) Sea ice type Sea ice cover Sea ice thickness	Long-time analysis and narrow coverage	Altika [25]	SRAL [24], [25]
			Available commercial platform (Non-exclusive)	
			SN-50 [21] Altair [20]	DAUNTLESS [85] SSTL-150 ESPA/-300/-600 [86,92,93] BCP-100 [87] SN-200 [94] Eagle-1M, LEOStar-2 BUS [90] TET-XL [13] LEOSTART-500XO [9] ELiTeBUS 1000 [10]
SAR Imager	Ocean surface currents Wind speed over sea surface Dominant wave direction Dominant wave period Significant wave Height Sea ice type Sea ice cover Sea ice thickness Iceberg tracking Sea ice drift Sea ice extent Ocean imagery and water leaving radiance	Narrow coverage	COSI [24]	Severjamin [25,83]
			Available commercial platform (Non-exclusive)	
			LEOStar-2 BUS [90]	LEOStar-2 BUS [90] EliTeBUS 1000 [10]
Lidar Altimeter	Sea ice thickness	Cloud sensitivity long time analysis narrow covarage	ATLAS [24]	GEDI lidar [24]
			Available commercial platform (Non-exclusive)	
			ELiTeBUS 1000 [10] LEOStart-2 BUS [90]	

The background color in the Table indicates the platform suitable for the instrument according to the power and mass requirements: light gray: micro-platform; gray:mini-platform.

5. Reference Instrument Selection

The main requests of any satellite monitoring mission can be summarized as follows: (1) that observations are acquired with the required revisit time; (2) preferably in all weather conditions (clouds, rain, haze, and fog) and in all illumination conditions; (3) with a large swath to reduce the revisit time; (4) with the required radiometric and spatial resolutions; (5) with low manufacturing and launch costs, and with minimum deployment time in case of failure; and (6) keeping these parameters in mind, the reference instruments can be selected. In this way, the identification of instruments is based on the state-of-the-art at the payload level and the need to fulfill the gaps of the current Copernicus infrastructure.

Reference instruments and small platforms have been selected in the previous chapter. In this way, it has as strategy been implemented a significant reduction of the development time and cost, thanks to the adoption of commercial technologies, but it requires that these have a good performance of the measurement capabilities. In this regard, the capability of the instrument technologies is evaluated according to the trends in the design for small satellites. For each instrument, the mass and power consumption constraints, and data quality (spatial resolution, swath, and accuracy) are taken as a reference. This chapter evaluates if the instruments selected to meet the requirements (defined in [6]) in terms of spatial resolution and accuracy. Table 8 summarizes the performance requirements over each instrument:

- SGR-ReSI instrument presents a good performance for sea ice cover [99] because it satisfies the minimum requirement for spatial resolution and accuracy. For ocean surface currents, and significant wave height measurements satisfy the minimum requirement of spatial resolution at 25 km [100]. For other measurements, such as sea ice thickness [46], soil moisture [101], and wind speed [80] present worse performance than the minimum spatial resolution and accuracy requirements.
- EON-MW is a satellite project under development and presents an approximate performance that the Advanced Technology Microwave Sounder (ATMS) [33], in this way, it will be expected that the instrument satisfies the minimum requirements for accuracy of 5% and spatial resolution at 23 km for atmospheric pressure over sea surface measurements (channel from 50 to 60 GHz).
- MIRAS instrument presents a coarse spatial resolution $\sim$35 $\times$ 50 km for horizontal- and vertical-polarization. This instrument has an accuracy of 0.04 m^3/m^3 for soil moisture measurements [102] that is worse than 0.01 m^3/m^3 required. For sea ice thickness, the accuracy is worse than the 1 cm required [103], but it can have an accuracy of 5% for sea ice cover.
- SSM/I using an antenna (inflatable) of 2.2 m from 600 km orbit altitude can obtain a spatial resolution of 10 km and satisfy the minimal spatial resolution requirement for wind speed, and sea ice cover measurements. The accuracy for wind speed measurement can be until 1.5 m/s [104], and for sea ice data from 10% to 20% [105].
- TMI in order to meet the minimal spatial resolution requirement of 10 km (at 10.65 GHz) was proposed the modification of the aperture size of the antenna at 3.4 m (inflatable antenna). The accuracy for SST is of 0.5 K [104]. The accuracy is between 10% and 20% for sea ice data [105].
- AVHRR/3 presents a spatial resolution $\sim$1 km, and computes an accuracy better than 0.1 K [106].
- EON-IR is expected to be better than 0.25 K and present, with spatial resolution at 13.5 km.
- SCAT—the accuracy for wind speed monitoring is 2 m/s, and for sea ice monitoring is 5% .
- SRAL in SAR mode has a spatial resolution of 300 m, the accuracy for wind speed measurements is of 2 m/s [107]; for significant wave height, the accuracy is between 2 cm to 8 cm [108].
- Severjamin has a spatial resolution from 400 m to 1 km depending on the operation mode can satisfy many minimal requirements for some measurements.
- GLAS acquires the geophysical variables with a vertical spatial resolution of 10 cm, which does not satisfy the user requirement for sea ice thickness measurements.

Table 8. Reference instruments selected to cover the measurements with gaps.

Instrument	Measurements	Requirements [109]	
		Accuracy	Spatial Resolution
SGR-ReSI [57]	Soil Moisture at the surface	$<0.01\ m^3/m^3$	10 km
	Sea ice thickness	1 cm	1 cm (vertical)
	Dominant wave direction	10°	1–15 km
	Wind speed over the sea surface	0.5 m/s	1–10 km
	Significant wave height	0.1 m	1–25 km
	Sea ice cover	5 %	12 km–10 m
	Ocean surface currents	0.5 m/s 10°	1–25 km
EON-Microwave [33] (Ka-, U-, D-bands) (22 channels)	Atmospheric pressure over sea surface	5 %	1–25 km
MIRAS [25,88] (L- band)	Soil Moisture at the surface	$<0.01\ m^3/m^3$	10 km
	Sea ice thickness	1 cm	1 cm (vertical)
	Crop grow & condition	-	2 km
	Sea ice cover	5 %	12 km–10 m
SSM/I [a] [84] (K, Ka, W)	Wind speed over sea surface	0.5 m/s	1–10 km
	Sea ice cover	5 %	12 km–10 m
	Sea ice type	0.25/classes	10 m
	Sea ice drift	0.5 m/s 10°	10 m
TMI [b] [24] (X, K, Ka, W)	Wind speed over sea surface	0.5 m/s	1–10 km
	Sea ice cover	5 %	12 km–10 m
	Sea ice type	0.25/classes	10 m
	Sea ice drift	0.5 m/s 10°	10 m
	Sea surface temperatture	0.3 K	1–10 km
AVHRR/3 [61] (VIS, NIR, MWIR, TIR)	Ocean chlorophyll concentration	$0.05\ mg/m^3$	1 km
	Ocean imagery and water leaving radiance	5%	1 km
	Color Dissolved Organic Mater (CDOM)	5%	1 km
	Sea Surface Temperature (SST)	0.3 K	1–10 km
	Detection of water stress in crops	5%	2–7 m
	Estimation of crop evapotranspiration	-	1–10 m
	Sea Ice Cover	5 %	12 km–10 m
COMIS [24] (VIS, NIR)	CDOM	5%	1 km
	Sea Ice Cover	5 %	12 km–10 m
EON-IR	Sea Surface Temperature (SST)	0.3 K	1–10 km
	Atmospheric pressure over sea surface	5 %	1 km–25 km
SCAT [24] (Ku-band)	Wind speed over the sea surface	0.5 m/s	1–10 km
	Sea ice extent	5%	12 km–10 m
	Sea ice cover	5 %	12 km–10 m

Table 8. *Cont.*

Instrument	Measurements	Requirements [109]	
		Accuracy	Spatial Resolution
SRAL[24,25] (C- & Ku-bands)	Ocean surface currents	0.5 m/s 10°	1–25 km
	Significant wave height	0.1 m	1–25 km
	Dominant wave direction	10°	1–15 km
	Sea ice type	0.25/classes	10 m
	Sea ice thickness	1 cm	1 cm (vertical)
	Sea ice cover	5 %	12 km–10 m
	Wind speed over the sea surface	0.5 m/s	1–10 km
Severjamin [25,83] (X-band)	Ocean surface currents	0.5 m/s 10°	1–25 km
	Iceberg tracking	5%	10 m
	Sea ice drift	0.5 m/s 10°	10 m
	Sea ice extent	5%	12 km–10 m
	Sea ice type	0.25/classes	10 m
	Sea ice cover	5 %	12 km–10 m
	Dominant wave direction	10°	1–15 km
	Significant wave height	0.1 m	1–25 km
	Sea ice thickness	1 cm	1 cm (vertical)
	Ocean Imagery and water leaving radiance	5%	1 km
	Wind speed over the sea surface	0.5 m/s	1–10 km
ATLAS [24] (VIS & NIR)	Sea ice thickness	1 cm	1 cm (vertical)

[a] antenna size of 2.2 m. [b] antenna size 3.4 m. The background color in the requirements denotes: Green: Requirement met or is better; Yellow: Minimum requirement met; Red: Have worst performance that the minimum requirement. The background color in the instrument indicates the platform suitable according to the power and mass requirements: Very light gray: Nano-platform; Light gray: Micro-platform; Gray: Mini-Platform.

6. Quantitative Method to Identify the Potential Technologies to Cover the Future Copernicus Gaps

In order to identify the potential technologies to cover future gaps over Copernicus infrastructure, a quantitative method has been defined starting from the perspective of the instrument technologies and the variables with gaps. The analysis is centered on the list of the top 10 use cases and 20 variables detected with gaps, and the potential instruments which have been proposed in Table 8. A quantitative method has been applied to rank the technologies suitable to measure the variables with gaps, and identify which technologies cover most of the requirements. The rank order weights used is based on the user requirements, and measurements priorities.

A weighting system for the instrument performance parameter has been implemented. First, it defined the numerical score for each instrument capability based on user requirements (Table 9). Then, these numerical scores are evaluated for each measurement with gaps and each factor. In this way, the numerical score for latency is assigned for measurement that required latency time <1 h; for spatial resolution, a high score is assigned for measurement that required spatial resolution <1 km; for the revisit time, a high score for geophysical variables that required <3 h is assigned; for accuracy, a high score for measurements that require accuracy better than the state of the art is assigned. For payload mass and power consumption, the corresponding score for mini and micro platform is assigned; the measurement relevance was assigned taking the following:

- High relevance measurements: ocean surface currents, wind speed over sea surface, dominant wave direction, and significant wave height measurements.

- Medium relevance measurements: sea ice cover, sea ice type, sea surface temperature, and atmospheric pressure over the sea surface.
- Low relevance measurements: Ocean chlorophyll concentration, ocean imagery and weather leaving radiance, CDOM, monitoring system vessels, sea ice extent, sea ice thickness, iceberg tracking, sea ice drift, estimation of crop evapotranspiration, detection of water stress in crops, crop grow and conditions.

Then, the weights for each factor (latency, revisit time, spatial resolution, accuracy, payload mass, payload power, and measurement relevance) are derived by the normalization of the average of the numerical score assigned for each measurement:

$$W_j = \frac{\frac{1}{n}\sum_i^n Numerical_{score_i}}{\sum_j^m \frac{1}{n}\sum_i^n Numerical_{score_i}}, \tag{1}$$

where i represents each measurement, and j represents each factor. In order to identify the potential technologies, new numerical scores are assigned based on the instrument capabilities to measure the variables with gaps and how those meet the user requirements. Instrument attributes are defined in Table 10. The requirements for the geophysical variables are evaluated in terms of seven criteria (factors) or instrument capabilities:

- Latency is referred the time to be processed the data to obtain the product.
- Swath is related to the ability of the instrument in order to cover an area, a wide swath indicates minor revisit time.
- Spatial resolution is evaluated for the reference instruments according to the user requirements for each measurement.
- Accuracy is a component of the data quality; it is evaluated according to it being closed to the user requirements for each reference instrument.
- Payload mass is evaluated for each reference instrument, giving priority to the instruments that are best suited to smaller platforms.
- Payload power is related to the power consumption of the payload; it also brings priority to the instruments that are best suited to smaller platforms.
- Data relevance is the potential of the sensor to provide the measure based on sensing constraints (e.g., long time to analyze the data, data limited by cloud cover, and daylight only)

This scoring method assigns a lower score to the technologies that require a large instrument (large mass and high power consumption), and the technologies that present low data quality (low coverage, low spatial resolution, high latency, low accuracy, and low relevance for specific measurement). The score for each instrument is expressed in the following equation:

$$Instrument_{score-by-mensurement} = \sum_j^m \left(\frac{Numerical_{score}}{3} * W_j\right), \tag{2}$$

where j represents each technology performances' parameters such as latency, spatial resolution, swath, accuracy, payload mass, payload power consumption, and data relevance for each potential instrument;

$Numerical_{score}$ is assigned to each instrument by measurement (0, 1, 2 or 3); and W_k, is the weight assigned for each factor obtained of Equation (1) (Table 9, second column).

Four critical use cases were evaluated, such as Marine for Weather forecast, Sea Ice Monitoring, Agriculture and Forestry: Hydric Stress, and Fishing Pressure (Table 11). Subsequently, high, medium, and low priority measurements were defined and its weights were assigned according to the use case to evaluate:

$$W_i = \frac{Numerical_{score_i}}{\sum_i^n Numerical_{score_i}}. \tag{3}$$

Table 9. Definition of the numerical score for the criteria and result of the weights.

Instrument Capabilities	Weight	Numerical Score		
		1	2	3
Latency	19.2%	>3 h	2–1 h	<1 h
Spatial Resolution	15.4%	>1 km	1 km	<1 km
Revisit time	15.4%	Revisit time >24 h	Revisit time: 3–24 h	Revisit time <3 h
Accuracy	14.1%	Worse that state of the art	Equal to state of the art	better that state of the art
Payload mass	12.8%	large	mini	nano-micro
Payload power Consumption	12.8%	large	mini	nano-micro
Measurements relevance	10.3%	Low	Medium	High

Table 10. Instrument technologies' attributes and related numerical scores.

Instrument Capabilities	Numerical Score			
	0	1	2	3
Latency	N/A	high	medium	low
Spatial Resolution	N/A	worse than required	minimun requirement met	requirement meet or better
Swath	N/A	Narrow swath <400 km	Moderate swath <1000 km	Wide swath >1000
Accuracy	N/A	Worse than required	Equal to requirement	Requirement meet or better
Payload mass	N/A	large	mini	nano-micro
Payload power Consumption	N/A	>150 W	25–150 W	≤25 W
Data relevance	N/A	Marginal	High	Primary

Table 11. The priority level of the measurement according to the use case priority.

Use Case Priority	Marine for Weather Forecast		Sea Ice Monitoring		Agriculture and Forestry: Hydric Stress		Fishing Pressure	
Measurements	**Priority Level**	**Weight [%]**	**Priority Level**	**Weight [%]**	**Priority Level**	**Weight [%]**	**Priority Level**	**Weight [%]**
Ocean Surface currents	H	9.375	M	5.000	L	3.570	M	5.410
Wind speed over sea surface	H	9.375	M	5.000	L	3.570	M	5.410
Dominant wave direction	H	9.375	M	5.000	L	3.570	M	5.410
Significant wave height	H	9.375	M	5.000	L	3.570	M	5.410
Sea Surface temperature	M	6.250	H	7.500	L	3.570	H	8.110
Atmospheric pressure over sea surface	M	6.250	L	2.500	L	3.570	M	5.410
Sea ice cover	M	6.250	H	7.500	L	3.570	L	2.700
Sea ice type	M	6.250	H	7.500	L	3.570	L	2.700
Sea ice thickness	L	3.125	H	7.500	L	3.570	M	5.410
Iceberg tracking	L	3.125	H	7.500	L	3.570	M	5.410
Sea ice drift	L	3.125	H	7.500	L	3.570	L	2.700
Sea ice extent	L	3.125	H	7.500	L	3.570	L	2.700
Surface soil moisture	L	3.125	L	2.500	H	10.710	L	2.700
Ocean chlorophyll concentration	L	3.125	L	2.500	L	3.570	H	8.110
Ocean imagery and weather leaving radiance	L	3.125	M	5.000	L	3.570	H	8.110
Color dissolved organic mater	L	3.125	L	2.500	L	3.570	H	8.110
Estimation of crop evapotranspiration	L	3.125	L	2.500	H	10.710	L	2.700
Detection of water stress in crops	L	3.125	L	2.500	H	10.710	L	2.700
Crop growth & condition	L	3.125	L	2.500	H	10.710	L	2.700
Monitoring system vessels	L	3.125	M	5.000	L	3.570	H	8.110

Priority level and numerical score: L: Low = 1; M: Medium = 2; H: High = 3.

When the instrument score by measurement is defined, the ranking of the instruments is obtained. The instrument ranking (Table 12) is computed as:

$$Ranking_{instrument} = \sum_{i}^{n} (instrument_{score-by-measurement} * W_i). \tag{4}$$

Table 12. Ranking results for each technology for each use case.

Instrument/Technology	Ranking Results [%]			
	Marine for Weather Forecast	Sea Ice Monitoring	Agriculture and Forestry: Hydric Stress	Fishing Pressure
Multispectral Radiometer	21.6	23.1	30.1	31.2
Hyperspectral Radiometer	11.2	10.7	19.9	11.8
Hyperspectral . Sounder (IR)	10.6	8.5	6.1	11.4
L-Microwave Radiometer	8.5	10.6	19.4	8.1
Ka, K, W-Microwave Radiometer	19.3	21.6	11.2	10.4
GNSS-R	39.4	29.6	24.8	25.3
X-, Ka, K, W-Microwave Radiometer	21.2	23.7	12.1	14.6
Ka-, U-, D-Microwave Sounder	5.9	2.37	3.4	5.1
Automatic Identification System (AIS)	3.1	5.0	3.6	8.1
Radar Scatterometer	11.9	12.7	6.8	6.8
Lidar	1.04	2.5	1.2	1.8
Synthetic Aperture Radar (SAR) Altimeter	27.9	21.4	12.7	16.3
X-SAR Imager	30.8	32.5	18.23	23.56

In order to evaluate the robustness of the methodology implemented, a sensitivity analysis at 25% has been performed to estimate the impact of the weights over the ranking of the technologies. Figure 3 shows the same trend in the rank of the technologies by varying randomly 100 times all weights at the same time for each use case prioritized. In this model, the priority level of the measurements and the number of measurements that can measure the sensors are the critical parameters to rank the technologies.

When the priority use case is Marine for Weather Forecast, the key technologies in ranked order are GNSS-R, X- band SAR imager, and Radar Altimeter with SAR processing (Table 12, columns 1 and 2). The sensitivity analysis is summarized in Figure 3a. The simultaneously random weights defined a clear trend in each technology. Columns 1 and 3 of the Table 12 shows the relevant technologies when selecting the Sea Ice Monitoring use case as the priority. They are X-band SAR, GNSS-R, X-, K-, Ka-, W-band MWIm, and Radar Altimeter (SAR). Figure 3b presents a similar tendency in the results when the weights are varying randomly.

The valuable technologies for the Agriculture- Hydric stress use case in ranked order are Multispectral sensors, GNSS-R, Hyperspectral, and L-band MW; the same distribution has been found in the sensitivity analysis (Figure 3c). Figure 3d shows the sensitivity analysis of the technology rank when the Fishing Pressure use case is the priority. The most important technology also is the Multispectral sensor.

In general, the prioritized list of the main technologies to ensure that the gaps are covered taking into account the priority level of different use cases in the time frame 2020–2030 are GNSS-R, imaging X-band SAR, with 1 km of spatial resolution, and Multispectral sensor. GNSS-R provides support to marine and land services of Copernicus and can collaborate with other technologies to improve the measurements. SAR can provide several data from the ocean and can collaborate with the land data. The best ranked optical payload to support multiple services of Copernicus program is a Multispectral sensor with bands in the VIS (442.5, 485, 490, 510, 560, 640, 660, 665 nm), NIR (1610 nm), MWIR (3.7, and 4.05 μm) and TIR (8.55, 11, and 12 μm).

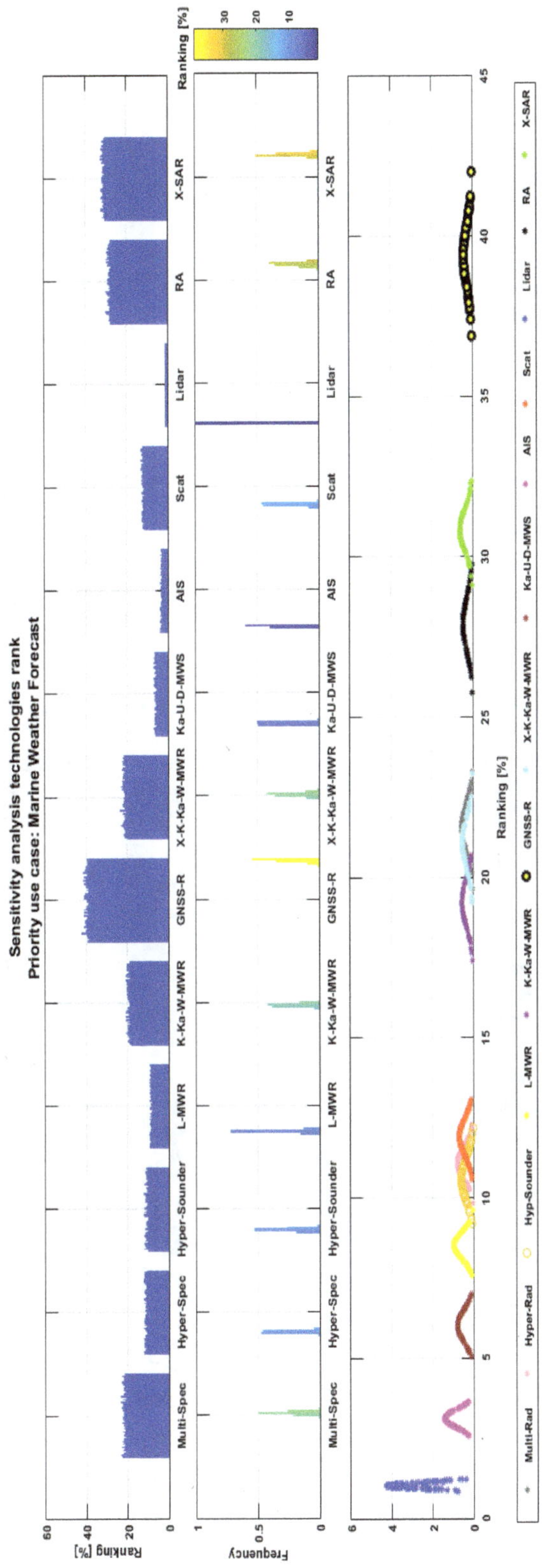

(a) High priority: Marine for Weather Forecast use case

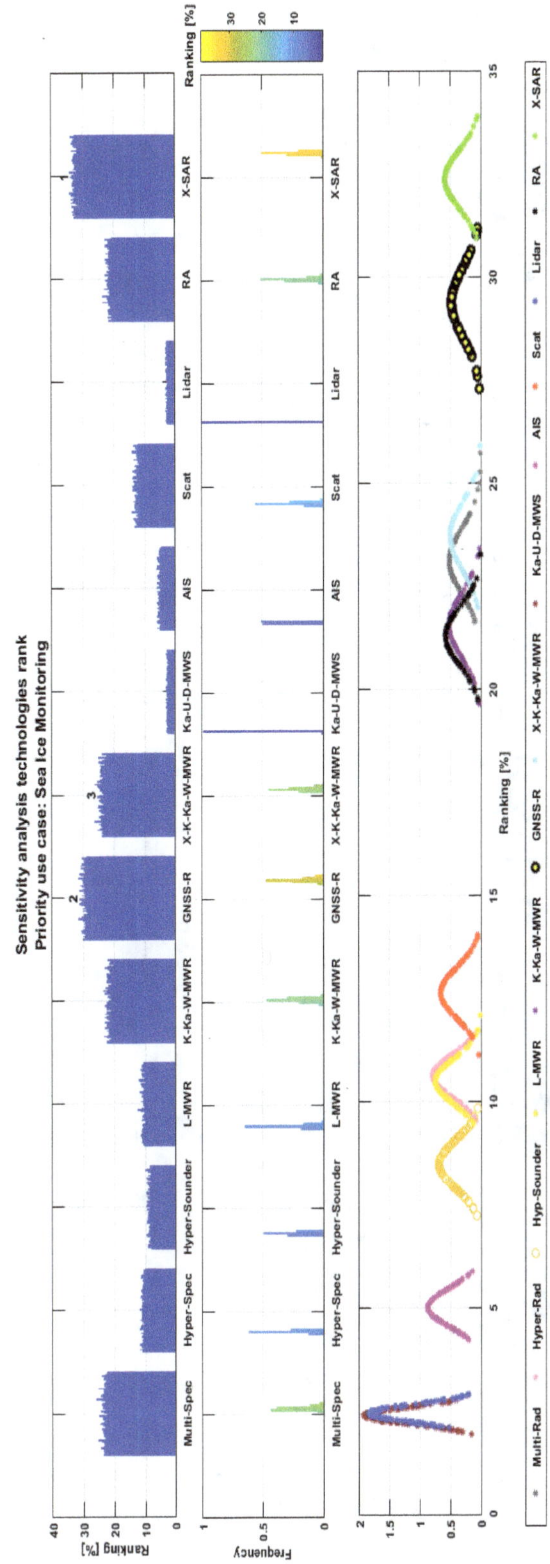

(b) High priority: Sea Ice Monitoring use case

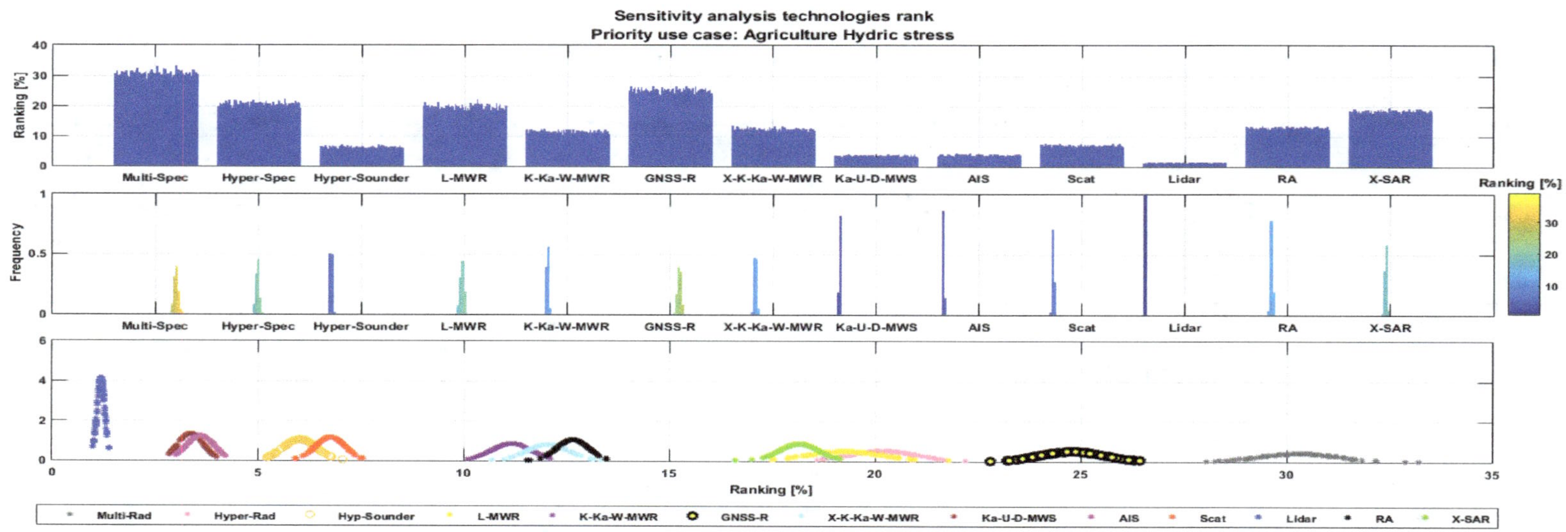

(c) High priority: Agriculture- Hydric Stress use case

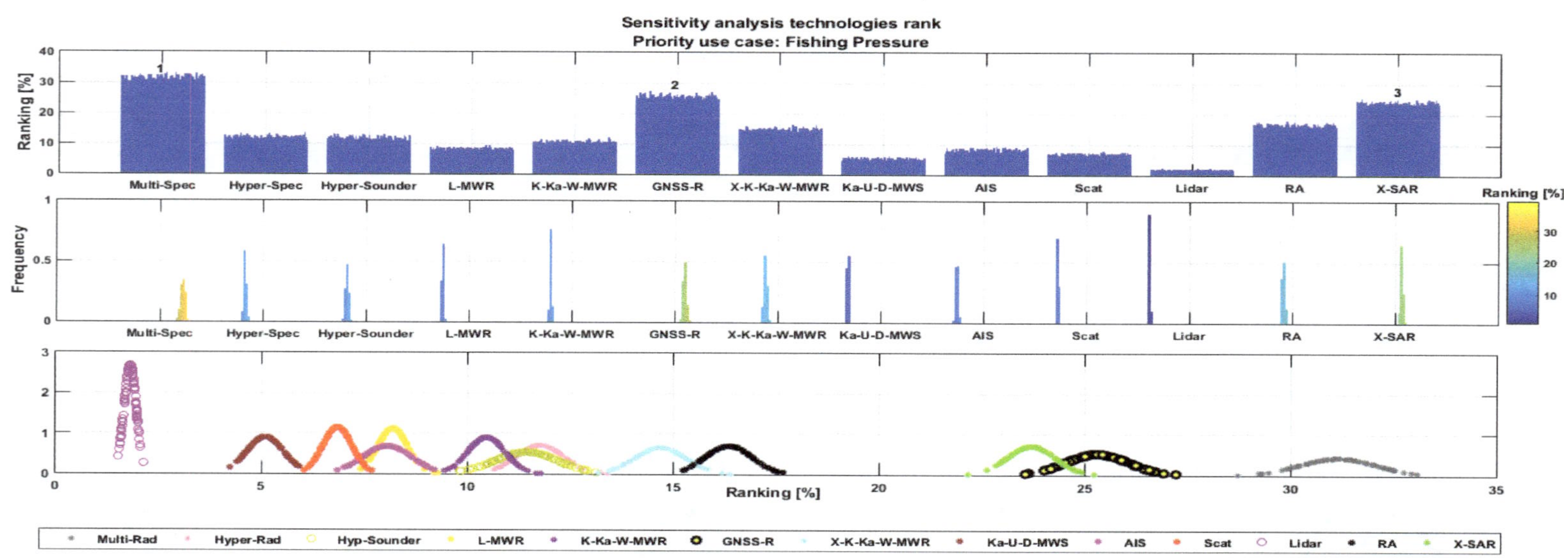

(d) High priority: Fishing Pressure use case

Figure 3. Sensitivity analysis at 25% for different use cases priorities. (**a**) Marine for Weather Forecast; (**b**) Sea Ice Monitoring; (**c**) Fishing Pressure; (**d**) Agriculture and Forestry: Hydric Stress.

7. Conclusions

This study has reviewed the state of the art in EO sensors and platforms and has presented a methodology to select the best instruments' technologies and platforms required to complement the Copernicus system in the time frame 2020–2030. Suitable instruments for small platforms have been analyzed using several attributes, and they have been ranked using a quantitative scoring method. Results show that the most relevant payloads capable of filling the measurements gaps are: GNSS-R at 10 km spatial resolution, X-band imaging SAR at 1 km spatial resolution, and multispectral Optical instrument with bands in the VIS (10 m of spatial resolution), NIR (10 m), MWIR (1 km), and TIR (1 km).

The high temporal resolution of one hour required can only be achieved if a sufficiently large number of spacecrafts are used; then, the architecture selection could be analyzed and optimized [31,71]. A distributed or Federated Satellite System (FSS) will help to reduce the temporal gaps. The possibility to create strategic alliances to establish distributed or federated architectures between different missions and agencies must be carefully evaluated to safe costs. Federated Satellite System (FSS) concepts could also be applied to future instrument technologies to cover the gaps, taking into account different satellites program and space agencies.

Author Contributions: E.L. performed the survey of the instruments and commercial platforms; E.L., A.C., H.P., P.R., S.T., J.C. and S.P. identified the potential instruments to cover the measurements gaps; E.L. developed the quantitative method; A.C. revised the quantitative method; E.L. wrote the paper; A.C., H.P., P.R., S.T., J.C. and S.P. revised it.

Appendix A

This section presents all the commercial LEO small platforms that have been considered in the survey with their corresponding references. Then, the commercial platforms are assessed in terms mass, power consumption, communications, pointing control, and knowledge. Tables A1–A3 summarize each platform and manufacturer with the available capability to support a wide range of available payload mass and power. These small platforms were categorized into three groups' nano-, micro-, and mini- platforms based on the criteria of the International Academy of Astronautics [110]. Nano-satellites have a mass smaller than 10 kg, micro-satellites have a mass between 10 kg and 100 kg, and mini-satellites have mass in the range from 100 kg to 1000 kg.

Table A1. Survey of available nano-satellite platforms for Earth Observation.

Product	Manufacturer	Total Mass [kg]	Size [cm]	Payload Mass [kg]	Payload Volume	Payload Power [W]	Pointing Control	Pointing Knowledge	Communication Downlink	Propulsion
THUNDER (3U) [85]	Space Flight Laboratory	3.5	10 × 10 × 34	1	1000 cm^3	1–2 average	2°	-	S-band 32 kbps–2 Mbps	Cold Gas
Endeavour-3U [18]	Tyvak NanoSatellite Technology Inc.	5.99	30 × 10 × 10	-	2U	12 average, 70 peak	3°	25 arcsec	UHF, S-band 10 Mbps	Cold gas
GRYPHON (GNB) [85]	Space Flight Laboratory	7	20 × 20 × 20	2	1700 cm^3	3–4 average, 6 peak	2°	-	S-band 32 kbps–2 Mbps	Cold gas
GOMX 1U [98]	GomSpace ApS	0.725	1U	-	0.4U	1.33 average	10°	5°	UHF, VHF	-
GOMX 2U [98]	GomSpace ApS	1.2	2U	-	1.4U	2.48 average	10°	5°	UHF, VHF	-
GOMX 3U [98]	GomSpace ApS	1.5	3U	-	2.3U	9.4 average	10°	5°	UHF, VHF, optional X-band	-
SMALL SAT 6U [12]	Nexeya	10	10 × 22 × 34	3	-	7 average, 100 peak	-	-	X-band 100 Mb	Available
XB-12 [19]	Blue Canyon Technologies LLC	-	12U	-	11U	-	1 arcsec	0.002°	UHF, S-band, X-band Up to 15 Mbps	Up to 7 thrusters
XB-3 [19]	Blue Canyon Technologies LLC	-	3U	-	2U	-	-	-	UHF, S-band, X-band Up to 15 Mbps	Up to 7 thrusters
XB-6 [19]	Blue Canyon Technologies LLC	-	6U	-	5U	-	1 arcsec	0.002°	UHF, S-band, X-band Up to 15 Mbps	Up to 7 thrusters
MAI-3000 [17]	Maryland Aerospace	8	10 × 10 × 30	3	1.5U	12 average	$0.1^{\circ\,a}$ or $1.1^{\circ\,b}$	$0.01^{\circ\,a}$ or $1^{\circ\,b}$	S-band Up to 2 Mbps, X-band available.	Compatible with existing 3U launch adapters

Table A2. Survey of available micro-satellite platforms for Earth Observation.

Product	Manufacturer	Total Mass [kg]	Size [cm]	Payload Mass [kg]	Payload Volume	Payload Power Average/Peak [W]	Pointing Control	Pointing Knowledge	Communication	Propulsion
MAI-6000 [23]	Maryland Aerospace	29	10 × 20 × 30	12	4U	20	0.1°	0.01°	S-band Up to 2 Mbps and X-band available.	Compatible with existing launch dispensers
SN-50 [21]	Sierra Nevada Corporation Space Systems	-	-	50	40 × 40 [a] cm	100	0.03°	0.024°	3.5 Mbps	Optional green propulsion capability
Altair [20]	Millennium Space Systems	-	30 × 30 × 30	50	-	90/250	20 arcsec	10 arcsec	S-Band—2 Mbps downlink	-
LEOS-30 [b] [95]	Berlin Space Technologies GmbH	30	30 × 30 × 50	8	-	15/60	-	-	S-Band—2 Mbps downlink	-
LEOS-50 [b] [95]	Berlin Space Technologies GmbH	60	50 × 50 × 30	15	-	20/140	-	-	X-band—100 Mbps downlink	-
NEMO [85]	Space Flight Laboratory	15	20 × 30 × 40	6	8000 cm^3	45	2	-	S-band 32 kbps–2 Mbps downlink	Cold gas, resistojet, monopropulsion
DEFIANT [85]	Space Flight Laboratory	20–30	30 × 30 × 40	6–10	11,000 cm^3	45	2	-	32 kbps–50 Mbps downlink	Cold gas, resistojet, monopropulsion
SMALL SAT 12U [12]	Nexeya	20	22 × 22 × 34	30	-	12/100	-	-	S-band 2.5 Mbps downlink, 256 kbps uplink, Optional X-band 100 Mbps downlink	Available
SMALL SAT 16U [12]	Nexeya	-	46 × 22 × 22	13	-	16/150	-	-	-	Available
SMALL SAT 27U [12]	Nexeya	40	35 × 35 × 34	25	-	30/200	-	-	-	Available
SSTL-12 [22]	Surrey Satellite Technology Limited	40–75	39 × 39 × 47	Up 45	39 × 39 × 37 cm^3	10–30	2°	0.007°	Up to 160 Mbps (X-band)	Available
SSTL-X50 Platform [91]	Surrey Satellite Technology Limited	75	-	Up 45	53 × 43 × 40 cm^3	35/85	0.07°	10 arcsec	-	Available
SSTL-100 [92]	Surrey Satellite Technology Limited	Up 100	-	15	32.1 × 30.3 × 24.6 [c] cm^3 17.9 × 21.6 × 39 [d] cm^3	24/48	2880 arcsec	2520 arcsec	Up 80 Mbps	Liquefied Butane Gas
XB Microsat [19]	Blue Canyon Technologies LLC	75	-	-	45 × 45 × 80 cm^3	-	0.002°	1 arcsec	UHF, S-band, X-band Up to 150 Mbps downlink	Up to 7 thrusters
BCP-50 [96]	Ball Aerospace Commercial Technologies Corp.	80	-	30	30 × 30 × 55 cm^3	30 [e], 100 [f]	0.03°–0.10°	0.03°	2 Mbps downlink,	-
LEOS-100 [f] [95]	Berlin Space Technologies GmbH	90	60 × 60 × 82.5	30	-	60/140	-	-	X-band—100 Mbps downlink	-

[a] Height limited by LV Fairing; [b] Integrated payload. Carry optical payload; [c] Main payload; [d] Secondary payload; [e] Worse case; [f] Best case.

Table A3. Survey of available mini-platforms for Earth Observation.

Product	Manufacturer	Total Mass [kg]	Size [cm^3]	Payload Mass [kg]	Payload Volume [cm^3]	Payload Power Average/Peak [W]	Pointing Control	Pointing Knowledge	Payload Data [Downlink]	Propulsion
NAUTILUS (NEMO-150) [85]	Space Flight Laboratory	Up 150	60 × 60 × 60	Up 70	Up 108000	50/500	2°	-	up to 50 Mbps	Cold gas, resistojet, monopropulsion, Hall thruster.
DAUNTLESS [85]	Space Flight Laboratory	Up 500	100 × 100 × 100	Up 250	Up 500000	200/1000	2°	-	up to 200 Mbps	Cold gas, resistojet, monopropulsion, Hall thruster.
SSTL-150 [92]	Surrey Satellite Technology Limited	Up 150	60 × 60 × 30	50	27.95 × 23.15 × 25.25	50 average, 100 peak.	36 arcsec	25 arcsec	80 Mbps	Hot gas Xenon resistojet.
SSTL-150 ESPA [86]	Surrey Satellite Technology Limited	-	60 × 60 × 80	65	47.5 × 50.5 × 21.1 41 × 54.7 × 24.4	120	1 arcmin	2.5 arcsec	2 Mbps	Available
SSTL-300 [92,93]	Surrey Satellite Technology Limited	368	89.9 × 81.5 × 106.1	150	27.95 × 23.15 × 25.25	140	360 arcsec	72 arcsec	S-Band	Hot gas Xenon resistojet
TET-1 [111]	Astro- und Feinwerktechnik Adlershof	120	67 × 58 × 88	50	460 × 460 × 428	20 to 80 average, 160 peak for 20 min	2 arcmin	10 arcsec	S-band—2.2 Mbps	-
BCP-100 [87]	Ball Aerospace Commercial Technologies Corp.	180	60.9 × 71.1 × 96.5	70	140,000	100–200	0.03°–0.10°	0.03°	2 Mbps for each payload[a]	Green Propellant, Hydrazine options
SN- 200 [94]	Sierra Nevada Corporation Space Systems	Up 355	-	200	-	200	0.1°	0.05°	274 Mbps (X-band)	Xenon HET (TacSat), 4.5
SSTL-600 [92]	Surrey Satellite Technology Limited	Up 429	190 × 140 × 47.6	200	90.1 × 90.8 × 26	386 average, 450 peak	605 arcsec	360 arcsec	500 Mbps (X-band)	Liquefied butane gas
Eagle-1M [90]	Northrop Grumman	-	-	>175	-	500 average, 1200 peak.	0.05°	90 arcsec	-	200 m/s modular
TET-X [13]	OHB	120	58 × 88 × 67	50	1700	Max. 80 , 160 peak for 25 min	-	10 arcsec	100 Mbit/s (X-Band)	Micro propulsion system
TET-XL [13]	OHB	200	80 × 84.5 × 80	80	900	Max. 150 , 460 peak for 25 min.	-	10 arcsec	400 Mbit/s (X-Band), or 1.2 Gbit/s (Ka-Band)	Micro propulsion system
LEOStar-2 BUS [90]	Northrop Grumman	150–500	-	210–550	1,388,000	up to 2k (optional)	15 arcsec	6 arcsec	2 Mbps (S-Band), 150 Mbps (X-band)	Blowdown monopropellant hydrazine;
LEOSTART-500XO [9]	Astrium	500–1000	-	150–600	-	250 average, 450 peak	0.35°	0.24 deg	1.6 Mbps (downlink),	Available
ELiTeBUS 1000 [10]	Thales Alenia Space	-	-	350	38 × 27.12 × 14.25	1000–1500	360 arcsec	22 arcsec	-	Mono-prop (N2H4)

Appendix B

This section presents the sensors that have been considered in the survey with their corresponding references.The sensors are assessed in terms mass, power consumption, data rate, and orbit altitude.

Table A4. Survey of microwave imagers (MWI).

Instrument [Mission]	Frequencies Bands [GHz]	Spatial Resolution [km]	Antenna Size [m]	Swath Width [km]	Mass [kg]	Power [W]	Data Rate [kbps]	Orbit Altitude [km]
Soil Moisture Active and Passive (SMAP) [SMAP] [25,89]	1.41	40	6	1000	356	448	40,000	685
Microwave Imaging Radiometer using Aperture Synthesis (MIRAS) [Soil Moisture and Ocean Salinity (SMOS)] [25,88]	1.41	<50	4 [a]	1000	355	511	89	755
WindSat (Coriolis) [25]	6.8, 10.7, 18.7, 23.8, 37	39 × 71 to 8 × 13	1.83	1200	341	350	256	838
AMSR (ADEOS-II) [25]	6.93, 10.65, 18.7, 23.8, 36.5, 50.3, 52.8 and 89	3 × 6 to 40 × 70	2	1600	320	400	130	812
AMSR-2 (GCOM) [25]	6.93, 7.3, 10.65, 18.7, 23.8, 36.5 and 89	5 to 50 [b]	2	1450	320	400	130	700
AMSR-E (Aqua) [112]	6.93, 7.3, 10.65, 18.7, 23.8, 36.5 and 89	3 × 5 to 35 × 62	2.4	1450	314	350	874	705
Aquarius (SAC-D) [25]	1.4 GHz	100	2.5	390	247	291	5	661
MWI (Metop-SG) [25]	18.7–183.31 (26 channels)	8 × 13 to 40 × 65	0.75	1700	220	250	160	817
MADRAS (Megha Tropiques) [25]	18.7, 23.8, 36.5, 89 and 157	40 × 60 to 6 × 9	0.65	1700	162	153	37	867
GMI (GPM) [25]	10.65, 18.7, 23.8, 36.5, 89, 166, 183.31	19 × 32 to 4.4 × 7.2	1.2	850	150	140	25	407
TMI (TRMM) [25]	10.65, 19.35, 21.3, 37, 85.5	37 × 63 to 5 × 7	0.61	790	65	50	8.8	402
SSM/I (DMSP) [84]	19.35, 23.235, 37, 85.5	45 × 68 to 11 × 16	0.61	1400	48.5	45	3.3	850

The background color in the table indicates the type of platform suitable for the instrument according to the power and mass requirements; Light gray: Micro-platform; Gray: Mini-Platform; [a] 3 arm size; [b] Resolution range for standard products.

Table A5. Survey of microwave sounders (MWS).

Instrument [mission]	Frequencies [GHz]	Spatial Resolution [km]	Antenna Size [m]	Swath Width [km]	Mass [kg]	Power (W)	Data Rate [kbps]	Orbit Altitude [km]
ATMS (SNPP, JPSS) [25]	23.8–183 (22 channels)	16, 32 and 75	-	2600	75	130	30	824
AMSU-A (NOAA-15/16/17/18/19, Metop A/B/C and Aqua) [25]	23 to 89 (15 Channels)	48	0.17 and 0.08 [a]	2100	104	99	3.4	817
Tri-band Microwave Radiometer (MiRaTA) [25,32]	52–58 175–191 203.8–206.8 (10 channels)	-	0.1	-	<4.5	6	10	400
Miniature microwave sounder (EON-MW) [33]	23/31, 50–60/88, 166/183 (22 channels)	44, 23, 7.5	0.11	1000	5	23	50	505

The background color in the table indicates the type of platform suitable for the instrument according to the power and mass requirements: Very light gray: Nano-platform; Light gray: Micro-platform; Gray: Mini-Platform; [a] This instrument has two antennas with different apertures.

Table A6. Survey of GNSS-R instruments.

Available Instruments	Frequencies & Signals	Spatial Resolution [km]	Swath [km]	Mass [Kg]	Power [W]	Data Rate [kbps]	Orbit Altitude [km]
SGR-ReSI (TechDemoSat-1 (TDS-1), CYGNSS) [57]	L1 C/A Code (Options: Galileo E1, GPS L2C, Glonass L1, GPS L5, Galileo E5)	20–50	740	1.4 [a]	<12	200	680
GEROS-ISS (GEROS-ISS) [80]	L1 C/A Code (Options: Galileo E1, GPS L2C, Glonass L1, GPS L5, Galileo E5, and QZSS)	30	~2000	376	395	1200	375–435
FMMPL-2 (FSSCAT) [49]	L1 C/A Code (Options: Galileo E1)	0.3	~350	1.5	>8.0	40	500–550

The background color in the table indicates the type of platform suitable for the instrument according to the power and mass requirements: Very light gray: Nano-platform; Light gray: Micro-platform; Gray: Mini-Platform; [a] Antenna mass doesn't include.

Table A7. Survey of Automatic Identification System (AIS) missions.

Missions	Satellite Mass [kg]	Size	Power Consumption [W]	Launch Date	Payload
Triton-2/E-SAIL [97]	100	60 × 60 × 70 cm	100	2018	AIS
Norsat-2/SAT-AIS [97]	1.5	51 × 140 × =168 mm	5	2016	AIS
AISSat [25]	14	1 U	15	2013	AIS
3CAT-4 [113]	9	6U	2	-	AIS + VIS/NIR camera
Canx-6 [25]	6.5	2U	5.6	2008	AIS
AISSat 1 [25]	6	-	0.97	2010	AIS
AISSat 2 [25]	6	-	0.97	2014	AIS
ZACube-2 [25]	4	3U	-	2017	AIS + imager
AAUSAT-4 [25]	0.88	1U	1.15	2016	AIS

The background color in the table indicates the type of platform suitable for the instrument according to the power and mass requirements: Very light gray: Nano-platform; Light gray: Micro-platform.

Table A8. Survey of optical radiometer instruments: multispectral and hyperspectral.

Instrument (Mission)	Classification	Wavelength [μm]	Aperture Size [m]	Spatial Resolution [km]	Swath Width [km]	Mass [kg]	Power [W]	Data Rate [Mbps]	Orbit Altitude [km]
MetImage [a] (MetOp-SG) [25]	Radiometer/ Multispectral resolution	[0.443–13.345] 20 spectral channels	0.17	0.25 to 0.5 or 1	2670	296	465	18	817
VIIRS [a] (NOAA-20) [25]	Radiometer/ Multispectral resolution	[0.4–12.5] 22 spectral channels	0.184	0.375 to 0.75	3000	275	240	5.9	825
Modis [a] (Terra/Aqua) [25]	Radiometer/ Multispectral resolution	[0.4–14] 36 spectral channels	0.178	0.25–1	2330	229	162.5	6.1	705
SLSTR [b] (Sentinel-3) [24,25]	Radiometer/ Multispectral resolution	[0.545–12.5] 11 spectral channels	-	0.5–1.0	1400	140	100	64	814.5
OLCI [c] (Sentinel-3) [25]	Radiometer/ Multispectral resolution	[0.55–10.85] 21 spectral channels	-	0.3	1270	150	124	5	814.5
AATSR [b] (Envisat) [25]	Radiometer/ Multispectral resolution	[0.4–15] 7 spectral channels	-	1	500	101	100	0.625	774
VIRS [b] (TRMM) [25]	Radiometer/ Multispectral resolution	[0.58–12.05]	-	2	833	34.5	40	0.05	402
AVHRR/3 [b] (Metop/ NOAA) [25,61]	Radiometer/ Multispectral resolution	[0.58–12.5] 6 spectral channels	0.21 × 0.295	1.1	2900	33	27	0.621	850
Naomi [b] (SPOT-6/7) [25]	Radiometer/ Multispectral resolution	[0.45–0.89] 5 spectral channels	-	0.08	25	18.5	-	60	695
CHRIS [b] (PROBA-1) [25]	Imager Spectrometer/ Hyperspectral resolution	[0.4–1.05] 63 spectral channels	0.12	0.036	14	14	8	1	615
COMIS [c] (STSat-3) [25]	Imager Radiometer/ Hyperspectral resolution	[0.4–1.05] 64 spectral channels	-	30 or 60	15 or 30	4.3	5	-	700
HyperScout/ FSSCAT, (3CAT 5/B) [49]	Imager/ Hyperspectral resolution	[0.4–1.0] 45 spectral bands	0.1	0.04	164	1.1	11	-	300
CIRC [c] (ALOS-2) [25]	Infrared radiometer	[8–12] Single TIR channel	0.08	200	128	3	<20	-	640

The background color in the table indicates the type of platform suitable for the instrument according to the power and mass requirements: Very light gray: Nano-platform; Light gray: Micro-platform; gray: mini-platform; [a] Instrument affordable for a wide range of geophysical variables, from cloud classification and properties, to aerosol main properties, land surface variables and sea surface variables. [b] Instrument affordable for cloud analysis, aerosol inference, land surface variables and sea surface variables. [c] Instrument affordable for Observation of land surface (e.g., vegetation), marine biology (e.g., ocean color), and cloud/aerosol.

Table A9. Survey of optical sounders instruments: multispectral and hyperspectral.

Instrument (Mission)	Classification	Wavelength [μm]	Aperture Size [m]	Spatial Resolution [km]	Swath Width [km]	Mass [kg]	Power [W]	Data Rate [Mbps]	Orbit Altitude [km]
IASI [b] (MetOp) [25]	Fourier Transform spectrometer b Radiometer/ Hyperspectral resolution	[3.62–15.5] 8461 spectral samples	1.1	25, 1–30	2052	236	210	1.5	827
AIRS [a] (Aqua) [25]	Infrared sounder/ Hyperspectral resolution	[0.4–15.4] spectral channel >2300	0.219	13.5, 1	1650	177	220	1.27	705
CrIS [a] (JPSS) [25]	Infrared Sounder/ Hyperspectral resolution	[3.92–15.38] 1345 spectral channels	0.8	14	2200	152	124	1.5	824
HIRS/4 [a] (MetOp, NOAA) [25]	Infrared sounder/ Multispectral resolution	[0.69–14.95] 20 spectral channels	0.15	10	2160	35	24	0.003	850
EON-IR [a] CIRAS [25,62]	Infrared Sounder/ Hyperspectral resolution	[4.08–5.13] 625 channels	0.15	3, 13.5	2200	2.5	15	2	450–600

The background color in the table indicates the type of platform suitable for the instrument according to the power and mass requirements: Very light gray: Nano-platform; Light gray: Micro-platform; gray: mini-platform; [a] Instrument affordable for cloud analysis, aerosol inference, land surface variables and sea surface variables.

Table A10. Survey of Radar Altimeter instruments.

Instrument/Mission	Frequency [GHz]	Antenna Size [m]	Spatial Resolution [km]	Mass [kg]	Power [W]	Data Rate [kbps]	Orbit Altitude [km]
Altika/SARAL [25]	23.8, 36.5, 35.75	1	10	40	85	43	800
SWIM/CFOSAT [25]	13.58	0.9	-	-	120	50	519
Altimeter/SWOT [24]	5.3, 13.58	1.2	25	70	78	22.5	891
Karin*/SWOT [24]	35.75	5 × 0.25	0.05 [a] 1 [b]	300	1100	320,000	891
RA-2/Envisat [24,25]	3.2, 13.6	1.5	20	110	161	100	774
SSALT/TOPEX- Poseidon [24,25]	13.65	1.5	25	24	49	-	1336

The background color in the table indicates the type of platform suitable for the instrument according to the power and mass requirements: light gray: micro-platform; gray: mini-platform. * Interferometry; [a] Spatial resolution over land; [b] Spatial resolution over ocean.

Table A11. Survey of scatterometer instruments.

Instrument (Mission)	Frequencies [GHz]	Spatial Resolution [km]	Swath Width [km]	Mass [kg]	Power [W]	Data Rate [kbps]	Orbit Altitude [km]
ASCAT (Metop) [25]	5.255	50, 25 and 12.5	550	260	215	42	817
RapidScat (ISS RapidScat) [24]	13.4	50, 25 and 12.5	900	200	220	40	407
SCA (Metop-SG-B1/B2/B3) [25]	5.3	17–25	550	600	540	5000	817
SCAT (CFOSAT) [25,82]	13.256	50, 10	>1000	<70	<200	220	500
WindRAD (FY-3E/3H) [25]	5.3 and 13.265	20 (C-band), and 10 (Ku-band)	1200	-	265	-	836

The background color in the table indicates the type of platform suitable for the instrument according to the power and mass requirements: gray: mini-platform; black: large-platform.

Table A12. Survey of Radar Altimeter instruments with SAR processing.

Instrument/Mission	Frequency [GHz]	Antenna Size [m]	Spatial Resolution [km]	Mass [kg]	Power [W]	Data Rate [kbps]	Orbit Altitude [km]
SIRAL/Cryosat-2 [24,25]	13.56	1.2	15 0.25 [a]	70	149	24,000	717
SRAL/Sentinel-3 [24,25]	5.3, 13.58	1.2	20 0.3 [a]	60	90	12,000	810
Poseidon-4/ Sentinel-6 [24]	5.3, 13.58	-	20 0.3 [a]	60	90	12,000	1336

The background color in the table indicates the type of platform suitable for the instrument according to the power and mass requirements: light gray: micro-platform; gray: mini-platform. [a] Along track resolution (SAR mode).

Table A13. Survey of SAR imager instruments.

Instrument/ Mission	Frequency [GHz]	Spatial Resolution [m] @ Swath [km]	Mass [kg]	Power [W]	Data Rate [Mbps]	Orbit Altitude [km]
L-band SAR/SAOCOM-2 [24]	1.275	10–100 @ 30–320	1500	-	300	620
X-Band SAR/TSX-NG [24]	9.65	1–16 @ 10–100	1230	2400	680	515
SAR/RISAT-1/1A/2 [24]	5.35	1–50 @ 10–220	950	3100	1478	546
C-Band SAR/Sentinel-1 [24]	5.405	9–50 @ 80–400	880	4400	600	693
SAR (CSA)/RADARSAT [24]	5.405	16–100 @ 20–500	705	1650	105	798
SAR RCM/RCM [24]	5.4	3–100 @ 20–500	600	1270	-	592
COSI/KOMPSAT-5 [24]	9.66	1–20 @ 5–100	520	600	310	550
Severjanin-M/Meteor-M N2 [25,83]	9.623	400–1000 @ 600	150	1000	10	830

The background color in the table indicates the type of platform suitable for the instrument according to the power and mass requirements: Gray: mini-platform; black: large-platform.

Table A14. Survey of the lidar instruments.

Type of Lidar	Instrument/ Mission	Wavelength [nm]	Mass [kg]	Power [W]	Data Rate [kbps]	Vertical Spatial Resolution [m]	Swath [m]	Orbit Altitude [km]
Doppler Lidar	ALADIN/ ADM-Aeolus [25]	355	500	840	11	250	50,000	405
Backscatter LIDAR	ATLID/ EarthCare [25]	354.8	230	320	820	100	100	394
	CALIOP/ CALIPSO [25]	532, 1064	156	124	332	30	333	705
	CATS/ ISS CATS [25]	355, 532, 1064	494	1000	2000	30	3500	407
LIDAR Altimeter	VCL/ DESDynl [24]	1064	225	336	800	1	25,000	400
	GEDI-Lidar/ ISS GEDI [24]	1064.5	230	516	2100	25	7000	407
	ATLAS/ ICESat-2 [24]	1064	298	300	0.45	0.1	170	478
	GLAS/ ICESat [24]	532, 1064	298	300	0.45	0.1	170	600
Differential Absorption Lidar (DIAL)	IPDA LIDAR/ MERLIN [24,25]	1645	32.5	57	150,000	100	0.1	506

The background color in the table indicates the type of platform suitable for the instrument according to the power and mass requirements: gray: mini-platform; black: large-platform.

References

1. Earth Explorer Missions. Available online: https://www.esa.int/Our_Activities/Observing_the_Earth/Earth_Explorers_an_overview (accessed on 19 December 2018).
2. European Space Agency: Two New Earth Explorer Concepts to Understand Our Rapidly Changing World. Available online: https://www.esa.int/Our_Activities/Observing_the_Earth/Two_new_Earth_Explorer_concepts_to_understand_our_rapidly_changing_world (accessed on 19 December 2018).
3. ESA Status Report. Available online: http://www.wmo.int/pages/prog/sat/meetings/ET-SAT-11/documents/ET-SAT-11_Doc_3.4_ESA%20EOP%20%20Status%20Report-%204%20April%202017.pdf (accessed on 19 December 2018).
4. Study to Examine the Socio-Economic Impact of Copernicus in the EU—Report on the Socio-Economic Impact of the Copernicus Programme. Available online: https://publications.europa.eu/en/publication-detail/-/publication/62e30ab0-aa54-11e6-aab7-01aa75ed71a1/language-en/format-PDF/source-search (accessed on 17 December 2018).
5. Matevosyan, H.; Lluch, I.; Poghosyan, A.; Golkar, A. A Value-Chain Analysis for the Copernicus Earth Observation Infrastructure Evolution: A Knowledgebase of Users, Needs, Services, and Products. *IEEE Geosci. Remote Sens. Mag.* **2017**, *5*, 19–35. [CrossRef]
6. Lancheros, E.; Camps, A.; Park, H.; Sicard, P.; Mangin, A.; Matevosvan, I.; Lluch, I. Gaps Analysis and Requirements Specification for the Evolution of Copernicus System for Polar Regions Monitoring: Addressing the Challenges in the Horizon 2020–2030. *Remote Sens.* **2018**, *10*, 1098. [CrossRef]
7. European Space Agency: Sentinels Satellites. Available online: http://www.esa.int/Our_Activities/Observing_the_Earth/Copernicus/Overview4 (accessed on 17 December 2018).
8. European Space Agency: Copernicus Contributing Missions. Available online: https://spacedata.copernicus.eu/web/cscda/missions (accessed on 17 December 2018).
9. Dribault, L.; Durteste, C.; Salvatori, A. LeoStart: Lessons Learnt and Perspectives. Small Satellite Conference. 2013. Available online: https://digitalcommons.usu.edu/smallsat/2000/All2000/22/ (accessed on 17 December 2018).
10. Thales Alenia Space: ELiTeBUS^{TM}1000 Platform. Available online: https://www.thalesgroup.com/sites/default/files/asset/document/20161212_pr_elitebus_rsdo_en.pdf (accessed on 7 June 2018).
11. Bowen, J.; Tsuda, A.; Abel, J.; Villa, M. CubeSat Proximity Operations Demonstration (CPOD) mission update. In Proceedings of the IEEE Aerospace Conference Proceedings, Big Sky, MT, USA, 7–14 March 2015. [CrossRef]
12. Nexeya: Small Satellites. Available online: http://www.nexeya.com/solutions/space-systems/nano-satellites/ (accessed on 16 January 2019)
13. OHB System AG: TET the Small Satellites Series. Available online: https://www.ohb-system.de/tl_files/system/images/mediathek/downloads/pdf/OHB_Messe_TET_2014_web.pdf (accessed on 7 June 2018).

14. Small Geostationary Satellite (SGEO) / Telecommunications & Integrated Applications / Our Activities / ESA. Available online: http://www.esa.int/Our_Activities/Telecommunications_Integrated_Applications/Small_Geostationary_Satellite_SGEO (accessed on 7 June 2018).
15. OHB System AG: SmallGeo. Available online: https://www.ohb-system.de/small-geo-luxor-english.html (accessed on 7 June 2018).
16. Sun, D.W.; Ellmers, F.; Winkler, A.; Schuff, H.; Sansegundo Chamarro, M.J. European small geostationary communications satellites. *Acta Astron.* **2011**, *68*, 802–810. [CrossRef]
17. Maryland Aerospace: Platform Products—MAI-3000—3U CubeSat. Available online: https://www.adcolemai.com/3u-cubesat-mai-3000 (accessed on 7 June 2018).
18. Macgillivray, S. Tyvak Nano-Satellite Systems LLC TM Endeavour: The Product Suite for Next Generation CubeSat Mission. Available online: http://mstl.atl.calpoly.edu/~bklofas/Presentations/SummerWorkshop2012/MacGillivray_Endevour.pdf (accessed on 12 June 2018).
19. Blue Canyon Technologies: Spacecraft Buses, Systems & Solutions—XB Spacecraft Buses. Available online: http://bluecanyontech.com/wp-content/uploads/2018/01/DataSheet_XBSpacecraft_11_F.pdf (accessed on 8 June 2018).
20. Millennium Space Systems—Altair Platforms. Available online: http://www.millennium-space.com/platforms.html#altair (accessed on 8 June 2018).
21. Sierra Nevada Corporation (SNC): Spacecraft Systems—SN-50 Nanosat. Available online: http://mediakit.sncorp.com/mediastore/mediakit/sncspace/146/snc's%20spacecraft%20systems%20overview%20brochure_sn-50_low%20res.pdf (accessed on 8 June 2018).
22. Surrey Satellite Technology Ltd.: SSTL-12 Platform—Mission Configurations. Available online: http://www.sst-us.com/downloads/datasheets/us_platform_sstl-12.pdf (accessed on 7 June 2018).
23. Maryland Aerospace: Platform Products—MAI-6000—6U CubeSat. Available online: https://www.adcolemai.com/6u-cubesat-mai-6000 (accessed on 7 June 2018).
24. World Meteorological Organization: Observing Systems Capability Analysis and Review Tool—Space-Based Capabilities. Available online: https://www.wmo-sat.info/oscar/spacecapabilities (accessed on 7 June 2018).
25. Earth Observation Portal Directory: Satellite Missions Database. Available online: https://directory.eoportal.org/web/eoportal/satellite-missions (accessed on 7 June 2018).
26. Zhang, L.; Shi, H.; Wang, Z.; Yu, H.; Yin, X.; Liao, Q. Comparison of Wind Speeds from Spaceborne Microwave Radiometers with In Situ Observations and ECMWF Data over the Global Ocean. *Remote Sens.* **2018**, *10*, 425. [CrossRef]
27. Tauro, C.B.; Hejazin, Y.; Jacob, M.M.; Jones, W.L. An Algorithm for Sea Surface Wind Speed from SAC-D/Aquarius Microwave Radiometer. *IEEE J. Sel. Top. Appl. Earth Obs. Remote Sens.* **2015**, *8*, 5485–5490. [CrossRef]
28. Kaleschke, L.; Tian-Kunze, X.; Maaß, N.; Mäkynen, M.; Drusch, M. Sea ice thickness retrieval from SMOS brightness temperatures during the Arctic freeze-up period. *Geophys. Res. Lett.* **2012**, *39*. [CrossRef]
29. Ricker, R.; Hendricks, S.; Kaleschke, L.; Tian-Kunze, X.; King, J.; Haas, C. A weekly Arctic sea-ice thickness data record from merged CryoSat-2 and SMOS satellite data. *Cryosphere* **2017**, *11*, 1607–1623. [CrossRef]
30. Ivanova, N.; Pedersen, L.T.; Tonboe, R.T.; Kern, S.; Heygster, G.; Lavergne, T.; Sørensen, A.; Saldo, R.; Dybkjær, G.; Brucker, L.; et al. Inter-comparison and evaluation of sea ice algorithms: Towards further identification of challenges and optimal approach using passive microwave observations. *Cryosphere* **2015**. [CrossRef]
31. Araguz, C.; Llaveria, D.; Lancheros, E.; Bou-Balust, E.; Camps, A.; Alarcón, E.; Lluch, I.; Matevosyan, H.; Golkar, A.; Tonetti, S.; et al. Architectural optimization results for a network of Earth-observing satellite nodes. In Proceedings of the 5th Federated and Fractionated Satellite Systems Workshop, Toulouse, France, 2–3 November 2017.
32. Blackwell, W.J. New Small Satellite Capabilities for Microwave Atmospheric Remote Sensing. In Proceedings of the IEEE MTT-S International Microwave Symposium, Honolulu, HI, USA, 4–9 June 2017.
33. Blackwell, W.J.; Pereira, J. *New Small Satellite Capabilities for Microwave Atmospheric Remote Sensing: The Earth Observing Nanosatellite-Microwave (EON-MW)*; American Meteorological Society: Boston, MA, USA, 2016.
34. Carreno-Luengo, H.; Camps, A.; Perez-Ramos, I.; Forte, G.; Onrubia, R.; Diez, R. 3Cat-2: A P(Y) and C/A GNSS-R experimental nano-satellite mission. In Proceedings of the International Geoscience and Remote Sensing Symposium (IGARSS), Melbourne, VIC, Australia, 21–26 July 2013; pp. 843–846. [CrossRef]

35. Semmling, A.M.; Beckheinrich, J.; Wickert, J.; Beyerle, G.; Schön, S.; Fabra, F.; Pflug, H.; He, K.; Schwabe, J.; Scheinert, M. Sea surface topography retrieved from GNSS reflectometry phase data of the GEOHALO flight mission. *Geophys. Res. Lett.* **2014**. [CrossRef]
36. Morris, M.; Ruf, C.S. Determining tropical cyclone surface wind speed structure and intensity with the CYGNSS satellite constellation. *J. Appl. Meteorol. Climatol.* **2017**. [CrossRef]
37. Clarizia, M.P.; Gommenginger, C.P.; Gleason, S.T.; Srokosz, M.A.; Galdi, C.; Di Bisceglie, M. Analysis of GNSS-R delay-Doppler maps from the UK-DMC satellite over the ocean. *Geophys. Res. Lett.* **2009**. [CrossRef]
38. Mashburn, J.; Axelrad, P.; Lowe, S.T.; Larson, K.M. Global Ocean Altimetry with GNSS Reflections from TechDemoSat-1. *IEEE Trans. Geosci. Remote Sens.* **2018**, *56*, 4088–4097. [CrossRef]
39. Clarizia, M.P.; Ruf, C.S.; Jales, P.; Gommenginger, C. Spaceborne GNSS-R minimum variance wind speed estimator. *IEEE Trans. Geosci. Remote Sens.* **2014**, *52*, 6829–6843. [CrossRef]
40. Wang, F.; Yang, D.; Zhang, B.; Li, W. Waveform-based spaceborne GNSS-R wind speed observation: Demonstration and analysis using UK TechDemoSat-1 data. *Adv. Space Res.* **2018**, *61*, 1573–1587. [CrossRef]
41. Camps, A.; Park, H.; Pablos, M.; Foti, G.; Gommenginger, C.P.; Liu, P.W.; Judge, J. Sensitivity of GNSS-R Spaceborne Observations to Soil Moisture and Vegetation. *IEEE J. Sel. Top. Appl. Earth Obs. Remote Sens.* **2016**, *9*, 4730–4742. [CrossRef]
42. Chew, C.; Shah, R.; Zuffada, C.; Hajj, G.; Masters, D.; Mannucci, A.J. Demonstrating soil moisture remote sensing with observations from the UK TechDemoSat-1 satellite mission. *Geophys. Res. Lett.* **2016**, *43*, 3317–3324. [CrossRef]
43. Alonso-Arroyo, A.; Camps, A.; Aguasca, A.; Forte, G.; Monerris, A.; Rudiger, C.; Walker, J.P.; Park, H.; Pascual, D.; Onrubia, R. Improving the accuracy of soil moisture retrievals using the phase difference of the dual-polarization GNSS-R interference patterns. *IEEE Geosci. Remote Sens. Lett.* **2014**, *11*, 2090–2094. [CrossRef]
44. Rodriguez-Alvarez, N.; Bosch-Lluis, X.; Camps, A.; Aguasca, A.; Vall-Llossera, M.; Valencia, E.; Ramos-Perez, I.; Park, H. Review of crop growth and soil moisture monitoring from a ground-based instrument implementing the Interference Pattern GNSS-R Technique. *Radio Sci.* **2011**, *46*. [CrossRef]
45. Larson, K.M.; Braun, J.J.; Small, E.E.; Zavorotny, V.U.; Gutmann, E.D.; Bilich, A.L. GPS Multipath and Its Relation to Near-Surface Soil Moisture Content. *IEEE J. Sel. Top. Appl. Earth Obs. Remote Sens.* **2010**, *3*, 91–99. [CrossRef]
46. Li, W.; Cardellach, E.; Fabra, F.; Rius, A.; Ribó, S.; Martín-Neira, M. First spaceborne phase altimetry over sea ice using TechDemoSat-1 GNSS-R signals. *Geophys. Res. Lett.* **2017**, *44*, 8369–8376. [CrossRef]
47. Alonso-Arroyo, A.; Zavorotny, V.U.; Camps, A. Sea Ice Detection Using U.K. TDS-1 GNSS-R Data. *IEEE Trans. Geosci. Remote Sens.* **2017**, *55*, 4989–5001. [CrossRef]
48. Foti, G.; Gommenginger, C.; Jales, P.; Unwin, M.; Shaw, A.; Robertson, C.; Rosellõ, J. Spaceborne GNSS reflectometry for ocean winds: First results from the UK TechDemoSat-1 mission. *Geophys. Res. Lett.* **2015**, *42*, 5435–5441. [CrossRef]
49. Copernicus Masters: FSSCat—Towards Federated EO Systems. Available online: https://www.copernicus-masters.com/winner/ffscat-towards-federated-eo-systems/ (accessed on 9 June 2018).
50. Castellví, J.; Camps, A.; Corbera, J.; Alamús, R. 3Cat-3/MOTS nanosatellite mission for optical multispectral and GNSS-R earth observation: Concept and analysis. *Sensors* **2018**, *18*, 140. [CrossRef] [PubMed]
51. Park, H.; Pascual, D.; Camps, A.; Martin, F.; Alonso-Arroyo, A.; Carreno-Luengo, H. Analysis of spaceborne GNSS-R delay-doppler tracking. *IEEE J. Sel. Top. Appl. Earth Obs. Remote Sens.* **2014**. [CrossRef]
52. Martín-Neira, M.; D'Addio, S.; Buck, C.; Floury, N.; Prieto-Cerdeira, R. The PARIS ocean altimeter in-Orbit demonstrator. *IEEE Trans. Geosci. Remote Sens.* **2011**, *49*, 2209–2237. [CrossRef]
53. Camps, A.; Park, H.; Valencia I Domenech, E.; Pascual, D.; Martin, F.; Rius, A.; Ribo, S.; Benito, J.; Andres-Beivide, A.; Saameno, P.; et al. Optimization and performance analysis of interferometric GNSS-R altimeters: Application to the PARIS IoD mission. *IEEE J. Sel. Top. Appl. Earth Obs. Remote Sens.* **2014**, *7*, 1436–1451. [CrossRef]
54. Martin-Neira, M. A Passive Reflectometry and Interferometry System(PARIS)- Application to ocean altimetry. *ESA J.* **1993**, *17*, 331–355.
55. Camps, A.; Park, H.; Sekulic, I.; Rius, J.M. GNSS-R altimetry performance analysis for the GEROS experiment on board the international space station. *Sensors* **2017**, *17*, 1583. [CrossRef]

56. Ribo, S.; Arco, J.C.; Oliveras, S.; Cardellach, E.; Rius, A.; Buck, C. Experimental results of an x-band PARIS receiver using digital satellite TV opportunity signals scattered on the sea surface. *IEEE Trans. Geosci. Remote Sens.* **2014**, *52*, 5704–5711. [CrossRef]
57. Surrey Satellite Technology: SGR-ReSI. Available online: http://www.sst-us.com/getfile/e4b5fba3-8d9a-499d-8e50-568ffa156bcd (accessed on 9 June 2018).
58. GOMSpace. GOMspace: Ship Tracking with Space based AIS Receiver. Available online: https://gomspace.com/Shop/payloads/ship-tracking.aspx (accessed on 18 September 2018).
59. Cooley, S.W.; Smith, L.C.; Stepan, L.; Mascaro, J. Tracking dynamic northern surface water changes with high-frequency planet CubeSat imagery. *Remote Sens.* **2017**, *9*, 1306. [CrossRef]
60. Selva, D.; Krejci, D. A survey and assessment of the capabilities of Cubesats for Earth observation. *Acta Astron.* **2012**, *74*, 50–68. [CrossRef]
61. Polar Operational Envoromental Satellites (POES): AVHRR/3. Available online: https://poes.gsfc.nasa.gov/avhrr3.html (accessed on 12 June 2018).
62. Pagano, T.; Abesamis, C.; Andrade, A.; Aumann, H.; Gunapala, S.; Heneghan, C.; Jarnot, R.; Johnson, D.; Lamborn, A.; Maruyama, Y.; et al. Design and development of the CubeSat Infrared Atmospheric Sounder (CIRAS). *SPIE In. Soc. Opt. Eng.* **2017**, *10402*. [CrossRef]
63. Dawson, G.J.; Bamber, J.L. Antarctic Grounding Line Mapping From CryoSat-2 Radar Altimetry. *Geophys. Res. Lett.* **2017**. [CrossRef]
64. Tournadre, J.; Bouhier, N.; Boy, F.; Dinardo, S. Detection of iceberg using Delay Doppler and interferometric Cryosat-2 altimeter data. *Remote Sens. Environ.* **2018**. [CrossRef]
65. Richard, J.; Enjolras, V.; Rys, L.; Vallon, J.; Nann, I.; Escudier, P. Space Altimetry from Nano-Satellites: Payload Feasibility, Missions and System Performances. In Proceedings of the IGARSS 2008, 2008 IEEE International Geoscience and Remote Sensing Symposium, Boston, MA, USA, 7–11 July 2008. [CrossRef]
66. Vogelzang, J.; Stoffelen, A. Scatterometer wind vector products for application in meteorology and oceanography. *J. Sea Res.* **2012**, *74*, 16–25. [CrossRef]
67. Reimer, A.C.; Melzer, B.T.; Kidd, C.R.; Wagner, D.W. Validation of the enhanced resolution ERS-2 scatterometer soil moisture product. In Proceedings of the International Geoscience and Remote Sensing Symposium (IGARSS), Munich, Germany, 22–27 July 2012; pp. 1208–1211. [CrossRef]
68. Brakke, T.W.; Kanemasu, E.T.; Steiner, J.L.; Ulaby, F.T.; Wilson, E. Microwave radar response to canopy moisture, leaf-area index, and dry weight of wheat, corn, and sorghum. *Remote Sens. Environ.* **1981**, *11*, 207–220. [CrossRef]
69. Foster, J.L.; Hall, D.K.; Eylander, J.B.; Kim, E.D.; Riggs, G.A.; Tedesco, M.; Nghiem, S.V.; Kelly, R.E.J.; Choudhury, B.; Reichle, R. Blended Visible, Passive Microwave and Scatterometer Global Snow Products. In Proceedings of the 64th Eastern Snow Conference, St. John's, NL, Canada, 29 May–1 June 2007; pp. 27–36.
70. Belmonte Rivas, M.; Verspeek, J.; Verhoef, A.; Stoffelen, A. Bayesian sea ice detection with the advanced scatterometer ASCAT. *IEEE Trans. Geosci. Remote Sens.* **2012**, *50*, 2649–2657. [CrossRef]
71. Alarcon, E.; Sanchez, A.A.; Araguz, C.; Barrot, G.; Bou-Balust, E.; Camps, A.; Cornara, S.; Cote, J.; Pena, A.G.; Lancheros, E.; et al. Design and optimization of a polar satellite mission to complement the copernicus system. *IEEE Access* **2018**. [CrossRef]
72. Gebert, N.; Domínguez, B.C.; Davidson, M.W.J.; Martin, M.D.; Silvestrin, P. SAOCOM-CS—A passive companion to SAOCOM for single-pass L-band SAR interferometry. In Proceedings of the European Conference on Synthetic Aperture Radar, Berlin, Germany, 3–5 June 2014; pp. 1251–1254.
73. Surrey Satellite Technology Ltd.: NovaSAR-S—The Small Satellite Approack to Synthetic Aperture Radar. Available online: https://www.sstl.co.uk/Downloads/Brochures/115184-SSTL-NovaSAR-Brochure-high-res-no-trims (accessed on 7 June 2018).
74. ICEYE Satellite Missions. Available online: https://www.iceye.com/resources/satellite-missions (accessed on 4 October 2018).
75. Slobbe, D.C.; Lindenbergh, R.C.; Ditmar, P. Estimation of volume change rates of Greenland's ice sheet from ICESat data using overlapping footprints. *Remote Sens. Environ.* **2008**. [CrossRef]
76. Lefsky, M.A.; Harding, D.J.; Keller, M.; Cohen, W.B.; Carabajal, C.C.; Del Bom Espirito-Santo, F.; Hunter, M.O.; de Oliveira, R. Estimates of forest canopy height and aboveground biomass using ICESat. *Geophys. Res. Lett.* **2005**. [CrossRef]

77. Winker, D.M.; Pelon, J.; Coakley, J.A.; Ackerman, S.A.; Charlson, R.J.; Colarco, P.R.; Flamant, P.; Fu, Q.; Hoff, R.M.; Kittaka, C.; et al. The Calipso Mission: A Global 3D View of Aerosols and Clouds. *Bull. Am. Meteorol. Soc.* **2010**. [CrossRef]
78. Illingworth, A.J.; Barker, H.W.; Beljaars, A.; Ceccaldi, M.; Chepfer, H.; Clerbaux, N.; Cole, J.; Delanoë, J.; Domenech, C.; Donovan, D.P.; et al. The earthcare satellite: The next step forward in global measurements of clouds, aerosols, precipitation, and radiation. *Bull. Am. Meteorol. Soc.* **2015**. [CrossRef]
79. Stoffelen, A.; Marseille, G.J.; Bouttier, F.; Vasiljevic, D.; de Haan, S.; Cardinali, C. ADM-Aeolus Doppler wind lidar Observing System Simulation Experiment. *Q. J. R. Meteorol. Soc.* **2006**. [CrossRef]
80. Wickert, J.; Cardellach, E.; Martin-Neira, M.; Bandeiras, J.; Bertino, L.; Andersen, O.B.; Camps, A.; Catarino, N.; Chapron, B.; Fabra, F.; et al. GEROS-ISS: GNSS REflectometry, Radio Occultation, and Scatterometry Onboard the International Space Station. *IEEE J. Sel. Top. Appl. Earth Obs. Remote Sens.* **2016**, *9*, 4552–4581. [CrossRef]
81. Camps, A.J.; Swift, C.T. A two-dimensional Doppler-radiometer for earth observation. *IEEE Trans. Geosci. Remote Sens.* **2001**. [CrossRef]
82. Hauser, D.; Tison, C.; Amiot, T.; Delaye, L.; Mouche, A.; Guitton, G.; Aouf, L.; Castillan, P. CFOSAT: A new Chinese-French satellite for joint observations of ocean wind vector and directional spectra of ocean waves. *Remote Sens.* **2016**, *98780T*, 1–20. [CrossRef]
83. Vnotchenko, S.; Dostovalov, M.; Dudukin, V.; Kovalenko, A.; Musinyants, T.; Riman, V.; Selyanin, A.; Smirnov, S.; Telichev, A.; Chernishov, V.; et al. Wide-swath spaceborne SAR system "Severyanin-M" for remote sensing: First results. In Proceedings of the 9th European Conference on Synthetic Aperture Radar, Nuremberg, Germany, 23–26 April 2012; pp. 422–425.
84. Special Sensor Microwave Imager (SSM/I) | National Snow and Ice Data Center. Available online: https://nsidc.org/data/pm/ssmi-instrument (accessed on 9 June 2018).
85. UTIAS Space Flight Laboratory | Satellite Platforms. Available online: http://www.utias-sfl.net/?page_id=89 (accessed on 14 June 2018).
86. Surrey Satellite Technology Ltd.: SSTL-150 ESPA Satellite Platform. Available online: http://www.sst-us.com/getfile/95b740f8-2873-4943-93e7-97a99655c798 (accessed on 15 June 2018).
87. Ball Aerospace & Technologies Corp.: Ball Configurable Platform (BCP)- 100. Available online: http://www.ball.com/aerospace/Aerospace/media/Aerospace/Downloads/D3072_BCP100-ds_1_14.pdf?ext=.pdf (accessed on 12 June 2018).
88. Camps, A.; Bosch-Lluis, X.; Ramos-Perez, I.; Marchán-Hernández, J.F.; Rodríguez, N.; Valencia, E.; Tarongi, J.M.; Aguasca, A.; Acevo, R. New passive instruments developed for ocean monitoring at the remote sensing lab-Universitat Politècnica de Catalunya. *Sensors* **2009**, *9*, 10171–10189. [CrossRef]
89. Jet Propulsion Labratory: Instrument | Observatory – SMAP. Available online: https://smap.jpl.nasa.gov/observatory/instrument/ (accessed on 8 June 2018).
90. Northrop Grumman: Spacecraft Buses. Available online: http://www.northropgrumman.com/Capabilities/SpacecraftBuses/Pages/default.aspx (accessed on 15 June 2018).
91. Surrey Satellite Technology Ltd.: SSTL-X50 Satellite Platform. Available online: http://www.sst-us.com/downloads/datasheets/sstl-x50.pdf (accessed on 15 June 2018).
92. Abbott, B. Surrey Satellite Technology US LLC. Available online: https://www.sprsa.org/sites/default/files/conference-presentation/2015_06_PayloadHostingBusiness%20Final.pdf (accessed on 14 June 2018).
93. SSTL-300 Satellite Platform. Available online: http://www.sst-us.com/getdoc/e39526e0-8336-4d5d-8d98-7e18f531126d/1353-sstl-300-s1-datasheet.pdf (accessed on 14 June 2018).
94. Sierra Nevada Corporation (SNC): Spacecraft Systems—SN-50 Nanosat. Available online: http://mediakit.sncorp.com/api/document/SN-200_FINAL_web.pdf (accessed on 20 December 2018).
95. Berlin Space Technologies: Small satellite Systems. Available online: https://www.berlin-space-tech.com/products/small-satellite-systems/ (accessed on 7 June 2018).
96. Ball Aerospace & Technologies Corp.: Ball Configurable Platform (BCP)- 50. Available online: http://www.ball.com/aerospace/Aerospace/media/Aerospace/Downloads/D3103_BC__50-ds_0714.pdf?ext=.pdf (accessed on 18 June 2018).
97. Sat-AIS: Satellite-Based Automatic Identification System. Available online: https://esamultimedia.esa.int/docs/telecom/SAT-AIS-factsheet{_}WEB.pdf (accessed on 9 June 2018).
98. GOMSPACE: Platforms. Available online: https://gomspace.com/platforms.aspx (accessed on 18 June 2018).

99. Yan, Q.; Huang, W. Sea ice detection from GNSS-R Delay-Doppler Map. In Proceedings of the 2016 17th International Symposium on Antenna Technology and Applied Electromagnetics, Montreal, QC, Canada, 10–13 July 2016. [CrossRef]
100. Schiavulli, D.; Nunziata, F.; Migliaccio, M.; Frappart, F.; Ramilien, G.; Darrozes, J. Reconstruction of the Radar Image from Actual DDMs Collected by TechDemoSat-1 GNSS-R Mission. *IEEE J. Sel. Top. Appl. Earth Obs. Remote Sens.* **2016**, *9*, 4700–4708. [CrossRef]
101. Camps, A.; Park, H.; Portal, G.; Rossato, L. Sensitivity of TDS-1 GNSS-R Reflectivity to Soil Moisture: Global and Regional Differences and Impact of Different Spatial Scales. *Remote Sens.* **2018**, *10*, 1856. [CrossRef]
102. Van der Schalie, R.; Parinussa, R.M.; Renzullo, L.J.; van Dijk, A.I.; Su, C.H.; de Jeu, R.A. SMOS soil moisture retrievals using the land parameter retrieval model: Evaluation over the mUrrumbidgee Catchment, southeast Australia. *Remote Sens. Environ.* **2015**. [CrossRef]
103. Patilea, C.; Heygster, G.; Huntemann, M.; Spreen, G. Combined SMAP/SMOS Thin Sea Ice Thickness Retrieval. *Cryosphere* **2017**. [CrossRef]
104. Robinson, I.S. *Measuring the Oceans from Space: The Principles and Methods of Satellite Oceanography*; Springer: Berlin/Heidelberg, Germany, 2004; p. 669.
105. Global Sea Ice Concentration Data Record Relase 1.2 (SSMI/SSMIS). Available online: http://www.osi-saf.org/?q=content/global-sea-ice-concentration-data-record-ssmissmis (accessed on 20 December 2018).
106. The Advanced Very High Resolution Radiometer (AVHRR) Multi-Purpose Imaging Instrument Is Used for Global Monitoring of Cloud Cover, Sea Surface Temperature, Ice, Snow and Vegetation Cover Characteristics. Available online: https://www.eumetsat.int/website/home/Satellites/CurrentSatellites/Metop/MetopDesign/AVHRR/index.html (accessed on 19 December 2018).
107. Sentinel-3 Altimetry: Wind Speed. Available online: https://sentinels.copernicus.eu/web/sentinel/user-guides/sentinel-3-altimetry/overview/geophysical-measurements/wind-speed (accessed on 19 December 2018).
108. Sentinel-3 Altimetry: Significant Wave Height. Available online: https://sentinels.copernicus.eu/web/sentinel/user-guides/sentinel-3-altimetry/overview/geophysical-measurements/significant-wave-height (accessed on 19 December 2018).
109. World Meteorological Organization Observational Requirements and Capabilities. Available online: https://www.wmo-sat.info/oscar/requirements (accessed on 21 June 2018).
110. Sandau, R. Status and trends of small satellite missions for Earth observation. *Acta Astron.* **2010**, *66*, 1–12. [CrossRef]
111. Astr-und Feinwerktechnik Adlershof GmbH: TET-1 Platform. Available online: http://www.astrofein.com/2728/dwnld/admin/Brochure_Satellite_TET-1.pdf (accessed on 7 June 2018).
112. AMSR-E Instrument Description | National Snow and Ice Data Center. Available online: https://nsidc.org/data/amsre/amsre-instrument (accessed on 9 June 2018).
113. Castelvi, J.; Lancheros, E.; Camps, A.; Park, H. Feasibility of Nano-Satellites Constellations for AIS Decoding and Fire detection. In Proceedings of the FSS Workshop 2016, Rome, Italy, 10 October 2016; pp. 1–6.

PERMISSIONS

All chapters in this book were first published in MDPI; hereby published with permission under the Creative Commons Attribution License or equivalent. Every chapter published in this book has been scrutinized by our experts. Their significance has been extensively debated. The topics covered herein carry significant findings which will fuel the growth of the discipline. They may even be implemented as practical applications or may be referred to as a beginning point for another development.

The contributors of this book come from diverse backgrounds, making this book a truly international effort. This book will bring forth new frontiers with its revolutionizing research information and detailed analysis of the nascent developments around the world.

We would like to thank all the contributing authors for lending their expertise to make the book truly unique. They have played a crucial role in the development of this book. Without their invaluable contributions this book wouldn't have been possible. They have made vital efforts to compile up to date information on the varied aspects of this subject to make this book a valuable addition to the collection of many professionals and students.

This book was conceptualized with the vision of imparting up-to-date information and advanced data in this field. To ensure the same, a matchless editorial board was set up. Every individual on the board went through rigorous rounds of assessment to prove their worth. After which they invested a large part of their time researching and compiling the most relevant data for our readers.

The editorial board has been involved in producing this book since its inception. They have spent rigorous hours researching and exploring the diverse topics which have resulted in the successful publishing of this book. They have passed on their knowledge of decades through this book. To expedite this challenging task, the publisher supported the team at every step. A small team of assistant editors was also appointed to further simplify the editing procedure and attain best results for the readers.

Apart from the editorial board, the designing team has also invested a significant amount of their time in understanding the subject and creating the most relevant covers. They scrutinized every image to scout for the most suitable representation of the subject and create an appropriate cover for the book.

The publishing team has been an ardent support to the editorial, designing and production team. Their endless efforts to recruit the best for this project, has resulted in the accomplishment of this book. They are a veteran in the field of academics and their pool of knowledge is as vast as their experience in printing. Their expertise and guidance has proved useful at every step. Their uncompromising quality standards have made this book an exceptional effort. Their encouragement from time to time has been an inspiration for everyone.

The publisher and the editorial board hope that this book will prove to be a valuable piece of knowledge for researchers, students, practitioners and scholars across the globe.

LIST OF CONTRIBUTORS

Emmanouil Psomiadis and Konstantinos X. Soulis
Department of Natural Resources Management and Agricultural Engineering, Agricultural University of Athens, 75 Iera Odos st., 11855 Athens, Greece

Michalis Diakakis
Faculty of Geology and Geoenvironment, National and Kapodistrian University of Athens, 15784 Panepistimioupoli Zografou, 10679 Athens, Greece

Nikos Charizopoulos
Department of Natural Resources Management and Agricultural Engineering, Agricultural University of Athens, 75 Iera Odos st., 11855 Athens, Greece

Nikolaos Efthimiou
Faculty of Environmental Sciences, Czech University of Life Sciences Prague, Kamýcká 129, 165 00 Prague-Suchdol, Czech Republic

Ioannis Charalampopoulos
Department of Crop Science, Agricultural University of Athens, 75 Iera Odos st., 11855 Athens, Greece

Sanjib Kumar Kar and Yuei-An Liou
Center for Space and Remote Sensing Research, National Central University, No. 300, Jhongda Rd., Jhongli Dist., Taoyuan City 32001, Taiwan

Francisco Navas-Guzmán and Lorena Moreira
Institute of Applied Physics, University of Bern, CH-3012 Bern, Switzerland

Klemens Hocke, Leonie Bernet and Christian Mätzler
Institute of Applied Physics, University of Bern, CH-3012 Bern, Switzerland
Oeschger Centre for Climate Change Research, University of Bern, CH-3012 Bern, Switzerland

Estefany Lancheros, Adriano Camps and Hyuk Park
Universitat Politècnica Catalunya—BarcelonaTech & IEEC, Campus Nord, 08034 Barcelona, Spain

Pierre Sicard and Antoine Mangin
ACRI Bâtiment Le Grand Large, Quai de la Douane—2 ème éperon, 29200 Brest, France

Hripsime Matevosyan and Ignasi Lluch
Skolkovo Institute of Science and Technology, Moscow 143026, Russia

M. Venkat Ratnam, Ghouse Basha and N. Narendra Reddy
National Atmospheric Research Laboratory, Gadanki 517112, India

Xiaojuan Bie
College of Earth Science, Chengdu University of Technology, Chengdu 610059, China

Dong He and Tingbin Zhang
College of Earth Science, Chengdu University of Technology, Chengdu 610059, China
The Engineering & Technical College of Chengdu University of Technology, Leshan 614000, China

Guihua Yi
College of Management Science, Chengdu University of Technology, Chengdu 610059, China

Jiaqing Miao
The Engineering & Technical College of Chengdu University of Technology, Leshan 614000, China
Department of Geophysical Engineering, Montana Tech of the University of Montana, Butte, MT 59801, USA

Jingji Li
Chengdu University of Technology, College of Environmental and Civil Engineering Institute, Chengdu 610059, China
Chengdu University of Technology, Institute of Ecological Resource and Landscape, Chengdu 610059, China

Lin Xiao
Key Laboratory of Remote Sensing of Gansu Province, Heihe Remote Sensing Experimental Research Station, Northwest Institute of Eco-Environment and Resources, Chinese Academy of Sciences, Lanzhou 730000, China
University of Chinese Academy of Sciences, Beijing 100049, China

Tao Che
Key Laboratory of Remote Sensing of Gansu Province, Heihe Remote Sensing Experimental Research Station, Northwest Institute of Eco-Environment and Resources, Chinese Academy of Sciences, Lanzhou 730000, China
Center for Excellence in Tibetan Plateau Earth Sciences, Chinese Academy of Sciences, Beijing 100101, China

Linling Chen
Nansen Environmental and Remote Sensing Center, N-5006 Bergen, Norway

Hongjie Xie
Laboratory for Remote Sensing and Geoinformatics, Department of Geological Sciences, The University of Texas at San Antonio, One UTSA Circle, San Antonio, TX 78249, USA

Liyun Dai
Key Laboratory of Remote Sensing of Gansu Province, Heihe Remote Sensing Experimental Research Station, Northwest Institute of Eco-Environment and Resources, Chinese Academy of Sciences, Lanzhou 730000, China
Jiangsu Center of Collaborative Innovation in Geographical Information Resource Development and Application, Nanjing 210023, China

Haibo Wang and Liying Geng
Key Laboratory of Remote Sensing of Gansu Province, Heihe Remote Sensing Experimental Research Station, Northwest Institute of Eco-Environment and Resources, Chinese Academy of Sciences, 320 Donggang West Road, Lanzhou 730000, China

Xin Li
Institute of Tibetan Plateau Research, Chinese Academy of Sciences, Beijing 100101, China
CAS Center for Excellence in Tibetan Plateau Earth Sciences, Chinese Academy of Sciences, Beijing 100101, China

Mingguo Ma
Chongqing Engineering Research Center for Remote Sensing Big Data Application, School of Geographical Sciences, Southwest University, Beibei, Chongqing 400715, China

Yeseul Kim and No-Wook Park
Department of Geoinformatic Engineering, Inha University, Incheon 22212, Korea

Kyung-Do Lee
National Institute of Agricultural Sciences, Rural Development Administration, Wanju 55365, Korea

Pedro Rodriguez
Thales Alenia Space, 28760 Madrid, (España)Madrid, Spain

Stefania Tonetti
Deimos Space S.L.U, 28760 Madrid, Spain

Judith Cote and Stephane Pierotti
Thales Alenia Space, 06150 Toulouse, France

Index

www.ingramcontent.com/pod-product-compliance
Lightning Source LLC
LaVergne TN
LVHW081025210926
848485LV00004B/954